COULOMB INTERACTIONS IN NUCLEAR AND ATOMIC FEW-BODY COLLISIONS

Finite Systems and Multiparticle Dynamics

Series Editors

Frank S. Levin, Brown University, Providence, Rhode Island
David A. Micha, University of Florida, Gainesville, Florida

COULOMB INTERACTIONS IN NUCLEAR AND ATOMIC FEW-BODY COLLISIONS
Edited by Frank S. Levin and David A. Micha

LONG-RANGE CASIMIR FORCES: Theory and Recent Experiments on Atomic Systems
Edited by Frank S. Levin and David A. Micha

A Continuation Order Plan is available for this series. A continuation order will bring delivery of each new volume immediately upon publication. Volumes are billed only upon actual shipment. For further information please contact the publisher.

COULOMB INTERACTIONS IN NUCLEAR AND ATOMIC FEW-BODY COLLISIONS

Edited by

FRANK S. LEVIN
Brown University
Providence, Rhode Island

and

DAVID A. MICHA
University of Florida
Gainesville, Florida

PLENUM PRESS • NEW YORK AND LONDON

Library of Congress Cataloging-in-Publication Data

Coulomb interactions in nuclear and atomic few-body collisions /
  edited by Frank S. Levin and David A. Micha.
      p.   cm. -- (Finite systems and multiparticle dynamics)
   Includes bibliographical references and index.
   ISBN 0-306-45149-2
   1. Few-body problem. 2. Coulomb excitation.   I. Levin, F. S.
 (Frank S.), 1933-   . II. Micha, David. III. Series.
 QC174.17.P7C685 1996
 539.7'57--dc20   95-51224
                                                               CIP

ISBN 0-306-45149-2

© 1996 Plenum Press, New York
A Division of Plenum Publishing Corporation
233 Spring Street, New York, N. Y. 10013

All rights reserved

10 9 8 7 6 5 4 3 2 1

No part of this book may be reproduced, stored in a retrieval system, or transmitted in any form or by any means, electronic, mechanical, photocopying, microfilming, recording, or otherwise, without written permission from the Publisher

Printed in the United States of America

Contributors

F. O. Alt · *Johannes-Gutenberg-Universität, Institute für Physik, 55099 Mainz, Germany*

J. S. Briggs · *Fakultät für Physik, Albert-Ludwigs-Universität, D-9104 Freiburg, Germany*

H. C. Bryant · *Department of Physics and Astronomy, University of New Mexico, Albuquerque, New Mexico 87131-1156*

J. L. Friar · *Theoretical Division, Los Alamos National Laboratory, Los Alamos, New Mexico 87545*

M. Halka · *Department of Physics, Portland State University, Portland, Oregon 97207-0751*

Helmut Kröger · *Département de Physique, Université Laval, Québec, Québec 61K 7P4, Canada*

G. L. Payne · *Department of Physics and Astronomy, University of Iowa, Iowa City, Iowa 52242*

W. Sandhas · *Physikalisches Institut der Universität Bonn, 53115 Bonn, Germany*

Editors' Foreword

This series, *Finite Systems and Multiparticle Dynamics*, is intended to provide timely reviews of current research topics, written in a style sufficiently pedagogic so as to allow a nonexpert to grasp the underlying ideas as well as understand technical details.

The series is an outgrowth of our involvement with three interdisciplinary activities, namely, those arising from the American Physical Society's Topical Group on Few-Body Systems and Multiparticle Dynamics, the series of Gordon Research Conferences first known by the title "Few-Body Problems in Chemistry and Physics" and later renamed "Dynamics of Simple Systems in Chemistry and Physics," and the series of Sanibel Symposia, sponsored in part by the University of Florida. The vitality of these activities and the enthusiastic response to them by researchers in various subfields of physics and chemistry have convinced us that there is a place—even a need—for a series of timely reviews on topics of interest not only to a narrow band of experts but also to a broader, interdisciplinary readership. It is our hope that the emphasis on pedagogy will permit at least some of the books in the series to be useful in graduate-level courses.

Rather than use the adjective "Few-Body" or "Simple" to modify the word "Systems" in the title, we have chosen "Finite." It better expresses the wide range of systems with which the reviews of the series may deal. On the "few-body" side these could include nucleons or deuterons, light nuclei, low-Z atoms, and small molecules; on the "many-body" side, possibilities include polyatomic molecules, heavier nuclei, clusters, and "few-body" dynamics in larger systems and at interfaces, although infinite systems like nuclear matter or the electron gas are excluded.

The series will emphasize theoretical, computational, and experimental methods. It is clear that researchers in fields such as molecular and nuclear sciences, for example, may develop theoretical methods similar to those of another field, often without either group being aware of it. These parallel developments are the result of efforts made to analyze complex physical systems in terms of their simpler constituents, and their complex dynamical properties in terms of the dynamics of subsystems and of collective modes. A similar comment holds with respect to experimental techniques. The pedagogic aspect of the reviews commissioned for *Finite Systems and Multiparticle Dynamics* is designed specifically to promote both learning and cross-disciplinary exchanges.

To give some notion of the scope we envisage for the series, we list below the general area of each of the first three volumes:

Vol. I.   Long-Range Casimir Forces
Vol. II. Coulomb Interactions in Nuclear and Atomic Few-Body Collisions
Vol. III. Recent Developments in Molecular and Nuclear Reaction Dynamics

Topics for further volumes include time-dependent methods for collisions and numerical and computational techniques for bound states of finite systems.

It is our hope and expectation that this series will encourage and enhance the interdisciplinary activities already fostered by the organizations and meetings noted at the start of this Foreword.

PREFACE

This second volume in the series *Finite Systems and Multiparticle Dynamics* is concerned with three-body systems where the Coulomb interaction plays an important role. The subject is interdisciplinary and so is this book. Of the various areas from which examples could have been drawn, we have chosen two: nuclear physics and atomic physics, each in the low-energy regime. The selection of only two fields has led to a volume of manageable size, although to achieve this, further selectivity was obviously required.

The emphasis is on collision phenomena. Such emphasis is almost a natural one in the nuclear case, since Coulomb effects tend to be much less important for nuclear bound states than for nuclear collision states. Correspondingly, the already complicated description of collisions involving three strongly interacting particles (Faddeev-type theories) is made even more so when Coulomb interactions are added to the strong nuclear forces.

In contrast to the nuclear case, there is essentially only one type of interaction in low-energy, atomic few-body systems, *viz*, the Coulomb interaction itself. It is as important for atomic bound states as for atomic collision states, and both types of states are discussed in detail by the authors of the chapters on atomic few-body systems.

The first three chapters deal with aspects of the nuclear case. Since the interpretation of the data depends so intimately on the theoretical analysis, these chapters are all written by theorists, who provide references to relevant experimental results, although they concentrate more on the theory needed to understand experiments. There are three different ways to formulate scattering theory, and each chapter deals with a different formulation. Thus, one can work wholly within a time-dependent framework or one can use one of the two time-independent, stationary-state approaches: a momentum-space analysis employing transition operators or a coordinate-space analysis involving wave functions.

Considered first are the stationary-state formalisms and their applications to the proton–deuteron system. Chapter 1, by Erwin Alt and Werner Sandhas, accentuates the direct calculation of collision amplitudes using momentum-space integral equations and the screening method for handling Coulomb effects. These authors have pioneered this approach and present the underlying ideas, starting with the two-body case, in an effective, pedagogic style. Chapter 2, by Jim Friar and Jerry Payne, emphasizes the coordinate-space, wave function analysis. Of great importance here is the proper incorporation of the asymptotic boundary conditions. Friar and Payne have been in the forefront of those researchers who have investigated and used these methods, and their treatment, like that of Chapter 1, stresses the pedagogic aspects. In addition to analyzing aspects of p–d collisions, they also briefly consider the p–d bound state ^{3}He and also p–d fusion reactions, particularly those involving muons.

Chapter 3, by Helmut Kröger, is devoted to the third approach, that based on time-dependent methods, in particular, on the time-evolution operator and its Hilbert-space approximation, known as the strong-operator approximation. Kröger and his collaborators have developed this particular method, and the chapter has been written in a manner again stressing pedagogy, essential for a method based on time-evolution operators.

In the second part of the book, the thrust shifts not only from nuclear to atomic physics, but also to reviews of experimental as well as theoretical work. Chapter 4, by Howard Bryant and Monica Halka, is on the H^- (e^-pe^-) system, and both bound state and continuum state aspects are discussed, including, e.g., resonances seen in e^- + H collisions. The beautiful experiments in which laser beams are crossed with a relativistic H^- beam is a perfect topic for inclusion in this volume; the description by Bryant (one of the pioneers of the method) and Halka is a notably clear contribution about an exciting field.

The final chapter, by John Briggs, is a theoretical one in which both atomic and molecular three-particle systems are examined. Bound and continuum states are considered, in particular processes at energies above the three-particle breakup (ionization) threshold. Briggs and his collaborators have, in recent years, made important contributions to both the "molecular" description of atomic bound states and to the theoretical analysis of ionization events, leading to final states of the e^-pe^- and $e^-\alpha e^-$ ($\alpha = {}^4$He nucleus) systems. These are among the variety of topics which his chapter so nicely portrays.

Frank S. Levin
David A. Micha

Providence, Rhode Island
Gainesville, Florida

Contents

CHAPTER 1

COLLISION THEORY FOR TWO- AND THREE-PARTICLE SYSTEMS INTERACTING VIA SHORT-RANGE AND COULOMB FORCES

E. O. Alt and W. Sandhas

1. Introduction	1
2. Elementary Scattering Theory for Two Charged Particles	5
2.1. Radial Wave Functions for a Short-Range Potential	5
2.2. Short-Range Scattering States and Amplitudes	7
2.3. Radial Wave Functions for the Coulomb Potential	8
2.4. Coulomb Scattering States and Amplitudes	9
2.5. Radial Wave Functions for a Short-Range plus Coulomb Potential	11
2.6. The Full Scattering States and Amplitudes	12
3. Screening Method	15
3.1. Screened Coulomb Phase Shifts: S- and T-Matrices	15
3.2. Screened Coulomb Wave Functions	17
3.3. Zero-Screening Limits of the Full Phase Shifts	19
3.4. Zero-Screening Limits of the Full S- and T-Matrices	20
3.5. The Full Screened Scattering Amplitude	21
3.6. Practical Procedure	22
4. Formal Theory of Two-Body Scattering with Coulomb Distortion	23
4.1. Scattering States	24
4.2. Explicit Representations	25
4.3. S- and T-Matrices	26
4.4. Operator Lippmann–Schwinger Equation	28
4.5. Short-Range plus (Screened) Coulomb Potential	28
4.6. The Two-Potential Formula	29
4.7. Zero-Screening Limit	30

4.8. Practical Construction of $\tilde{T}^{SC}$ as Zero-Screening Limit	32
4.9. Integral Equation for T^{SC}	34
5. Three-Body AGS Formalism for Short-Range Potentials	35
5.1. Hamiltonians and Channel States	35
5.2. S-Matrices and Transition Amplitudes	38
5.3. Some More Details	40
5.4. The AGS Equations	42
5.5. Breakup Reactions	43
6. Three-Body Collision Theory for Charged Particles	45
6.1. Potentials, Hamiltonians, and Resolvents	46
6.2. The Two-Potential Formula	48
6.3. Details on the ε-Limit	49
6.4. Zero-Screening Limit	51
6.5. Practical Construction of $\tilde{T}^{SC}_{\beta m,\alpha n}$ as Zero-Screening Limit	52
6.6. Integral Equation for $U^{SC}_{\beta\alpha}$	53
6.7. Three-Body Breakup with Two Charged Outgoing Particles	56
6.8. Equivalence with the Dollard Amplitudes for Rearrangement Scattering	57
6.9. Equivalence with the Dollard Amplitudes for Breakup Processes	59
7. Effective Two-Body Formalism	61
7.1. Rearrangement Collisions	62
7.2. Breakup Reactions	65
7.3. Specialization to Proton–Deuteron Scattering	66
7.4. Justification of the Screening and Renormalization Method within the Effective Two-Body Formalism	68
8. Numerical Results for Nuclear Three-Body Systems	70
8.1. General Remarks	70
8.2. Results for the Proton–Deuteron System	73
8.3. Applications to Deuteron–Nucleus Reactions and Related Processes	88
9. Some Further Developments	88
9.1. Coulomb Effects on Analytic Properties of Particle Transfer Amplitudes	88
9.2. Long-Range Behavior of the Effective Potential between a Charged Elementary and a Charged (or Neutral) Composite Particle	89
References	90

CHAPTER 2

PROTON–DEUTERON SCATTERING AND REACTIONS

J. L. Friar and G. L. Payne

1. Introduction	97
2. Strong-Interaction Dynamics	101
3. Two-Nucleon Problem	105
3.1. Non-Coulomb Case	105
3.2. Coulomb Case	108
3.3. Exact Forms	113
4. Variational Estimates	118
5. Three-Nucleon Problem	121
5.1. Geometric Aspects	121
5.2. Boundary Conditions	123
5.3. Faddeev–Noyes Equations	125
5.4. Three-Nucleon Forces	127
5.5. Coulomb-Modified Faddeev Equations	128
6. Boundary Conditions	129
6.1. Elastic Scattering	130
6.2. Breakup Scattering	132
6.3. Boundary Conditions Redux	133
7. Coulomb Polarization Potential	135
7.1. Rayleigh Scattering	135
7.2. Polarization Potential	136
7.3. Boundary Conditions for $1/r$- plus $1/r^4$-Potentials	138
8. Practical Three-Nucleon Calculations	142
8.1. Spin–Isospin–Orbital Coupling Schemes	142
8.2. Scales	143
8.3. Three-Nucleon Channels and Partial Waves	144
8.4. Partial-Wave Faddeev Equations	146
8.5. Wave Function Normalization	149
8.6. Numerical Modeling	150
9. The ^{3}He Bound State	151
10. Nucleon–Deuteron Scattering	155
10.1. Threshold Scattering: Nucleon–Deuteron Scattering Lengths	155
10.2. Proton–Deuteron Scattering below Breakup Threshold	158
10.3. Scattering above Breakup Threshold	159

11. Proton–Deuteron Fusion Reactions 162
    11.1. Muon-Catalyzed Fusion 162
    11.2. Muon Internal Conversion 163
    11.3. Wolfenstein–Gershtein Effect 164

12. Summary and Prognosis 165

References 166

CHAPTER 3

TIME-DEPENDENT SCATTERING IN COULOMBIC FEW-BODY SYSTEMS AND THE STRONG OPERATOR APPROXIMATION METHOD

Helmut Kröger

1. Introduction 169

2. Time-Dependent Scattering Theory: One-Body Scattering . . 172
    2.1. Short-Range Potential 172
    2.2. Coulomb Potential 177

3. Coulomb-like Potentials: Scattering in Charged Two-Body and Three-Body Systems 187
    3.1. Classes of Asymptotic Situations 187
    3.2. One-Particle Scattering with Coulomb-like Potentials . . . 187
    3.3. Two-Particle Scattering with Coulomb-like Potentials . . . 188
    3.4. Three-Particle Scattering with Coulomb-like Potentials . . 189
    3.5. Wave Operator for Class $C1$ 191
    3.6. Wave Operator for Class $C2$ 191
    3.7. Wave Operator for Class $C3$ 192

4. Hilbert Space Approximation Method 193
    4.1. Construction of Finite-Dimensional Operators 193
    4.2. Approximation Parameters 197

5. Strong Resolvent Convergence 200

6. Numerical Results 207
    6.1. Proton–Proton Scattering 207
    6.2. Proton–Deuteron Breakup Scattering 212

7. Summary 216

References 216

CHAPTER 4

H⁻ SPECTROSCOPY

H. C. Bryant and M. Halka

1. Introduction 221

2. H⁻ Structure 222
    2.1. Simple Considerations 222
    2.2. Electron Correlations 225

3. Experimental Techniques 226
    3.1. Relativistic Kinematics 227
    3.2. An Experimental Application of Time Dilation 229
    3.3. Aberration of Light 229
    3.4. The Electromagnetic Tensor 230
    3.5. Application to Photoabsorption: Doppler Tuning . . . 231
    3.6. Energy Resolution 233
    3.7. Transit Time Broadening (an Illusion?) 234

4. One-Electron Detachment Thresholds 235
    4.1. Ground State Detachment in Zero Field 235
    4.2. Ground State Detachment in dc Fields 235
    4.3. $N = 2$ Threshold 240
    4.4. Higher Thresholds in Zero Field 240
    4.5. Higher Thresholds in Applied dc Fields 242
    4.6. Two-Electron Detachment Threshold 245

5. Doubly-Excited States 246
    5.1. Introduction 246
    5.2. Two Very Different Types of Resonances: Feshbach and Shape 250
    5.3. Convenient Coordinate Representations 251
    5.4. H⁻ ($n = 2$) Resonances 252
    5.5. Field Effects on H⁻ ($n = 2$) Resonances 256
    5.6. Higher n Resonances 257

6. Angular Distribution of Photoelectrons 263

7. Multiphoton Studies 264

8. Foil Transmission Studies 270

9. Summary and Outlook 272

Appendix A. Hyperspherical Coordinates 273

Appendix B. Prolate Spheroidal Coordinates 275

Appendix C. Propensity Rules 276
References 276

CHAPTER 5

COULOMB FORCES IN THREE-PARTICLE ATOMIC AND MOLECULAR SYSTEMS

J. S. Briggs

1. Introduction 281
2. Below the Three-Body Breakup Threshold 284
   2.1. The Adiabatic Approximation 285
   2.2. The Correlation Diagram 290
   2.3. Symmetries of Doubly-Excited States 296
   2.4. Nodal Structure of Doubly-Excited States 301
   2.5. The Semiclassical Helium Atom 303
3. The Three-Body Threshold—and Beyond 307
   3.1. Ionization Mechanisms near Threshold 309
   3.2. Symmetries of Electron Impact Ionization at Threshold . 312
   3.3. Three-Body Continuum Wave Functions 314
   3.4. The Triply-Differential Cross Section at Near-Threshold
        and Intermediate Energies 317
   3.5. Comparison with Experimental Data 320
   3.6. Double Photoionization 325
   3.7. The High-Energy Continuum 328
4. Conclusions 336
References 338

INDEX 341

CHAPTER 1

COLLISION THEORY FOR TWO- AND THREE-PARTICLE SYSTEMS INTERACTING VIA SHORT-RANGE AND COULOMB FORCES

E. O. ALT AND W. SANDHAS

1. INTRODUCTION

In two- and three-particle reactions with light nuclei, a rich body of precise experimental data exists in which both projectile and target and/or the fragments occurring in the final state are charged. In order to make optimal use of these data for extracting physically interesting information about the nuclear interactions, the effects of the Coulomb force must be separated out in a reliable manner. For this purpose the mastering of the intricacies of charged-particle scattering theory is of vital importance.*

The scattering of *two elementary charged particles* has been investigated from a variety of possible points of view and all of its facets can be considered as being fully understood. The basic reason is that in the pure two-body Coulomb case, the coordinate-space Schrödinger equation, as well as the corresponding radial equation, can be solved analytically. Also in momentum space all quantities of interest, such as the scattering wave

---

*The systems for which the methods described in this chapter were developed, and to which they have been applied, are those noted in the first sentence above. We point out that this methodology could also be applicable to collisions involving a pair of molecular ions, where the analog of the short-range nuclear interactions would be the molecular potential energy surfaces. Since such applications have not yet been made, the concretization of the general theory is stated in the language of nuclear physics.

E. O. ALT • Johannes-Gutenberg-Universität, Institut für Physik, 55099 Mainz, Germany.
W. SANDHAS • Physikalisches Institut der Universität Bonn, 53115 Bonn, Germany.

Coulomb Interactions in Nuclear and Atomic Few-Body Collisions, edited by Frank S. Levin and David A. Micha. Plenum Press, New York, 1996.

function, the resolvent, the off-shell T-matrix, or the (on-shell) scattering amplitude, are given explicitly. However, as is well known, due to the long range of the Coulomb potential in coordinate space, or, equivalently, due to its singular behavior in momentum space, the definitions of these quantities, and their relations among one another, do not coincide with what is known to hold for short-range potential scattering.*

For instance, neither the usual coordinate-space asymptotic condition for the scattering wave function (plane wave plus outgoing spherical wave) is satisfied, nor does the partial-wave expansion of the scattering amplitude converge. The latter, however, has been shown to exist as a distribution.[13] This, in particular, implies that the standard construction of the scattering amplitude for short-range plus Coulomb potentials, presented in most textbooks via purely formal partial-wave manipulations, has a mathematically rigorous basis. These questions are discussed in detail in Section 2.

Furthermore, the question of the validity of the Lippmann–Schwinger (LS) equation in the Coulomb case requires careful investigation.[14–22] But for short-range plus Coulomb scattering this problem can be circumvented by splitting the full amplitude into the explicitly known pure Coulomb amplitude and a Coulomb-distorted short-range part. The latter does satisfy a well-behaved LS-type equation (compare Section 4.9). Also, with some modifications, time-dependent scattering theory can be extended to the Coulombic case. As shown by Dollard,[23] the nonexistence of the Coulomb Møller operators as strong time limits in the conventional sense can be cured by choosing appropriately modified time evolution operators. In this way the scattering by Coulomb-like potentials is fully incorporated into the rigorous framework of the time-dependent theory.

Another approach to the two-body Coulomb scattering problem, which is both physically and mathematically appealing, is provided by the "screening and renormalization" method.[13,24] When the Coulomb potential is made short-range by a screening function, all techniques of short-range theory can be applied for the calculation of the corresponding "screened" scattering states and amplitudes. In the zero-screening limit, i.e., for the screening radius going to infinity, neither of these quantities has a limit. However, each can be shown to converge toward the states and amplitudes, respectively, of the unscreened theory, after being multiplied by renormalization factors which compensate for their divergencies (see Sections 3 and 4). It is the generalization of this approach which has led to the practical three-body technique reviewed in the second part of this chapter.

As compared to the two-body case, the situation for *three-charged-particle scattering* is rather different. No three-body Coulomb quantity is known explicitly. Therefore, most of the special techniques developed for

---

*These peculiarities are addressed, at least partly, in textbooks on quantum mechanical scattering theory such as References (1–10), and in considerable detail in van Haeringen[11] (see also Chen and Chen[12]).

two charged particles cannot be taken over directly. Rather, only the general concepts and strategies can be expected to have their three-body extensions.

A generalization of the two-body asymptotic condition in position space has been proposed, e.g., by Merkuriev.[25] This approach is not incorporated in our presentation as it has already been reviewed, though fairly sketchily, in a monograph.[26] Furthermore, several attempts to generalize the above-mentioned idea of deriving well-defined LS equations, if not for the full then at least for Coulomb-modified scattering states or amplitudes, have been made. They are based on splitting off either the full or the center-of-mass Coulomb potential from the total interaction between the colliding fragments. We describe some of these attempts in Section 6.

Dollard's time-dependent theory[23] applies to arbitrary multichannel collisions, and thus also covers the three-body case. Though primarily designed as a mathematically rigorous basis, it has also been used for numerical construction of charged-particle scattering amplitudes. This application is presented in Chapter 3 of this book (see also Kröger[27]). We recall its underlying idea only when showing the equivalence of Dollard's definitions of transition amplitudes with those following from the alternative screening method.

The first reliable three-nucleon results with Coulomb distortion were obtained by means of the screening and renormalization method of Alt, Sandhas, and Ziegelmann (ASZ).[28] It is based on momentum-space integral equations with screened, i.e., short-range Coulomb potentials, the screening radius being successively increased until the resulting, appropriately renormalized, solutions reach their zero-screening limits. One of the decisive advantages of this procedure is that all successful practical methods of modern three-body theory for short-range potentials can immediately be applied to it. In this context the Alt–Grassberger–Sandhas (AGS) equations,[29] especially their reduction to effective two-body equations, turn out to be particularly well suited. In our presentation of the three-charged-particle theory we therefore concentrate primarily on this approach. Our presentation of the subject is as follows.

In Section 2 we describe the basic aspects of the standard coordinate-space scattering theory of two particles interacting via a short-range plus Coulomb potential. Recalling this elementary approach enables us to exhibit the characteristic features and complications of the Coulomb scattering problem in an explicit manner.

In Section 3 we introduce the important concept of the "screening and renormalization" method within the coordinate-space formalism. This allows us to explicitly read off the renormalization factors, which are needed when going over from the screened to the unscreened quantities. All these results have been proved in the literature on a rigorous, though sometimes rather formal basis. We have made an effort to demonstrate their validity in a manner which is convincing without being buried under all of the technical details of a mathematically exact consideration.

The screening approach is translated in Section 4 into the language of formal scattering theory. The procedure chosen emphasizes the algebraic structures and singularity properties of the relevant resolvent and T-operator relations more pointedly than is usually done in textbooks. Such an access to formal scattering theory is of interest in itself, and is especially well suited to the extension of the two-body to the general few-body problem.

As a kind of intermediate step we recall in Section 5 the AGS three-body theory.[29] In doing so the main emphasis is on working out the structural similarity of the relevant three-body to the corresponding two-body relations,[30] which makes this approach a natural starting point for generalizing two-body techniques to three-body scattering in an elegant and structurally transparent manner.

The extension along this line of the two-body "screening and renormalization" method to three-charged-particle scattering is presented in Section 6. It is based on the splitting of the channel interaction into a (screened) center-of-mass Coulomb potential, which acts between the total charges of the colliding fragments, and the short-range plus remaining Coulomb parts. In this way the appropriate renormalization factors, which guarantee the existence of the zero-screening limits of the correspondingly renormalized amplitudes, are found.[31] This investigation is carried through both for rearrangement and breakup scattering (in the latter case, however, only for one neutral and two charged particles). The equivalence of the amplitudes obtained hereby with the ones following from Dollard's time-dependent theory is shown.[32]

In Section 7 the conversion of the general theory into a practical integral equation approach for determining the screened amplitudes is performed. There we recapitulate the AGS reduction of the original three-body relations into equivalent effective-two-body LS equations, thereby recalling the original formulation of the ASZ screening and renormalization approach.[28,33] As a special example we consider proton–deuteron elastic and breakup scattering.

In Section 8 we first discuss qualitatively some of the merits and disadvantages of the coordinate- and the momentum-space formulations of three-charged-particle scattering as proposed and applied up to now. Then we summarize typical results which have been obtained for the proton–deuteron system by using the effective two-body formulation of the theory described above. This includes scattering lengths, elastic scattering phase parameters, and differential cross sections, as well as breakup cross sections. At the end, attention is directed to applications to other nuclear three-body problems.

Finally, in Section 9, some new developments which shed additional light on the pecularities of charged-composite-particle scattering are briefly touched upon.

We note that part of this material, as well as other approaches to charged-composite-particle scattering not taken up here, is the subject of several review articles.[34-36]

2. ELEMENTARY SCATTERING THEORY FOR TWO CHARGED PARTICLES

In this introductory section we review the conventional coordinate-space approach to the scattering of two particles interacting via a Coulomb-like potential. In particular, we recall the partial-wave construction of the scattering states and amplitudes first for the pure short-range, then the pure Coulomb, and finally the combined case. This allows us to exhibit the characteristic complications of the Coulomb scattering problem explicitly, e.g., the peculiar asymptotic behavior of the wave functions or the problems of convergence of the partial wave series. This latter important point will be discussed more thoroughly than is done in most textbooks. Furthermore, we introduce the explicit representations of the two-body Coulomb states and amplitudes needed throughout the following.

2.1. Radial Wave Functions for a Short-Range Potential

In the elementary coordinate-space approach[1-10,37] scattering states for a local, central potential are constructed as superpositions of the solutions of the radial Schrödinger equation (μ is the reduced mass, p the relative momentum, and r the relative coordinate):

$$-u''_{pl}(r) + \left[\frac{l(l+1)}{r^2} + 2\mu V(r)\right] u_{pl}(r) = p^2 u_{pl}(r) \qquad (1.2.1)$$

Here and in what follows natural units are chosen, i.e., $\hbar = c = 1$.

Let us first discuss the solution of this equation for a short-range potential $V(r) = V^S(r)$. To avoid difficulties which are not relevant in the present context, $V^S(r)$ is assumed to be continuous and less singular at the origin than r^{-2}. When solving (1.2.1) we therefore can impose the regularity condition

$$u_{pl}(r) \underset{r \to 0}{\sim} \alpha_l r^{l+1} \qquad (1.2.2)$$

which makes (1.2.1) a well-defined eigenvalue equation of a self-adjoint Hamiltonian. That is, this condition restricts the infinity of solutions of (1.2.1) to the physically relevant ones.

For potentials of finite range, which vanish identically for $r \geq r_0$, the general solution of (1.2.1) is given in the external region as a superposition of two linearly independent solutions of the free Schrödinger equation

$$u_{pl}(r) = A_l[\cos \delta_l \hat{j}_l(pr) - \sin \delta_l \hat{n}_l(pr)] \qquad \text{for} \quad r \geq r_0 \qquad (1.2.3)$$

chosen here as the Riccati–Bessel and the Riccati–Neumann functions $\hat{j}_l$ and $\hat{n}_l$. Due to the regularity condition (1.2.2) the solution of (1.2.1) is fixed except for an overall multiplicative factor. The constant δ_l introduced in (1.2.3) is thus uniquely determined (modulo π). In practice, it can be calculated by means of the logarithmic derivative $\gamma_l(r) = p^{-1} u'_{pl}(r)/u_{pl}(r)$, whose continuity at the matching point r_0 implies the well-known relation

$$\tan \delta_l = \frac{\hat{j}'_l(pr_0) - \gamma_l(r_0)\hat{j}_l(pr_0)}{\hat{n}'_l(pr_0) - \gamma_l(r_0)\hat{n}_l(pr_0)} \qquad (1.2.4)$$

The overall factor in $u_{pl}(r)$ may be fixed by prescribing the constants either in (1.2.2) or in (1.2.3). The choice $\alpha_l = 1$, e.g., defines the "regular solution," while $A_l = \exp(i\delta_l)$ yields the "physical solution." Taking into account the asymptotic behavior $\hat{j}_l(x) \sim \sin(x - l\pi/2)$ and $\hat{n}_l(x) \sim (-)\cos(x - l\pi/2)$ for $x \gg l$, the physical solution is seen to be characterized by the asymptotic condition

$$u_{pl}(r) \underset{r \to \infty}{\sim} e^{i\delta_l} \sin(pr - l\pi/2 + \delta_l) \qquad (1.2.5)$$

For potentials of short but not finite range, the value r, beyond which the motion of the particles can be considered as being practically free, is somewhat arbitrary. This suggests introducing the "variable phase" $\delta_l(r)$ by keeping the matching point open, i.e., by replacing the fixed value r_0 in (1.2.4) by the variable r. For such large values of r that the action of $V^S(r)$ can be neglected, $\delta_l(r)$ goes over into the true phase shift δ_l.

Instead of determining $\delta_l(r)$ via the logarithmic derivative of the regular solution of the radial Schrödinger equation, we can obtain it by solving the first-order, albeit nonlinear, differential equation[38]

$$\delta'_l(r) = -\frac{1}{p} 2\mu V^S(r)[\cos \delta_l(r) \hat{j}_l(pr) - \sin \delta_l(r) \hat{n}_l(pr)]^2 \qquad (1.2.6)$$

This relation represents a particularly simple means for calculating the phase shifts. Moreover, its integrated form allows us to check that the limit

$$\delta_l(r) \xrightarrow[r \to \infty]{} \delta_l$$

exists, provided $V^S(r)$ decreases for larger r more rapidly than the Coulomb potential. The validity of the asymptotic condition (1.2.5) is thus guaranteed even for potentials of fairly long, but slightly less than Coulombic range.

2.2. Short-Range Scattering States and Amplitudes

The (three-dimensional) scattering states $\psi_{\mathbf{p}}^{(+)}(\mathbf{r})$ are well known to be given by the partial-wave summation[1–10]

$$\psi_{\mathbf{p}}^{(+)}(\mathbf{r}) = \frac{1}{pr} \sum_{l=0}^{\infty} i^l (2l+1) u_{pl}(r) P_l(\cos \vartheta) \qquad (1.2.7)$$

of the physical solutions $u_{pl}(r)$. Here, $P_l(\cos \vartheta)$ are the Legendre polynomials, ϑ being the angle between $\mathbf{r}$ and $\mathbf{p}$. Making use of (1.2.5) one easily shows that (1.2.7) satisfies the conventional asymptotic condition

$$\psi_{\mathbf{p}}^{(+)}(\mathbf{r}) \underset{r \to \infty}{\sim} e^{i\mathbf{p}\cdot\mathbf{r}} + f(\vartheta) \frac{e^{ipr}}{r} \qquad (1.2.8)$$

with the scattering amplitude $f(\vartheta)$ being given by

$$f(\vartheta) = \frac{1}{p} \sum_{l=0}^{\infty} (2l+1) t_l P_l(\cos \vartheta) \qquad (1.2.9)$$

The partial-wave T-matrix $t_l = t_l(p)$ is related to the phase shift and the partialwave-S-matrix s_l via

$$t_l = \frac{e^{2i\delta_l} - 1}{2i} = \frac{s_l - 1}{2i} \qquad (1.2.10)$$

The representation (1.2.7) not only provides a practical means for constructing the scattering states, but also constitutes a simple proof that, as characterized by the asymptotic condition (1.2.8), they do in fact exist. This is true at least for potentials which decrease asymptotically as $r^{-3-\varepsilon}$, with $\varepsilon > 0$. The phase shifts can then be shown[10,37] to fall off as $l^{-2-\varepsilon}$ for $l \to \infty$, which implies the convergence of the sum in (1.2.9) for all angles.

In the purely nuclear two-body problem the interaction decreases exponentially for sufficiently large r. The effective interactions between composite particles, on the other hand, contain polarization terms which may decrease[4-6] only as $1/r^4$, i.e., only slightly more rapidly than required above. This gives at least an indication that partial-wave expansions will also converge in the two-cluster collisions studied later on, although much more slowly than in two-nucleon scattering.

2.3. Radial Wave Functions for the Coulomb Potential

Let us now consider the solutions of (1.2.1) for $V(r) = V^C(r) = e_1 e_2/r$, with e_1 and e_2 being the charges of the colliding particles. As noted earlier, in this case the variable phase $\delta_l(r)$ cannot be expected to converge toward a constant phase shift. On the contrary, Eq. (1.2.6) indicates that $\delta_l(r)$ diverges logarithmically for $r \to \infty$.

This expectation is confirmed by looking at the asymptotic behavior of the radial wave functions, which are explicitly known for the Coulomb potential. The solutions $F_{pl}(r)$ of (1.2.1) subject to the regularity condition (1.2.2) are given by the confluent hypergeometric function $_1F_1(a|b|x)$:

$$F_{pl}(r) = c_l(\eta) e^{ipr} (pr)^{l+1} {}_1F_1(l + 1 + i\eta|2l + 2|-2ipr) \qquad (1.2.11)$$

with the Coulomb parameter $\eta = \mu e_1 e_2/p$. If one chooses the arbitrary constant c_l in this expression as [$\Gamma(\cdot)$ is the gamma function]

$$c_l(\eta) = \frac{2^l e^{-\frac{1}{2}\pi\eta} |\Gamma(l + 1 + i\eta)|}{(2l + 1)!} \qquad (1.2.12)$$

then asymptotically[5,6,11]

$$F_{pl}(r) \underset{r \to \infty}{\sim} \sin[pr - l\pi/2 + \sigma_l - \eta \ln(2pr)] \qquad (1.2.13)$$

where

$$\sigma_l = \arg \Gamma(l + 1 + i\eta) \qquad (1.2.14)$$

is conventionally called the Coulomb phase shift. As compared to (1.2.5) the whole (variable) phase in (1.2.13),

$$\delta_l(r) = \sigma_l - \eta \ln(2pr) \qquad (1.2.15)$$

shows the expected logarithmic r-dependence. In other words, the Coulomb solution never becomes asymptotically free.

The analog of the "physical solution" (1.2.5) of the short-range case is defined by, and behaves asymptotically as

$$u_{pl}^C(r) = e^{i\sigma_l} F_{pl}(r) \underset{r \to \infty}{\sim} e^{i\sigma_l} \sin[pr - l\pi/2 + \sigma_l - \eta \ln(2pr)] \quad (1.2.16)$$

2.4. Coulomb Scattering States and Amplitudes

With the radial solutions (1.2.16) the three-dimensional Coulomb scattering wave function is introduced as in (1.2.7):

$$\psi_{\mathbf{p}}^{C(+)}(\mathbf{r}) = \frac{1}{pr} \sum_{l=0}^{\infty} i^l (2l+1) u_{pl}^C(r) P_l(\cos \vartheta) \quad (1.2.17)$$

On using (1.2.16), it is found that this expression behaves asymptotically as

$$\psi_{\mathbf{p}}^{C(+)}(\mathbf{r}) \underset{r \to \infty}{\sim} \psi_{\mathbf{p}}^{(i)} + f^C(\vartheta) \frac{1}{r} \exp\{i[pr - \eta \ln(2pr)]\} \quad (1.2.18)$$

i.e., roughly as (1.2.8). However, the outgoing spherical wave is modified by the typical logarithmic term in the Coulomb asymptotics (1.2.13), and the (asymptotic) incident part

$$\psi_{\mathbf{p}}^{(i)}(\mathbf{r}) = \frac{1}{2ipr} \sum_{l=0}^{\infty} (-1)^{l+1} (2l+1) \exp\{-i[pr - \eta \ln(2pr)]\} P_l(\cos \vartheta) \quad (1.2.19)$$

evidently cannot be summed up to a plane wave.

The scattering amplitude, on the other hand, is practically of the form (1.2.9),

$$f^C(\vartheta) = \frac{1}{p} \sum_{l=0}^{\infty} (2l+1) \frac{1}{2i} s_l^C P_l(\cos \vartheta) \quad (1.2.20)$$

with the only difference that now the partial-wave S-matrix

$$s_l^C = \exp(2i\sigma_l) \quad (1.2.21)$$

divided by $2i$ occurs instead of the corresponding partial-wave T-matrix:

$$t_l^C = \frac{s_l^C - 1}{2i} \quad (1.2.22)$$

Although in principle quite remarkable, this difference is of no relevance in the present context. In fact, the additional term $-1/2i$ would yield in the sum (1.2.20) only a δ-function specifying the forward direction:

$$\sum_{l=0}^{\infty} (2l + 1)P_l(\cos\vartheta) = 2\delta(1 - \cos\vartheta)$$

(this is a special case of the completeness relation of the Legendre polynomials). At $\vartheta = 0$, however, $f^C(\vartheta)$ is not defined (see below). In other words, for $\vartheta \neq 0$ the representation (1.2.20) could equally well be written with t_l^C in place of $s_l^C/2i$.

It is one of the characteristic features of the Coulomb problem that the regular solutions not only of the radial but also of the three-dimensional Schrödinger equation can be represented explicitly by means of the confluent hypergeometric function $_1F_1$. In most introductions to the Coulomb scattering problem this three-dimensional wave function,

$$\psi_\mathbf{p}^{C(+)}(\mathbf{r}) = \Gamma(1 + i\eta)e^{-\frac{1}{2}\pi\eta}e^{i\mathbf{p}\cdot\mathbf{r}}{}_1F_1(-i\eta|1|i(pr - \mathbf{p}\cdot\mathbf{r})) \qquad (1.2.23)$$

is chosen as the starting point. Using the asymptotic properties of $_1F_1$, one finds[5,6,11]

$$\psi_\mathbf{p}^{C(+)}(\mathbf{r}) \xrightarrow[r|1-\cos\vartheta|\to\infty]{} \exp\{i[\mathbf{p}\cdot\mathbf{r} + \eta\ln(pr - \mathbf{p}\cdot\mathbf{r})]\}$$

$$+ f^C(\vartheta)\frac{1}{r}\exp\{i[pr - \eta\ln(2pr)]\} \qquad (1.2.24)$$

with

$$f^C(\vartheta) = -\frac{\eta}{2p\sin^2\vartheta/2}\exp[-i\eta\ln(\sin^2\vartheta/2) + 2i\sigma_0] \qquad (1.2.25)$$

This asymptotic behavior is evidently equivalent to (1.2.18). In the present formulation, however, the partial-wave summations of both the incident wave (1.2.19) and, more importantly, of the Coulomb scattering amplitude (1.2.20) are expressed as simple functions. It should be noted that, starting from (1.2.23) and (1.2.25), the equivalence of these expressions with the representations (1.2.17) and (1.2.20) can be verified directly via formal use of standard Legendre polynomial expansions, a procedure conventionally adopted in the literature.

For a mathematically rigorous proof of equivalence it is, of course, essential to clarify the sense in which the convergence of the partial-wave series is to be understood. This question has been investigated, in particular, by Taylor,[13] and the following discussion is based upon that work. Let us concentrate on the amplitude. Using properties of the Γ-function, one derives $\sigma_l - \sigma_{l-1} = \arctan \eta/l$. This shows that the Coulomb phase shifts diverge as

$$\sigma_l \underset{l \to \infty}{\sim} \eta \ln l \tag{1.2.26}$$

a result not unexpected in view of the connection between the large-r behavior of the potential and the large-l behavior of δ_l discussed after Eq. (1.2.10). Consequently, $\exp(2i\sigma_l)$ moves around the unit circle indefinitely as l increases. Moreover, since $(2l + 1)P_l(\cos \vartheta)$ diverges as $l^{1/2}$ for $\varepsilon \leqslant \vartheta \leqslant \pi - \varepsilon$, $\varepsilon > 0$, and even grows as l for $\vartheta = 0$ and π, it is plausible, and can be shown rigorously,[13] that the partial-wave series (1.2.20) diverges for all angles, and not only in the forward direction where the amplitude (1.2.25) is itself infinite. However, after integration over smooth functions of ϑ, the partial-wave series has been shown to converge.[13] In this distribution theoretical sense it is therefore justified to write

$$f^C(\vartheta) = \frac{1}{2ip} \sum_{l=0}^{\infty} (2l + 1) \exp(2i\sigma_l) P_l(\cos \vartheta)$$

$$= -\frac{\eta}{2p \sin^2 \vartheta/2} \exp[-i\eta \ln(\sin^2 \vartheta/2) + 2i\sigma_0] \tag{1.2.27}$$

A thorough investigation shows that the partial-wave expansion (1.2.17) of (1.2.23) is also to be understood as a distribution,[11,39] at least asymptotically.

As long as we restrict ourselves to the pure Coulomb case, where all relevant three-dimensional quantities are known explicitly, it appears irrelevant that they can be represented as partial-wave series. However, this possibility will turn out to be crucial when constructing the amplitude for scattering by a Coulomb plus short-range potential.

2.5. Radial Wave Functions for a Short-Range plus Coulomb Potential

We now consider the collision of two elementary particles interacting via $V(r) = V^S(r) + V^C(r)$. For simplicity we assume again $V^S(r)$ to be of finite range, i.e., to vanish identically for $r \geqslant r_0$. In this external region the particles move exclusively under the influence of the Coulomb potential, which means that the solutions of (1.2.1) can be represented as superpositions of two linearly independent Coulomb wave functions:

$$u_{pl}(r) = A_l[\cos \delta_l^{SC} F_{pl}(r) + \sin \delta_l^{SC} G_{pl}(r)] \quad \text{for} \quad r \geqslant r_0 \tag{1.2.28}$$

Here we have especially chosen to work with the regular solution (1.2.11) and the irregular solution, denoted conventionally by $G_{pl}(r)$, which behaves asymptotically as

$$G_{pl}(r) \underset{r \to \infty}{\sim} \cos[pr - l\pi/2 + \sigma_l - \eta \ln(2pr)] \tag{1.2.29}$$

In analogy to the treatment in Section 2.1, the coefficients multiplying F_{pl} and G_{pl} have been written as the cosine and sine of the "Coulomb-modified short-range phase shift" δ_l^{SC}. The latter is determined by the analog of (1.2.4):

$$\tan \delta_l^{SC} = \frac{F'_{pl}(r_0) + \gamma_l(r_0) F_{pl}(r_0)}{G'_{pl}(r_0) + \gamma_l(r_0) G_{pl}(r_0)} \tag{1.2.30}$$

where $\gamma_l(r)$ is again the logarithmic derivative of the regular solution of Eq. (1.2.1), which now however takes into account the action of both V^S and V^C in the inner region. Note that the same value of $\gamma_l(r_0)$ would be obtained for V^S plus a Coulomb potential $V^C(r)\Theta(r_0 - r)$ cut off at $r = r_0$. The fact that $V_C(r)$ also acts in the external region is taken care of in (1.2.30) by the Coulomb functions F_{pl} and G_{pl}. The Coulomb-modified short-range phase shift therefore arises primarily from the interplay of the short-range and the Coulomb potential in the internal region, although the matching of the internal motion to a Coulombic rather than a free wave is also reflected in them.

Making use of (1.2.13) and (1.2.29), we infer from (1.2.28) that $u_{pl}(r)$ behaves asymptotically as

$$u_{pl}(r) \underset{r \to \infty}{\sim} A_l \sin[pr - l\pi/2 + \sigma_l + \delta_l^{SC} - \eta \ln(2pr)] \tag{1.2.31}$$

That is, $u_{pl}(r)$ shows the typical Coulomb asymptotics (1.2.13), with σ_l being replaced by the total phase shift

$$\delta_l = \sigma_l + \delta_l^{SC} \tag{1.2.32}$$

2.6. The Full Scattering States and Amplitudes

All considerations concerning construction and asymptotic behavior of the full (three-dimensional) scattering states can be taken over from the pure Coulomb case almost without any modification. Choosing $A_l = \exp(i\delta_l)$ in (1.2.31) we introduce the "physical solution" $u_{pl}(r)$ characterized by the asymptotic behavior

$$u_{pl}(r) \underset{r \to \infty}{\sim} e^{i\delta_l} \sin[pr - l\pi/2 + \delta_l - \eta \ln(2pr)] \qquad (1.2.33)$$

This allows us to define in complete analogy to (1.2.7) or (1.2.17) the full scattering states

$$\psi_{\mathbf{p}}^{(+)}(\mathbf{r}) = \frac{1}{pr} \sum_{l=0}^{\infty} i^l (2l+1) u_{pl}(r) P_l(\cos \vartheta) \qquad (1.2.34)$$

Employing (1.2.33) we reproduce the asymptotic form (1.2.18), with the scattering amplitude being given as in (1.2.20), viz.,

$$f(\vartheta) = \frac{1}{p} \sum_{l=0}^{\infty} (2l+1) \frac{1}{2i} s_l P_l(\cos \vartheta) \qquad (1.2.35)$$

The only difference is that the pure Coulomb phase shift in the S-matrix (1.2.21) is now replaced by the full phase shift,

$$s_l = \exp(2i\delta_l) = \exp[2i(\sigma_l + \delta_l^{SC})] \qquad (1.2.36)$$

According to (1.2.26), the pure Coulomb contribution σ_l in $\delta_l = \sigma_l + \delta_l^{SC}$ diverges logarithmically for $l \to \infty$. As discussed in the context of $f^C(\vartheta)$, the right-hand side of (1.2.35) can therefore exist only as a distribution.[24] To see this in more detail we write (1.2.36) in the form

$$s_l = s_l^C + 2it_l^{SC} s_l^C \qquad (1.2.37)$$

Here, s_l^C is the pure Coulomb S-matrix defined in (1.2.21), while

$$t_l^{SC} = \frac{\exp(2i\delta_l^{SC}) - 1}{2i} \qquad (1.2.38)$$

represents the T-matrix associated with the phase shift δ_l^{SC}. Inserting (1.2.37) in (1.2.35) we obtain the splitting

$$f(\vartheta) = f^C(\vartheta) + f^{SC}(\vartheta) \qquad (1.2.39)$$

of the full amplitude into the Coulomb amplitude $f^C(\vartheta)$, given by the partial-wave series (1.2.20), and the Coulomb-modified short-range amplitude f^{SC} defined by the series

$$f^{SC}(\vartheta) = \frac{1}{p} \sum_{l=0}^{\infty} (2l + 1) t_l^{SC} s_l^C P_l(\cos \vartheta) \qquad (1.2.40)$$

which converges as in the short-range case (see below).

The distribution-theoretical convergence of (1.2.20) toward the explicitly known expression (1.2.25) thus provides us with

$$f(\vartheta) = -\frac{\eta}{2p \sin^2 \vartheta/2} \exp[-i\eta \ln(\sin^2 \vartheta/2) + 2i\sigma_0] + f^{SC}(\vartheta) \qquad (1.2.41)$$

This important equation represents the basis of the conventional treatment of Coulomb distortion in nuclear reaction theory: solving the radial Schrödinger equation in the interior region, one obtains, e.g., via (1.2.30) the phase shifts δ_l^{SC} and then, by means of (1.2.40), the amplitude $f^{SC}(\vartheta)$ which, when inserted in (1.2.41), yields the full scattering amplitude.

It is worth pointing out that proceeding along this line requires about the same numerical effort as in the pure short-range case. All Coulomb modifications in the calculation of δ_l^{SC} are easily taken into account by employing the explicitly known functions F_{pl} and G_{pl} rather than the free solutions $\hat{j}_l$ and $\hat{n}_l$. In the final step, leading to $f(\vartheta)$ via (1.2.41), only the analytically given Coulomb amplitude $f^C(\vartheta)$ is needed.

Up to now we have restricted ourselves to potentials $V_S(r)$ which vanish identically for $r \geq r_0$. The above conclusions, however, remain valid under the much weaker condition

$$V^S(r) \underset{r \to \infty}{\sim} r^{-3-\varepsilon}$$

already assumed in the pure short-range case (ε can be any positive number). By keeping the matching point r_0 in (1.2.30) unspecified, we may introduce as in Section 2.1 a variable phase $\delta_l^{SC}(r)$ which approaches δ_l^{SC} for $r \to \infty$. Moreover, it can be shown[24,40] that

$$|t_l^{SC} s_l^C| = |t_l^{SC}| \underset{l \to \infty}{\sim} l^{-2-\varepsilon} \qquad (1.2.42)$$

which justifies the partial-wave construction (1.2.40) of $f^{SC}(\vartheta)$ [cf. also Reference (41)].

In practice this construction is most easily carried out for potentials of finite or rather short range since then, because of (1.2.42), only a few of the

t_l^{SC} differ essentially from zero. The method is, in other words, particularly well suited for proton–proton scattering or the collision of tightly bound nuclei with their short-range (effective) interactions. For loosely bound systems, e.g., the deuteron in proton–deuteron collisions, there are long-range polarization effects. In the few-body theory developed later on, the analog of the above construction of f^{SC} therefore requires in general the incorporation of quite a large number of partial-wave contributions. However, convergence, albeit slow, is also to be expected in this case, provided that the polarization potential decreases as r^{-4}, a behavior usually assumed or shown under some approximations (see, e.g., McDowell and Coleman[5] and Mott and Massey[6] and in particular Section 9.2).

3. SCREENING METHOD

The approach recapitulated in the preceding section is an extension of standard short-range scattering theory to the Coulomb problem. It is conceptually transparent and simple to use in practical applications. On the other hand, the special coordinate-space technique employed in this context is essentially restricted to the two-body case. When going over to multiparticle collisions, where alternative short-range methods, e.g., momentum-space integral equations, have turned out to be particularly efficient, a more general approach to the Coulomb problem is needed.

Such an approach is provided by the screening method. The relevant observables appear therein as zero-screening limits of the corresponding quantities obtained for screened Coulomb potentials. Although designed for general composite-particle reactions, the ingredients of this method can be transparently exhibited within the elementary two-body formalism.[13,24]

3.1. Screened Coulomb Phase Shifts: S- and T-Matrices

For convenience in what follows exponential screening is used almost exclusively. That is, our screened Coulomb potential is given by

$$V^R(r) = \frac{e_1 e_2}{r} e^{-r/R} = V^C(r) e^{-r/R} \tag{1.3.1}$$

For such a short-range potential the formalism of Section 2.1 is valid. According to (1.2.5), the asymptotic form of the physical solution is

$$u_{pl}^R(r) \underset{r \to \infty}{\sim} e^{i\sigma_l^R} \sin(pr - l\pi/2 + \sigma_l^R) \tag{1.3.2}$$

the R-dependent phase shift being denoted by σ_l^R. Comparison of the arguments of the sine in (1.3.2) and in (1.2.16) shows that for $R \to \infty$, i.e., in the zero-screening limit, σ_l^R will not simply go over into the Coulomb phase shift σ_l. What is to be expected is a behavior

$$\sigma_l^R \underset{R \to \infty}{\sim} \sigma_l + \phi_R \tag{1.3.3}$$

with ϕ_R diverging logarithmically. A careful analysis of this limit yields, in fact, the asymptotic representation (1.3.3) with[13]

$$\phi_R = -\eta[\ln(2pR) - C] + 0(R^{-1}) \tag{1.3.4}$$

Here, $C = 0.5772...$ is the Euler number. It is important to realize that Eq. (1.3.3) holds for all σ_l^R with the same l-independent ϕ_R.

We emphasize that the special form of ϕ_R depends on the screening function. The result (1.3.4) is characteristic for exponential screening. For a screening function $\theta(R - r)$, i.e., for a sharp cut-off, ϕ_R equals $-\eta \ln(2pR)$, a result which can immediately be read off by comparing (1.3.2) and (1.2.16) at $r = R$. For completeness we note the general formula[13]

$$\phi_R = -\eta \int_{(2p)^{-1}}^{\infty} dr \, (1/r) g_R(r) \tag{1.3.5}$$

valid for fairly arbitrary screening functions $g_R(r)$ in $V^R(r) = V^C(r) g_R(r)$.

From (1.3.3) we infer that the screened partial-wave S-matrix, multiplied by the inverse of the "renormalization factor"

$$Z_R = e^{2i\phi_R} \tag{1.3.6}$$

goes over into the Coulomb S-matrix (1.2.21) according to

$$s_l^R Z_R^{-1} = e^{2i\sigma_l^R} Z_R^{-1} \xrightarrow[R \to \infty]{} e^{2i\sigma_l} = s_l^C \tag{1.3.7}$$

The scattering amplitude (1.2.9) reads in the present case

$$f^R(\vartheta) = \frac{1}{p} \sum_{l=0}^{\infty} (2l + 1) t_l^R P_l(\cos \vartheta) \tag{1.3.8}$$

with the partial-wave T-matrix being given by

$$t_l^R = \frac{s_l^R - 1}{2i} \quad (1.3.9)$$

The presence of the term $-1/2i$ in t_l^R prevents $t_l^R Z_R^{-1}$ from approaching a well-defined limit for $R \to \infty$. However, as discussed in the context of (1.2.22), this term is irrelevant in the scattering amplitude as long as we restrict ourselves to scattering angles $\vartheta \neq 0$. For such angles the validity of the limit (1.3.7), together with (1.2.27), thus implies

$$\lim_{R \to \infty} f^R(\vartheta) Z_R^{-1} = \frac{1}{p} \sum_{l=0}^{\infty} (2l + 1) \frac{1}{2i} \left[\lim_{R \to \infty} s_l^R Z_R^{-1} \right] P_l(\cos \vartheta)$$

$$= \frac{1}{p} \sum_{l=0}^{\infty} (2l + 1) \frac{1}{2i} s_l^C P_l(\cos \vartheta)$$

$$= f^C(\vartheta) \quad (1.3.10)$$

It was pointed out in Section 2.4 that the sum over the unscreened S-matrices on the right-hand side of (1.3.10) converges only as a distribution. The present argumentation therefore guarantees the existence of the zero-screening limit of $f^R(\vartheta) Z_R^{-1}$, at least in the distribution-theoretical sense. A more careful analysis, which avoids the interchange of limit and summation typical for (1.3.10), shows that for exponential screening, and for a somewhat wider class of smooth screening functions, the limit

$$\lim_{R \to \infty} f^R(\vartheta) Z_R^{-1} = f^C(\vartheta) \quad \text{for} \quad \vartheta \neq 0 \quad \text{and} \quad p \neq 0 \quad (1.3.11)$$

also exists in the ordinary sense.[42,43] This is an additional argument in favor of the convenient exponential screening.[44-48]

3.2. Screened Coulomb Wave Functions

Besides the behavior of $f^R(\vartheta)$, the zero-screening property of the wave function $u_{pl}^R(r)$ is of considerable relevance for the general scattering formalism developed in the following. As will be seen later, this limit of $u_{pl}^R(r)$ has to be performed in expressions of the type

$$\int_0^\infty \varphi(r) V^S(r) u_{pl}^R(r) \, dr = \int_0^\infty \chi(r) u_{pl}^R(r) \, dr \quad (1.3.12)$$

In other words, what we are interested in is the distribution-theoretical limit of $u_{pl}^R(r)$, the test functions $\chi(r)$ being either of short or finite range depending on the range of $V^S(r)$.

Let us restrict ourselves for the moment to potentials V^S of finite range r_0, and a screened Coulomb potential V^R with a sharp cut-off at $r = R$, where $R > r_0$. The integral (1.3.12) therefore extends to $r = r_0$ only. Within this region $V^R(r)$ equals $V^C(r)$, and hence $u_{pl}^R(r)$ differs from $u_{pl}^C(r)$ only by a multiplicative factor corresponding to the different values of the constants α_l^R and α_l^C in the regularity condition (1.2.2). This implies that

$$\frac{\int_0^\infty \chi(r) u_{pl}^R(r)\, dr}{\int_0^\infty \chi(r) u_{pl}^C(r)\, dr} = \frac{\alpha_l^R}{\alpha_l^C} \tag{1.3.13}$$

On the other hand, in the physical solutions the overall factors are fixed by the corresponding asymptotic conditions. The quotient α_l^R/α_l^C is, therefore, given by the quotient of $u_{pl}^R(r)$ and $u_{pl}^C(r)$ at the value $r = R$ chosen so large that these functions have assumed their asymptotic forms (1.3.2) and (1.2.16), respectively, i.e.,

$$\frac{\alpha_l^R}{\alpha_l^C} = \frac{e^{i\sigma_l^R}\sin(pR - l\pi/2 + \sigma_l^R)}{e^{i\sigma_l}\sin[pR - l\pi/2 + \sigma_l - \eta\ln(2pR)]} \tag{1.3.14}$$

As noted after (1.3.4), in the simple cut-off case σ_l^R behaves asymptotically as $\sigma_l - \eta\ln(2pR)$. Hence, the relation (1.3.14) reduces to

$$\frac{\alpha_l^R}{\alpha_l^C} \underset{R\to\infty}{\sim} e^{-i\eta\ln(2pR)} = e^{i\phi_R} = Z_R^{1/2} \tag{1.3.15}$$

which implies for the zero-screening limit of (1.3.13)

$$\lim_{R\to\infty} \int_0^\infty \chi(r) u_{pl}^R(r)\, dr\, Z_R^{-1/2} = \int_0^\infty \chi(r) u_{pl}^C(r)\, dr \tag{1.3.16}$$

(for another derivation of this result see van Haeringen[11]).

The above considerations were based on a sharp cut-off and test functions $\chi(r)$ of finite domain. It is almost evident however that (1.3.16) is not restricted to these special cases. In particular, this limit exists for exponential screening, the only difference being that in the renormalization factor Z_R, the function ϕ_R has to be chosen according to (1.3.4).

We finally note that the above result implies an analogous distribution-theoretical limit,

$$\psi_p^{R(+)} Z_R^{-1/2} \xrightarrow[R \to \infty]{} \psi_p^{C(+)} \qquad (1.3.17)$$

for the screened three-dimensional scattering states.[49-51]

3.3. Zero-Screening Limits of the Full Phase Shifts

Let us now consider scattering with a potential V^S of finite range r_0 and a screened Coulomb potential V^R,

$$V^{(R)} = V^S + V^R \qquad (1.3.18)$$

Outside the interaction region of V^S the regular solution $u_{pl}^{(R)}(r)$ can be represented in the form (1.2.28),

$$u_{pl}^{(R)}(r) = A_l[\cos \delta_l^{SR} F_{pl}^R(r) + \sin \delta_l^{SR} G_{pl}^R(r)] \quad \text{for} \quad r \geqslant r_0 \qquad (1.3.19)$$

the regular and irregular Coulomb functions being replaced by the corresponding screened solutions F_{pl}^R and G_{pl}^R. Their asymptotic behavior is given by (1.2.13) and (1.2.29), however, with σ_l^R rather than $\sigma_l - \eta \ln(2pr)$. Correspondingly we have, instead of (1.2.31),

$$u_{pl}^{(R)}(r) \underset{r \to \infty}{\sim} A_l \sin(pr - l\pi/2 + \sigma_l^R + \delta_l^{SR}) \qquad (1.3.20)$$

with the total phase shift given by

$$\delta_l^{(R)} = \sigma_l^R + \delta_l^{SR} \qquad (1.3.21)$$

The Coulomb-modified short-range phase shift δ_l^{SR} introduced in (1.3.19) is determined by the logarithmic derivative of $u_{pl}^{(R)}$ and by the screened Coulomb solutions at $r = r_0$. In the case of a screening function sharply cut off at radius R larger than r_0, all these quantities agree with their unscreened counterparts, and consequently

$$\delta_l^{SR} = \delta_l^{SC} \quad \text{for} \quad R > r_0 \qquad (1.3.22)$$

This is made most explicit by comparing the right-hand sides of (1.2.30) and of its analog for the corresponding screened quantities.

For arbitrary screening functions these right-hand sides become identical for sufficiently large values of R, so that

$$\delta_l^{SR} \xrightarrow[R\to\infty]{} \delta_l^{SC} \qquad (1.3.23)$$

The total phase shift (1.3.21) thus contains, besides the logarithmically diverging Coulomb phase σ_l^R, this converging contribution. In other words,

$$\lim_{R\to\infty} (\delta_l^{(R)} - \sigma_l^R) = \delta_l^{SC} \qquad (1.3.24)$$

These considerations suggest the following practical procedure. Instead of calculating δ_l^{SC} directly by means of (1.2.30), we may determine $\delta_l^{(R)}$ and σ_l^R by any method of short-range theory. This is to be repeated for increasing values of R until their difference approaches a constant value, the Coulomb-modified short-range phase shift δ_l^{SC}. All further steps can be done as in the elementary theory. In fact, with the resulting values of δ_l^{SC} we obtain the amplitude $f^{SC}(\vartheta)$ via (1.2.38) and (1.2.40) and then, via (1.2.41), the full amplitude $f(\vartheta)$.

3.4. Zero-Screening Limits of the Full S- and T-Matrices

According to (1.3.3) and (1.3.23), the phase shift (1.3.21) behaves asymptotically as

$$\delta_l^{(R)} \underset{R\to\infty}{\sim} \sigma_l + \delta_l^{SC} + \phi_R = \delta_l + \phi_R \qquad (1.3.25)$$

That is, $\delta_l^{(R)}$ approaches the total phase shift $\delta_l = \sigma_l + \delta_l^{SC}$ in the same way as σ_l^R approaches σ_l. The full S-matrix $s_l^{(R)}$ associated with $\delta_l^{(R)}$ must therefore be renormalized in the same manner as the pure Coulomb S-matrix s_l^R. In fact, analogously to (1.3.7),

$$s_l^{(R)} Z_R^{-1} = e^{2i\delta_l^{(R)}} Z_R^{-1} \underset{R\to\infty}{\sim} e^{2i(\delta_l + \phi_R)} e^{-2i\phi_R}$$

$$= e^{2i\delta_l}$$

$$= s_l \qquad (1.3.26)$$

In the three-body case the corresponding proof will be based on a splitting of the full S- or T-matrices into a pure Coulomb part and a Coulomb-distorted term. To illustrate this procedure, we introduce the

Coulomb-modified short-range T-matrix by

$$t_l^{SR} = \frac{1}{2i}(e^{2i(\delta_l^{(R)} - \sigma_l^R)} - 1) = \frac{1}{2i}(e^{2i\delta_l^{SR}} - 1) \tag{1.3.27}$$

in order to decompose the S-matrix according to

$$s_l^{(R)} = s_l^R + 2it_l^{SR} s_l^R \tag{1.3.28}$$

Because of (1.3.24), for $R \to \infty$ the amplitude (1.3.27) approaches a well-defined limit without renormalization, while according to (1.3.7) the Coulomb S-matrix s_l^R must be multiplied by Z_R^{-1} before performing the zero-screening limit. Hence, after renormalization the right-hand side of (1.3.28) converges toward the representation (1.2.37) of the full S-matrix obtained in the elementary theory. That is, we have as an alternative proof of (1.3.26),

$$s_l^{(R)} Z_R^{-1} = s_l^R Z_R^{-1} + 2it_l^{SR} s_l^R Z_R^{-1} \xrightarrow[R \to \infty]{} s_l^C + 2it_l^{SC} s_l^C = s_l \tag{1.3.29}$$

It is instructive to study this limit for the T-matrices $t_l^{(R)} = (s_l^{(R)} - 1)/2i$ and $t_l^R = (s_l^R - 1)/2i$, i.e., to consider instead of (1.3.28) the splitting

$$t_l^{(R)} = t_l^R + t_l^{SR} s_l^R \tag{1.3.30}$$

Due to the terms $-1/2i$ in $t_l^{(R)}$ and t_l^R, neither the renormalized left-hand side nor the renormalized first term on the right-hand side of (1.3.30) converges pointwise. However, for their difference we have

$$(t_l^{(R)} - t_l^R) Z_R^{-1} = t_l^{SR} s_l^R Z_R^{-1} \xrightarrow[R \to \infty]{} t_l^{SC} s_l^C \tag{1.3.31}$$

This formulation of the zero-screening properties in the two-body problem is most closely related to the one adopted for the corresponding few-body quantities.

3.5. The Full Screened Scattering Amplitude

Let us finally consider the zero-screening limit of the full screened scattering amplitude. Partial-wave summation of (1.3.30) yields

$$f^{(R)}(\vartheta) = f^R(\vartheta) + f^{SR}(\vartheta) \tag{1.3.32}$$

Multiplied by Z_R^{-1} the Coulomb amplitude $f^R(\vartheta)$ converges toward $f^C(\vartheta)$ in the sense discussed in the context of (1.3.11). For the Coulomb-modified short-range amplitude we find

$$\lim_{R\to\infty} f^{SR}(\vartheta)Z_R^{-1} = \frac{1}{p}\sum_{l=0}^{\infty}(2l+1)\lim_{R\to\infty} t_l^{SR}s_l^R Z_R^{-1}P_l(\cos\vartheta)$$

$$= \frac{1}{p}\sum_{l=0}^{\infty}(2l+1)t_l^{SC}s_l^C P_l(\cos\vartheta)$$

$$= f^{SC}(\vartheta) \qquad (1.3.33)$$

Here we have used the fact that, according to (1.2.40), the final partial-wave series converges in the ordinary (pointwise) sense, provided that $V^S(r)$ decreases faster than $r^{-3-\varepsilon}$, $\varepsilon > 0$. The existence of these limits[24,42] implies that (for $\vartheta \neq 0$ and $p \neq 0$)

$$f^{(R)}(\vartheta)Z_R^{-1} = f^R(\vartheta)Z_R^{-1} + f^{SR}(\vartheta)Z_R^{-1} \xrightarrow[R\to\infty]{} f^C(\vartheta) + f^{SC}(\vartheta)$$

$$= f(\vartheta) \qquad (1.3.34)$$

Similarly to (1.3.31), this may also be written in the form

$$[f^{(R)}(\vartheta) - f^R(\vartheta)]Z_R^{-1} = f^{SR}(\vartheta)Z_R^{-1} \xrightarrow[R\to\infty]{} f^{SC}(\vartheta) \qquad (1.3.35)$$

Note that here the convergence is ensured for all angles, including the forward direction.

3.6. Practical Procedure

The preceding suggests the following alternative to the practical treatment described in the context of (1.3.24). Instead of first calculating the phase shifts δ_l^{SC} as the zero-screening limits of the differences $(\delta_l^{(R)} - \sigma_l^R)$ and then constructing the corresponding amplitude $f^{SC}(\vartheta)$ by partial-wave summation, we may determine $f^{(R)}(\vartheta)$ and $f^R(\vartheta)$ by any method of short-range theory for increasing values of R until their renormalized difference approaches the limit $f^{SC}(\vartheta)$.

The standard method for constructing $f^{(R)}$ and f^R would again be to sum up the corresponding (screened) partial-wave amplitudes. For finite R this is a converging procedure. However, when R is increased so that $\sigma_l^R - \phi_R$ more and more closely approaches σ_l^C, the divergence of the latter quantity for $l \to \infty$ would require a growing number of partial-wave contributions. It is therefore absolutely essential not to perform the zero-screening limit individually for $f^{(R)}$ and f^R, but only for their difference, which is governed by the, for increasing l decreasing, phase shifts δ_l^{SC}. (Compare the comments on this question at the end of Section 2.) In other words, in this treatment one carries out essentially the same limit as in (1.3.24), although less explicitly because of the partial-wave summation. That is, as long as partial-wave techniques are employed, nothing is gained by working with (1.3.35) instead of (1.3.24). On the contrary, it is evidently more convenient and reliable to study the convergence in (1.3.24) and the l-dependence of the resulting phase shifts δ_l^{SC} separately before summing up the corresponding partial-wave contributions.

Nevertheless, the result (1.3.35) is of considerable interest. It would be the basis when making use of direct three-dimensional constructions of $f^{(R)}$ and f^R. It also exhibits the content of the screening approach in a condensed form. Moreover, in the three-body case one arrives quite naturally at a generalization of (1.3.35), although the partial-wave projected version of this result, i.e., the three-body analog of (1.3.31) or (1.3.24), is generally employed in practice.

Let us finally emphasize once more that there is no need to go through the screening procedure in the present simple case, where the essential quantity δ_l^{SC} is given directly by (1.2.30). In the few-body problem, however, where one has to rely on more sophisticated short-range methods such as momentum-space integral equations, the screening technique turns out to be a particularly useful tool.

4. FORMAL THEORY OF TWO-BODY SCATTERING WITH COULOMB DISTORTION

The extension of the above screening technique to few-body collisions is performed most naturally and reliably within the framework of formal multichannel scattering theory. As a first step toward this goal we describe in this section the transition from elementary to formal theory for the same two-body model as considered in the preceding sections. That is, we translate the screening formalism into a more general language. The additional complications due to the higher complexity of the multichannel situation will be dealt with in subsequent sections.

Instead of simply taking over the basic definitions from any modern textbook on collision theory, we start with a brief recapitulation of the time-dependent approach.[8-10,52,53] This allows us to present the formalism in a form most appropriate for the subsequent applications and generalizations. Moreover, to introduce the fundamental quantities as time-limits represents the appropriate basis for proving the equivalence of the screening method and Dollard's time-dependent approach, although this equivalence will be described only in the three-body case.

4.1. Scattering States

Let the Hamiltonian be given by

$$H = \frac{P^2}{2\mu} + V = H_0 + V \qquad (1.4.1)$$

with **P** being the relative momentum operator and V the local short-range interaction between the colliding particles. (We do not yet specify V; in the following applications it may be either V^S or V^R or their sum.)

Incoming and outgoing scattering states $|\mathbf{p}^{(+)}\rangle$ and $|\mathbf{p}^{(-)}\rangle$ are associated with the free (plane wave) state $|\mathbf{p}\rangle$ via

$$|\mathbf{p}^{(\pm)}\rangle = \Omega^{(\pm)}|\mathbf{p}\rangle \qquad (1.4.2)$$

the Møller operators $\Omega^{(\pm)}$ being defined as strong limits in Hilbert space,

$$\Omega^{(\pm)} = \underset{t\to\mp\infty}{\text{s-lim}}\, e^{iHt}e^{-iH_0 t} \qquad (1.4.3)$$

These limits can be shown to exist for potentials which decrease more rapidly than the Coulomb potential. This is not only a sufficient but also a necessary condition. For, according to Dollard,[23] an additional time-dependent term has to be added to H_0 in order to achieve strong convergence in the Coulomb case. We will come back to this point in Section 6.8. Here and in the following two sections, we restrict ourselves, as in Section 3, to potentials which either are of short range or are made short range by screening functions. We also recall that the time limit in (1.4.3) can be replaced by the Abelian limit[1,8,9,52,53]

$$\Omega^{(\pm)} = \underset{\varepsilon\to 0}{\text{s-lim}}\,(\pm\varepsilon)\int_{\mp\infty}^{0} dt\, e^{\pm\varepsilon t}e^{iHt}e^{-iH_0 t} \qquad (1.4.4)$$

From the definition (1.4.3) one easily infers that the $\Omega^{(\pm)}$ are isometric operators. The scattering states thus possess the same δ-function normalization as the corresponding momentum states,

$$\langle \mathbf{p}'^{(\pm)}|\mathbf{p}^{(\pm)}\rangle = \langle \mathbf{p}'|\mathbf{p}\rangle = \delta(\mathbf{p}' - \mathbf{p}) \qquad (1.4.5)$$

Another important consequence of the existence of the strong limit (1.4.3) is the intertwining relation

$$H\Omega^{(\pm)} = \Omega^{(\pm)} H_0 \tag{1.4.6}$$

which implies that the scattering states are eigenvectors of H,

$$H|\mathbf{p}^{(\pm)}\rangle = E_p|\mathbf{p}^{(\pm)}\rangle = \frac{p^2}{2\mu}|\mathbf{p}^{(\pm)}\rangle \tag{1.4.7}$$

belonging to the same energy as the eigenvector $|\mathbf{p}\rangle$ of H_0.

4.2. Explicit Representations

The definitions (1.4.3) and (1.4.4) of $\Omega^{(\pm)}$ are equivalent as long as they are understood as limits in Hilbert space, i.e., when acting on wave packets. After being introduced in this well-defined manner, the resulting operators $\Omega^{(\pm)}$ can also be applied on the improper vectors $|\mathbf{p}\rangle$, providing us with (1.4.2). However, the limiting procedure can be performed on such states only in its Abelian form (1.4.4), leading to the representation

$$|\mathbf{p}^{(\pm)}\rangle = \lim_{\varepsilon \to 0} (\pm i\varepsilon) G(E_p \pm i\varepsilon)|\mathbf{p}\rangle \tag{1.4.8}$$

Here, $G(z) = (z - H)^{-1}$ is the resolvent of the full Hamiltonian. Making use of the fact that the full resolvent and the free resolvent $G_0(z) = (z - H_0)^{-1}$ satisfy the resolvent equation

$$G(z) = G_0(z) + G_0(z)VG(z) = G_0(z) + G(z)VG_0(z) \tag{1.4.9}$$

we infer from the defining relation (1.4.8) the integral equation

$$|\mathbf{p}^{(\pm)}\rangle = |\mathbf{p}\rangle + G_0(E_p \pm i0)V|\mathbf{p}^{(\pm)}\rangle \tag{1.4.10}$$

In coordinate representation this is the Lippmann–Schwinger (LS) equation

$$\psi_\mathbf{p}^{(\pm)}(\mathbf{r}) = (2\pi)^{-3/2} e^{i\mathbf{p} \cdot \mathbf{r}} + \int \langle \mathbf{r}|G_0(E_p \pm i0)|\mathbf{r}'\rangle V(\mathbf{r}')\psi_\mathbf{p}^{(\pm)}(\mathbf{r}') \, d^3r' \tag{1.4.11}$$

Apart from the additional factor $(2\pi)^{-3/2}$, the solution $\psi_\mathbf{p}^{(+)}(\mathbf{r})$ of (1.4.11) shows the asymptotic behavior (1.2.8), the scattering amplitude being given by

$$f(\mathbf{p}', \mathbf{p}) = -4\pi^2 \mu \langle \mathbf{p}'|V|\mathbf{p}^{(+)}\rangle = -4\pi^2 \mu T(\mathbf{p}', \mathbf{p}; E_p) \quad (1.4.12)$$

where $\mathbf{p}' = p\mathbf{r}/r$ and therefore $p'^2/2\mu = p^2/2\mu = E_p$. This establishes the relationship between the elementary approach of Section 2 and the present formal definitions.

According to Eq. (1.4.11), Eq. (1.4.10) is to be understood as an integral equation in coordinate space. In momentum space the first term $\langle \mathbf{p}'|\mathbf{p}\rangle = \delta(\mathbf{p}' - \mathbf{p})$ and the free Green's function $2\mu/(p^2 \pm i0 - p'^2)$ in the second term on the r.h.s. are distributions which require integrating over a momentum function. Choosing the potential as this function, i.e., multiplying (1.4.10) from the left by V, we end up with a well-defined momentum-space integral equation for an off-shell extension of the T-matrix introduced in (1.4.12),

$$T(\mathbf{p}', \mathbf{p}; E_p) = V(\mathbf{p}', \mathbf{p}) + 2\mu \int V(\mathbf{p}', \mathbf{k}) \frac{1}{p^2 - k^2 + i0} T(\mathbf{k}, \mathbf{p}; E_p) \, d^3k \quad (1.4.13)$$

4.3. S- and T-Matrices

The probability amplitude for the transition from an initial state $|\mathbf{p}\rangle$ to a final state $|\mathbf{p}'\rangle$ is given by the S-matrix

$$S(\mathbf{p}', \mathbf{p}) = \langle \mathbf{p}'^{(-)}|\mathbf{p}^{(+)}\rangle = \langle \mathbf{p}'|\Omega^{(-)\dagger}\Omega^{(+)}|\mathbf{p}\rangle = \langle \mathbf{p}'|S|\mathbf{p}\rangle \quad (1.4.14)$$

To give this expression a more explicit form, we use for $\Omega^{(-)}$ the definition (1.4.3), and for $\Omega^{(+)}|\mathbf{p}\rangle$ the representation (1.4.8). This leads to

$$S(\mathbf{p}', \mathbf{p}) = \lim_{t \to \infty} \lim_{\varepsilon \to 0} e^{i(E_{p'} - E_p)t} i\varepsilon \langle \mathbf{p}'|G(E_p + i\varepsilon)|\mathbf{p}\rangle \quad (1.4.15)$$

which shows that the S-matrix is completely determined by the momentum representation of the full resolvent.

In order to perform the ε- and t-limits we write $G(z)$ in the form

$$G(z) = G_0(z) + G_0(z)T(z)G_0(z) \quad (1.4.16)$$

introducing in this way the transition operator $T(z)$. The momentum representation of (1.4.16),

$$\langle \mathbf{p}'|G(z)|\mathbf{p}\rangle = \frac{\delta(\mathbf{p}' - \mathbf{p})}{z - E_p} + \frac{1}{z - E_{p'}} \langle \mathbf{p}'|T(z)|\mathbf{p}\rangle \frac{1}{z - E_p} \quad (1.4.17)$$

shows that by this definition the kinematical singularities $(z - E_{p'})^{-1}$ and $(z - E_p)^{-1}$, corresponding to the free motion of the incoming and outgoing particles, have been extracted from $\langle \mathbf{p}'|G(z)|\mathbf{p}\rangle$. It is just the occurrence of these singularities which make the limits in (1.4.15) nonvanishing. In fact, inserting (1.4.17) for $z = E_p + i\varepsilon$ in (1.4.15), we see that one of the singularities is cancelled by the factor ε, while the other one leads via the distribution-theoretical formula[52,54]

$$\lim_{t \to \pm\infty} \lim_{\varepsilon \to 0} \frac{e^{ixt}}{x - i\varepsilon} = \begin{cases} 2\pi i \delta(x) \\ 0 \end{cases} \quad (1.4.18)$$

to the final result

$$S(\mathbf{p}', \mathbf{p}) = \delta(\mathbf{p}' - \mathbf{p}) - 2\pi i \delta(E_{p'} - E_p)\langle \mathbf{p}'|T(E_p + i0)|\mathbf{p}\rangle \quad (1.4.19)$$

In this derivation it has tacitly been assumed that $\langle \mathbf{p}'|T(z)|\mathbf{p}\rangle$ is well-behaved when going with z to the real axis. To verify this, we make use of the fact that the T-operator can also be written in the form

$$T(z) = V + VG(z)V \quad (1.4.20)$$

a representation most easily obtained by inserting the two right-hand sides of (1.4.9) into each other and comparing with (1.4.16). In $\langle \mathbf{p}'|T(z)|\mathbf{p}\rangle$ the resolvent occurs sandwiched between vectors $V|\mathbf{p}\rangle$ and $\langle \mathbf{p}'|V$, which are normalizable for square-integrable potentials. Between such proper Hilbert states, $G(z)$ is known to be an analytic function with poles in the discrete and cuts in the continuous spectrum of H (dynamical singularities). This property means that the transition to the positive real axis, i.e., to the cut is justified in $\langle \mathbf{p}'|T(z)|\mathbf{p}\rangle$ under weak conditions[1] on V.

Let us add some comments on our derivation of the fundamental relation (1.4.19). It is based on splitting the free resolvents (kinematical singularities) off from the full resolvent. This is a general construction principle employed, in particular, in quantum field theory. Equation (1.4.17) is indeed a reduction formula of the type known from the Lehmann–Symanzik–Zimmermann (LSZ) formalism (see in this context Sandhas[30]). Moreover, in the few-body case this concept leads directly to the highly symmetric AGS equations. Also in our subsequent considerations of scattering by Coulombic potentials, the relevant relations are obtained in the same manner, i.e., by splitting off the kinematical singularities in the corresponding resolvent equations.

Finally we note that insertion of (1.4.16) in (1.4.8) yields the representation

$$|\mathbf{p}^{(\pm)}\rangle = [1 + G_0(E_p \pm i0)T(E_p \pm i0)]|\mathbf{p}\rangle \tag{1.4.21}$$

of the scattering states. In other words, the action (1.4.2) of the Møller operators on the momentum eigenstates is explicitly given by

$$\Omega^{(\pm)}(E_p) = [1 + G_0(E_p \pm i0)T(E_p \pm i0)] \tag{1.4.22}$$

4.4. Operator Lippmann–Schwinger Equation

It is instructive to discuss the derivation of the momentum-space LS equation (1.4.13) from the present point of view. According to (1.4.17) the resolvent, and consequently the resolvent equations (1.4.9), are highly singular in momentum space. To get rid of these singularities, we insert (1.4.16) in (1.4.9) and drop the outer free resolvents; this then leads to the T-operator LS equations

$$T(z) = V + VG_0(z)T(z) = V + T(z)G_0(z)V \tag{1.4.23}$$

The momentum space version of the first relation is

$$T(\mathbf{p}', \mathbf{p}; z) = V(\mathbf{p}', \mathbf{p}) + \int V(\mathbf{p}', \mathbf{k}) \frac{1}{z - k^2/2\mu} T(\mathbf{k}, \mathbf{p}; z) \, d^3k \tag{1.4.24}$$

As an operator identity, Eq. (1.4.23) and thus also (1.4.24) are fully off-shell relations. That is, the variables $\mathbf{p}'$, $\mathbf{p}$, and z are not restricted by any on-energy-shell condition. For $z = E_p + i0$, i.e., in half-on-shell restriction, Eq. (1.4.24) goes over to (1.4.13). In other words, the present S-matrix approach leads to a natural off-shell extension of the half-on-shell formalism developed within the wave function approach of Section 4.2.

Note that by comparing (1.4.16) with (1.4.9) one reads off the useful relations

$$VG(z) = T(z)G_0(z) \tag{1.4.25}$$

$$G(z)V = G_0(z)T(z) \tag{1.4.26}$$

4.5. Short-Range plus (Screened) Coulomb Potential

Let us now consider within the present formalism the scattering of two particles interacting via a potential $V^{(R)}$ which is composed of a short-range part V^S and a screened Coulomb potential V^R. The corresponding full resolvent, the (screened) Coulomb resolvent, and the free resolvent are given by, respectively,

$$G^{(R)}(z) = (z - H_0 - V^S - V^R)^{-1} \tag{1.4.27}$$

$$G^R(z) = (z - H_0 - V^R)^{-1} \tag{1.4.28}$$

$$G_0(z) = (z - H_0)^{-1} \tag{1.4.29}$$

Following the argumentation of Section 4.3, we define the screened full and Coulomb T-operators $T^{(R)}$ and T^R by factoring out the free resolvent from $G^{(R)}$ and G^R, respectively. Moreover, extracting the Coulomb resolvent from $G^{(R)}$, we introduce the Coulomb-modified short-range operator T^{SR}. That is,

$$G^{(R)} = G_0 + G_0 T^{(R)} G_0 \tag{1.4.30}$$

$$G^R = G_0 + G_0 T^R G_0 \tag{1.4.31}$$

$$G^{(R)} = G^R + G^R T^{SR} G^R \tag{1.4.32}$$

Above we have emphasized that by splitting off the free from the full resolvent, the resulting T-operator is well defined in momentum space even for values of z on the positive real axis. This was made explicit by going over from (1.4.16) to the equivalent definition (1.4.20). In the present case the analog of this step yields

$$T^{(R)} = V^{(R)} + V^{(R)} G^{(R)} V^{(R)} \tag{1.4.33}$$

$$T^R = V^R + V^R G^R V^R \tag{1.4.34}$$

$$T^{SR} = V^S + V^S G^{(R)} V^S \tag{1.4.35}$$

Similarly to what has been mentioned in this context, the transition between these equivalent definitions of the T-operators is most easily realized by inserting the various resolvent equations involving $G^{(R)}$, G^R, and G_0 into each other. Generalizing the considerations of Section 4.4, we obtain, as a further equivalent definition of the T-operators, the LS equations

$$T^{(R)} = V^{(R)} + V^{(R)} G_0 T^{(R)} = V^{(R)} + T^{(R)} G_0 V^{(R)} \tag{1.4.36}$$

$$T^R = V^R + V^R G_0 T^R = V^R + T^R G_0 V^R \tag{1.4.37}$$

$$T^{SR} = V^S + V^S G^R T^{SR} = V^S + T^{SR} G^R V^S \tag{1.4.38}$$

4.6. The Two-Potential Formula

The T-operators introduced in the preceding subsection are related via

$$T^{(R)}(z) = T^R(z) + [1 + T^R(z) G_0(z)] T^{SR}(z) [1 + G_0(z) T^R(z)] \tag{1.4.39}$$

This identity can be derived from any of the above three sets of definitions of $T^{(R)}$, T^R, and T^{SR}. A particularly transparent derivation consists of inserting (1.4.30) and (1.4.31) into (1.4.32) and dropping the superfluous free resolvents. The advantage of proceeding in this manner may appear to be a technical point of minor importance. However, in the much less trivial three-body case it is precisely the analog of this procedure which will provide the natural generalization of (1.4.39) most directly.

In view of the above considerations concerning the pole and cut structure of the T-matrices, it is evident that the on-shell limit of the momentum representation of (1.4.39) exists. Thus, bearing in mind that, according to (1.4.21), the (screened) Coulomb scattering states can be written as

$$|\mathbf{p}_R^{(\pm)}\rangle = [1 + G_0(E_p \pm i0)T^R(E_p \pm i0)]|\mathbf{p}\rangle \qquad (1.4.40)$$

we infer from (1.4.39) the important result

$$\langle \mathbf{p}'|T^{(R)}(E_p + i0)|\mathbf{p}\rangle = \langle \mathbf{p}'|T^R(E_p + i0)|\mathbf{p}\rangle + \langle \mathbf{p}_R'^{(-)}|T^{SR}(E_p + i0)|\mathbf{p}_R^{(+)}\rangle \qquad (1.4.41)$$

or in shorter notation

$$T^{(R)}(\mathbf{p}', \mathbf{p}) = T^R(\mathbf{p}', \mathbf{p}) + \langle \mathbf{p}_R'^{(-)}|T^{SR}(E_p + i0)|\mathbf{p}_R^{(+)}\rangle \qquad (1.4.42)$$

This relation is a particularly symmetric version[7,28,31] of what is known as the two-potential formula.[4,5,10,11] The first term on the r.h.s. is the scattering amplitude corresponding to the Coulombic part of the potential. The second term, although dominated by the short-range potential V^S, also reflects its interference with V^R.

4.7. Zero-Screening Limit

The limit $R \to \infty$ can be performed without any problem in T^{SR}. For, according to (1.4.35), the momentum representation of this operator contains the resolvent sandwiched between the states $V^S|\mathbf{p}\rangle$ and $\langle \mathbf{p}'|V^S$, which are normalizable provided V^S is square-integrable. Between such states not only the resolvent $G^{(R)}$, but also its zero-screening limit $G(z) = (z - H_0 - V^S - V^C)^{-1}$, are well-defined even on the positive real axis. The transition

$$T^{SR} = V^S + V^S G^{(R)} V^S \xrightarrow[R \to \infty]{} V^S + V^S G V^S = T^{SC} \qquad (1.4.43)$$

is therefore justified in any representation.

For all other quantities on the r.h.s. of (1.4.42) this limit exists only after renormalization. As discussed in Sections 3.1 and 3.2, the screened on-shell Coulomb amplitude converges toward the unscreened amplitude via

$$T^R(\mathbf{p}', \mathbf{p})Z_R^{-1}(p) \xrightarrow[R \to \infty]{} T^C(\mathbf{p}', \mathbf{p}) \qquad (1.4.44)$$

while for the screened Coulomb scattering states we have

$$|\mathbf{p}_R^{(\pm)}\rangle Z_R^{\mp 1/2}(p) \xrightarrow[R \to \infty]{} |\mathbf{p}_C^{(\pm)}\rangle \qquad (1.4.45)$$

[compare Eqs. (1.3.11) and (1.3.17)].

We recall that the latter limit is to be understood in the distribution-theoretical sense, i.e., for wave functions $\langle \mathbf{r}|\mathbf{p}_R^{(\pm)}\rangle = \psi_\mathbf{p}^{R(\pm)}(\mathbf{r})$ integrated over localized test functions $\chi(\mathbf{r})$ [compare the argumentation following (1.3.12)]. This, however, is all we need when studying the zero-screening limit of the second term in (1.4.42). For, in this expression the states $|\mathbf{p}_R^{(\pm)}\rangle$ sandwich the quantity T^{SR} and consequently are multiplied by the short-range functions V^S occurring in this operator.

From these considerations we can draw the following conclusion: After renormalization with $Z_R^{-1}(p)$ both terms on the r.h.s. of (1.4.42), and hence the renormalized full amplitude

$$T^{(R)}(\mathbf{p}', \mathbf{p})Z_R^{-1}(p) = T^R(\mathbf{p}', \mathbf{p})Z_R^{-1}(p)$$
$$+ Z_R^{-1/2}(p')\langle \mathbf{p}_R^{'(-)}|T^{SR}(E_p + i0)|\mathbf{p}_R^{(+)}\rangle Z_R^{-1/2}(p) \qquad (1.4.46)$$

(recall $p' = p$) converge for $R \to \infty$ toward their unscreened counterparts, in this way defining the scattering amplitude for short-range plus unscreened Coulomb potentials:

$$T(\mathbf{p}', \mathbf{p}) = T^C(\mathbf{p}', \mathbf{p}) + \langle \mathbf{p}_C^{'(-)}|T^{SC}(E_p + i0)|\mathbf{p}_C^{(+)}\rangle$$
$$= T^C(\mathbf{p}', \mathbf{p}) + \tilde{T}^{SC}(\mathbf{p}', \mathbf{p}) \qquad (1.4.47)$$

This, however, is nothing but the result (1.2.39) or (1.3.34) of the elementary approach. There we have already seen that the zero-screening behavior of the full amplitude is completely determined by the behavior of the Coulomb amplitude and the Coulomb wave functions. Within the present formalism this property appears as an immediate consequence of the fact that in the two-potential formula (1.4.42) the screened Coulomb amplitude and the screened Coulomb wave functions, with their diverging zero-screening behavior, are clearly separated from the well-behaved short-range operator T^{SR}.

In other words, the interplay of Coulomb and short-range forces is most clearly exhibited on the basis of the above general relationships, especially of the representation (1.4.42). On the other hand, the zero-screening properties of the pure Coulomb quantities were simply taken over from the explicit solutions found in the elementary approach. The insights gained by the above considerations, though quite instructive in the present two-body case, are primarily of interest with respect to their few-body generalizations, i.e., with respect to the treatment of processes such as proton–deuteron scattering or the collision of two charged composite particles.

4.8. Practical Construction of $\tilde{T}^{SC}$ as Zero-Screening Limit

As discussed in Section 3, the validity of the zero-screening limit of (1.4.46) immediately suggests the following practical procedure. Calculate the amplitudes $T^{(R)}(\mathbf{p}', \mathbf{p})$ and $T^R(\mathbf{p}', \mathbf{p})$ for increasing values of R by any method, e.g., by means of the LS equations (1.4.36) and (1.4.37). Their renormalized difference then yields the Coulomb-modified short-range amplitude

$$[T^{(R)}(\mathbf{p}', \mathbf{p}) - T^R(\mathbf{p}', \mathbf{p})]Z_R^{-1}(p) \xrightarrow[R \to \infty]{} \langle \mathbf{p}_C'^{(-)}|T^{SC}(E_p + i0)|\mathbf{p}_C^{(+)}\rangle \equiv \tilde{T}^{SC}(\mathbf{p}', \mathbf{p})$$

(1.4.48)

Adding the explicitly known Coulomb amplitude $T^C(\mathbf{p}', \mathbf{p})$ to this result, we obtain the desired full T-matrix (1.4.47).

Let us emphasize once more (cf. the corresponding discussion at the end of Section 3) that it is mandatory in this procedure to perform the limit for the difference of the full and the pure Coulomb T-matrices. Both these amplitudes are built up by an increasing, and for $R = \infty$ an infinite, number of partial-wave contributions, while in their difference only those partial waves contribute which determine the short-range operator T^{SC}.

In practice it is convenient to perform the limit (1.4.48) in each partial wave separately. Going over from $T^{(R)}(\mathbf{p}', \mathbf{p})$ and $T^R(\mathbf{p}', \mathbf{p})$ to the partial-wave amplitudes

$$t_l^{(R)} = \frac{e^{2i\delta_l^{(R)}} - 1}{2i}$$

(1.4.49)

and

$$t_l^R = \frac{e^{2i\sigma_l^R} - 1}{2i}$$

(1.4.50)

the limit (1.4.48) takes the form

$$(t_l^{(R)} - t_l^R)Z_R^{-1} = \frac{1}{2i}(e^{2i\delta_l^{(R)}} - e^{2i\sigma_l^R})Z_R^{-1} \xrightarrow[R\to\infty]{} \tilde{T}_l^{SC} \qquad (1.4.51)$$

with $\tilde{T}_l^{SC}$ denoting the partial-wave projection of $\tilde{T}^{SC}$. Its existence implies that the full phase shift $\delta_l^{(R)}$, as σ_l^R, must asymptotically contain the diverging phase ϕ_R in order to compensate for the same function in the renormalization factor $Z_R = \exp(2i\phi_R)$. In other words, besides

$$\sigma_l^R \underset{R\to\infty}{\sim} \sigma_l + \phi_R \qquad (1.4.52)$$

we must have

$$\delta_l^{(R)} \underset{R\to\infty}{\sim} \delta_l + \phi_R \qquad (1.4.53)$$

This behavior, in fact, guarantees the convergence of the difference of these phase shifts toward an R-independent value denoted by δ_l^{SC}:

$$\delta_l^{(R)} - \sigma_l^R \underset{R\to\infty}{\sim} \delta_l - \sigma_l = \delta_l^{SC} \qquad (1.4.54)$$

With these properties the limit (1.4.51) yields the explicit representation of the partial-wave T-matrix

$$\tilde{T}_l^{SC} = \frac{e^{2i\sigma_l}(e^{2i\delta_l^{SC}} - 1)}{2i} \qquad (1.4.55)$$

These relations are in complete agreement with the ones obtained in Section 3 by studying the asymptotics of the radial wave functions. However, in the present approach the phase shifts $\delta_l^{(R)}$ are defined by (1.4.49), and their large-R behavior is deduced from the existence of the zero-screening limit (1.4.48) of the two-potential formula (1.4.46).

In practice the amplitudes $t_l^{(R)}$ and t_l^R, and thus the corresponding phase shifts, may be determined by solving the partial-wave projected LS equations (1.4.36) and (1.4.37) in momentum space. This is to be done for increasing R until the difference of $\delta_l^{(R)}$ and σ_l^R approaches the limit δ_l^{SC}. Partial-wave summation of the resulting amplitude (1.4.55) yields the second term on the r.h.s. in (1.4.47) and then, on adding the explicitly known Coulomb amplitude $T^C(\mathbf{p}', \mathbf{p})$, the full amplitude $T(\mathbf{p}', \mathbf{p})$. This procedure is, of course, conceptually identical to the one proposed after (1.3.24) within the framework of the elementary approach.

4.9. Integral Equation for T^{SC}

The present formalism also provides a direct means for determining the Coulomb-modified short-range amplitude, i.e., the second term on the r.h.s. in (1.4.47). According to (1.4.43), the screened operator T^{SR} converges for $R \to \infty$ toward T^{SC}, and correspondingly the integral equation (1.4.38) goes over into its unscreened counterpart,

$$T^{SC}(z) = V^S + V^S G^C(z) T^{SC} \qquad (1.4.56)$$

The occurrence of the short-range potential V^S in its kernel guarantees that this equation is well-behaved, a fact most easily verified by applying the Rollnik[55-58] idea. It consists of replacing the original kernel by $|V^S|^{1/2} G^C(z) |V^S|^{1/2}$, i.e., by an expression which for sufficiently short-range potentials V^S satisfies the usual requirements of Fredholm theory [e.g., the existence of the Schmidt norm (Scadron et al.[56])].

In case of a repulsive Coulomb potential, inserting the spectral representation of $G^C(z)$ in (1.4.56) leads to an integral equation for the Coulomb-modified short-range amplitude

$$\langle \mathbf{p}_C'^{(-)} | T^{SC}(E + i0) | \mathbf{p}_C^{(+)} \rangle = \langle \mathbf{p}_C'^{(-)} | V^S | \mathbf{p}_C^{(+)} \rangle$$
$$+ \int d^3k \langle \mathbf{p}_C'^{(-)} | V^S | \mathbf{k}_C^{(-)} \rangle \frac{1}{E + i0 - k^2/2\mu} \langle \mathbf{k}_C^{(-)} | T^{SC}(E + i0) | \mathbf{p}_C^{(+)} \rangle \qquad (1.4.57)$$

This treatment represents in a sense the complement of the coordinate-space approach to the direct determination of δ_l^{SC} by means of (1.2.30). Both methods make use of the fact that the Coulomb wave functions are explicitly known. In (1.2.30) they are needed at the edge $r = r_0$ of the finite-range potential, while in the matrix elements of V^S, which enter (1.4.57) in the inhomogeneity and in the kernel, only their internal part (i.e., $r \leq r_0$) contributes. However, in contrast to (1.4.57), use of (1.2.30) is not restricted to repulsive Coulomb potentials.

Let us add some technical remarks. The matrix elements $\langle \mathbf{p}_C'^{(-)} | V^S | \mathbf{p}_C^{(\pm)} \rangle$ are in general most easily calculated in coordinate space by employing the known Coulomb wave functions. For potentials given in momentum space, such as the effective potentials of the few-body theory described below, this necessitates unpleasant Fourier transformations, a difficulty which considerably reduces the advantage of working with the unscreened relation (1.4.57). Only for special, in particular separable, potentials can the above matrix elements in the Coulomb representation be evaluated directly in momentum space. (There exists a vast literature on this subject which can, at least partly, be traced from van Haeringen;[11] see also Dreissigacker et al.[59]) Equation (1.4.57), however, may be used with profit for the calculation of the large-l partial-wave amplitudes $\tilde{T}_l^{SC}$ when the inhomogeneous term, i.e., the Born approximation, gives a sufficiently accurate result.

The screening procedure, on the other hand, requires time-consuming successive solutions of integral equations for increasing R. But it has the advantage that these equations contain in their kernel and inhomogeneity the original momentum-space nuclear potentials. These advantages and disadvantages become particularly relevant in the three-body case. There it is primarily the generalization of the screening procedure which has been successfully applied up to now. The proof that this screening method leads to the same transition amplitudes as Dollard's theory is deferred to the three-body case in Section 6.

5. THREE-BODY AGS FORMALISM FOR SHORT-RANGE POTENTIALS

Our main subject is the generalization of the foregoing two-body screening and renormalization technique to three-body scattering with charged particles, especially to proton–deuteron scattering. Due to the conceptual and structural equivalence of the Alt–Grassberger–Sandhas (AGS) formalism[29] with the T-operator approach presented in the preceding section, this extension is performed most naturally within the AGS framework. We therefore review in the present section the basic definitions and relations of the AGS theory, following closely the presentation given in Sandhas.[30] Its application to the Coulomb distortion problem will be described in subsequent sections.

5.1. Hamiltonians and Channel States

Let the Hamiltonian H be translation invariant. This suggests introducing the familiar Jacobi coordinates, i.e., the relative coordinate $\mathbf{r}_\alpha$ of particles β and γ,

$$\mathbf{r}_\alpha = \mathbf{x}_\beta - \mathbf{x}_\gamma \tag{1.5.1}$$

and the coordinate $\boldsymbol{\rho}_\alpha$ of particle α relative to the center of mass of the $(\beta\gamma)$-subsystem,

$$\boldsymbol{\rho}_\alpha = \mathbf{x}_\alpha - \frac{m_\beta \mathbf{x}_\beta + m_\gamma \mathbf{x}_\gamma}{m_\beta + m_\gamma} \tag{1.5.2}$$

The corresponding momentum operators are given by

$$\mathbf{P}_\alpha = (1/i)\mathbf{\nabla}_{r_\alpha} \tag{1.5.3}$$

and

$$\mathbf{Q}_\alpha = (1/i)\mathbf{\nabla}_{\rho_\alpha} \tag{1.5.4}$$

We assume that the three particles interact via (not necessarily local) two-body potentials $V_\alpha(\mathbf{r}'_\alpha, \mathbf{r}_\alpha)$, where, consistent with (1.5.1), the complementary notation V_α has been used for the interaction of particles β and γ. (We only mention that three-body potentials can be included by slight generalizations of the formalism.[29,60,61]) In the center-of-mass system the total Hamiltonian reads

$$H = H_0 + V = Q_\alpha^2/2M_\alpha + P_\alpha^2/2\mu_\alpha + \sum_{\gamma=1}^{3} V_\gamma \tag{1.5.5}$$

the reduced masses being given by

$$\mu_\alpha = \frac{m_\beta m_\gamma}{m_\beta + m_\gamma} \quad \text{and} \quad M_\alpha = \frac{m_\alpha(m_\beta + m_\gamma)}{m_\alpha + m_\beta + m_\gamma}$$

Let us consider the collision of particle α, having relative momentum $\mathbf{q}_\alpha$, with a bound state of particles β and γ described by the wave function $|\psi_{\alpha n}\rangle$. Its quantum numbers are collectively denoted by "n". The corresponding "channel state" is defined as the product

$$|\Phi_{\alpha n, \mathbf{q}_\alpha}\rangle = |\psi_{\alpha n}\rangle |\mathbf{q}_\alpha\rangle \tag{1.5.6}$$

or, in coordinate space, by

$$\langle \boldsymbol{\rho}_\alpha, \mathbf{r}_\alpha | \Phi_{\alpha n, \mathbf{q}_\alpha}\rangle = \psi_{\alpha n}(\mathbf{r}_\alpha) e^{i\mathbf{q}_\alpha \cdot \boldsymbol{\rho}_\alpha}/(2\pi)^{3/2} \tag{1.5.7}$$

This state is an (improper) eigenvector of the channel Hamiltonian

$$H_\alpha = H_0 + V_\alpha = Q_\alpha^2/2M_\alpha + P_\alpha^2/2\mu_\alpha + V_\alpha \tag{1.5.8}$$

i.e., it satisfies the eigenvalue equation

$$H_\alpha |\Phi_{\alpha n, \mathbf{q}_\alpha}\rangle = (q_\alpha^2/2M_\alpha + \hat{E}_{\alpha n})|\Phi_{\alpha n, \mathbf{q}_\alpha}\rangle = E_{\alpha n}|\Phi_{\alpha n, \mathbf{q}_\alpha}\rangle \tag{1.5.9}$$

with $\hat{E}_{\alpha n}$ denoting the binding energy of the pair in state $|\psi_{\alpha n}\rangle$. The full Hamiltonian (1.5.5) can be split into the channel Hamiltonian and the channel interaction $\bar{V}_\alpha$,

$$H = H_\alpha + \bar{V}_\alpha \tag{1.5.10}$$

The latter represents the full interaction between the fragments in channel α, i.e., it consists of the two potentials acting between particle α and each of the particles β and γ of the bound subsystem:

$$\bar{V}_\alpha = V - V_\alpha = \sum_{\gamma=1}^{3} \bar{\delta}_{\alpha\gamma} V_\gamma \tag{1.5.11}$$

Here and throughout the following, the conventional notation $\bar{\delta}_{\alpha\gamma} = (1 - \delta_{\alpha\gamma})$ is used for the anti-Kronecker symbol.

The channel states $|\Phi_{\alpha n, \mathbf{q}_\alpha}\rangle$ evidently play the role of free states in the present two-fragment case, and H_α is the corresponding "free" Hamiltonian. In complete analogy to (1.4.2) scattering states

$$|\Psi_{\alpha n, \mathbf{q}_\alpha}^{(\pm)}\rangle = \Omega_\alpha^{(\pm)} |\Phi_{\alpha n, \mathbf{q}_\alpha}\rangle \tag{1.5.12}$$

are associated with the channel states by means of the channel Møller operators

$$\Omega_\alpha^{(\pm)} = \operatorname*{s-lim}_{t \to \mp \infty} e^{iHt} e^{-iH_\alpha t} \mathbb{P}_\alpha \tag{1.5.13}$$

Here,

$$\mathbb{P}_\alpha = \sum_n \int d^3 q_\alpha |\Phi_{\alpha n, \mathbf{q}_\alpha}\rangle \langle \Phi_{\alpha n, \mathbf{q}_\alpha}|$$

denotes the projector onto the corresponding channel space. As in (1.4.4) the time limits can be replaced by the ε-limits

$$\Omega_\alpha^{(\pm)} = \operatorname*{s-lim}_{\varepsilon \to \infty} (\pm \varepsilon) \int_{\mp \infty}^{0} dt \, e^{\pm \varepsilon t} e^{iHt} e^{-iH_\alpha t} \mathbb{P}_\alpha \tag{1.5.14}$$

From (1.5.13) one easily infers the analog of (1.4.6),

$$H\Omega_\alpha^{(\pm)} = \Omega_\alpha^{(\pm)} H_\alpha \qquad (1.5.15)$$

i.e., the intertwining relation between the full and the channel Hamiltonians. It implies, as in the two-body case, that the scattering states (1.5.12) are eigenvectors of H belonging to the energy $E_{\alpha n}$ of the asymptotically free fragments,

$$H|\Psi_{\alpha n,\mathbf{q}_\alpha}^{(\pm)}\rangle = E_{\alpha n}|\Psi_{\alpha n,\mathbf{q}_\alpha}^{(\pm)}\rangle \qquad (1.5.16)$$

The three-body counterpart of the normalization property (1.4.5) reads ("asymptotic orthogonality")

$$\langle \Psi_{\beta m,\mathbf{q}_\beta'}^{(\pm)}|\Psi_{\alpha n,\mathbf{q}_\alpha}^{(\pm)}\rangle = \delta_{\beta\alpha}\langle \Phi_{\alpha m,\mathbf{q}_\alpha'}|\Phi_{\alpha n,\mathbf{q}_\alpha}\rangle$$
$$= \delta_{\beta\alpha}\delta_{mn}\delta(\mathbf{q}_\alpha' - \mathbf{q}_\alpha) \qquad (1.5.17)$$

5.2. S-Matrices and Transition Amplitudes

The three-body generalization of (1.4.14), i.e., the probability amplitude for the transition from an incident state $|\Phi_{\alpha n,\mathbf{q}_\alpha}\rangle$ to an outgoing asymptotic configuration $|\Phi_{\beta m,\mathbf{q}_\beta'}\rangle$, is

$$S_{\beta m,\alpha n}(\mathbf{q}_\beta', \mathbf{q}_\alpha) = \langle \Psi_{\beta m,\mathbf{q}_\beta'}^{(-)}|\Psi_{\alpha n,\mathbf{q}_\alpha}^{(+)}\rangle \qquad (1.5.18)$$

or, according to (1.5.12),

$$S_{\beta m,\alpha n}(\mathbf{q}_\beta', \mathbf{q}_\alpha) = \langle \Phi_{\beta m,\mathbf{q}_\beta'}|\Omega_\beta^{(-)\dagger}\Omega_\alpha^{(+)}|\Phi_{\alpha n,\mathbf{q}_\alpha}\rangle$$
$$= \langle \Phi_{\beta m,\mathbf{q}_\beta'}|S_{\beta\alpha}|\Phi_{\alpha n,\mathbf{q}_\alpha}\rangle \qquad (1.5.19)$$

This representation clearly exhibits the characteristic difference between two- and three-body scattering. Since any one of particles 1, 2, or 3 can be the unbound one before or after the collision, we have a 3×3 matrix of S-operators. Consequently, the single two-body T-operator will be replaced by a 3×3 transition-operator matrix, and the LS equation (1.4.23) by a set of 3×3 coupled relations.

To give the definition (1.5.19) a more explicit form, we utilize for $\Omega_\alpha^{(\pm)}$ the representation (1.5.14), which, when inserted in (1.5.12), yields

$$|\Psi_{\alpha n,\mathbf{q}_\alpha}^{(\pm)}\rangle = \lim_{\varepsilon \to \infty} \pm i\varepsilon G(E_{\alpha n} \pm i\varepsilon)|\Phi_{\alpha n,\mathbf{q}_\alpha}\rangle \qquad (1.5.20)$$

As in the analogous two-body formula (1.4.8), $G(z) = (z - H)^{-1}$ is the resolvent of the full (three-body) Hamiltonian. With the definition (1.5.13) for $\Omega_\beta^{(-)}$ and the property (1.5.16), the S-matrix then takes the form

$$S_{\beta m, \alpha n}(\mathbf{q}'_\beta, \mathbf{q}_\alpha) = \lim_{t \to \infty} \lim_{\varepsilon \to 0} e^{i(E'_{\beta m} - E_{\alpha n})t} i\varepsilon \langle \Phi_{\beta m, \mathbf{q}'_\beta} | G(E_{\alpha n} + i\varepsilon) | \Phi_{\alpha n, \mathbf{q}_\alpha} \rangle \quad (1.5.21)$$

which is the three-body generalization of the two-body result (1.4.15).

As discussed in Section 4.3, we need to extract from $G(z)$ the kinematical singularities corresponding to the free motion of the colliding fragments. Thus, in analogy to (1.4.16) we write

$$G(z) = \delta_{\beta\alpha} G_\alpha(z) + G_\beta(z) U_{\beta\alpha}(z) G_\alpha(z) \quad (1.5.22)$$

defining in this way the AGS transition operators $U_{\beta\alpha}$. Of course, the role of the free resolvent is now played by the channel resolvents G_α and G_β, where

$$G_\alpha(z) = (z - H_\alpha)^{-1} \quad (1.5.23)$$

Instead of one T-operator we have a set of 3×3 operators $U_{\beta\alpha}$, a fact anticipated in the context of Eq. (1.5.19). If (1.5.22) is sandwiched between the channel states, the kinematical singularities are explicitly seen to be split off from the full resolvent,

$$\langle \Phi_{\beta m, \mathbf{q}'_\beta} | G(z) | \Phi_{\alpha n, \mathbf{q}_\alpha} \rangle = \delta_{\beta\alpha} \delta_{mn} \delta(\mathbf{q}'_\alpha - \mathbf{q}_\alpha) \frac{1}{z - E_{\alpha n}}$$
$$+ \frac{1}{z - E'_{\beta m}} \langle \Phi_{\beta m, \mathbf{q}'_\beta} | U_{\beta\alpha}(z) | \Phi_{\alpha n, \mathbf{q}_\alpha} \rangle \frac{1}{z - E_{\alpha n}} \quad (1.5.24)$$

Inserting this representation into (1.5.21) and repeating the steps which led to (1.4.19), we end up with the three-body generalization of that fundamental relation, viz,

$$S_{\beta m, \alpha n}(\mathbf{q}'_\beta, \mathbf{q}_\alpha) = \delta_{\beta\alpha} \delta_{mn} \delta(\mathbf{q}'_\alpha - \mathbf{q}_\alpha) - 2\pi i \delta(E'_{\beta m} - E_{\alpha n}) \langle \Phi_{\beta m, \mathbf{q}'_\beta} | U_{\beta\alpha}(E_{\alpha n} + i0) | \Phi_{\alpha n, \mathbf{q}_\alpha} \rangle$$
$$(1.5.25)$$

The transition amplitudes from an incoming α-channel to an outgoing β-channel configuration are thus given by the matrix elements

$$T_{\beta m, \alpha n}(\mathbf{q}'_\beta, \mathbf{q}_\alpha; E_{\alpha n}) = \langle \mathbf{q}'_\beta | \langle \psi_{\beta m} | U_{\beta\alpha}(E_{\alpha n} + i0) | \psi_{\alpha n} \rangle | \mathbf{q}_\alpha \rangle \quad (1.5.26)$$

They are defined, and enter the integral equations given below, for arbitrary values of the momentum variables and the energy parameter. In the S-matrix, and hence in physical three-body observables, the momenta $\mathbf{q}'_\beta$ and $\mathbf{q}_\alpha$ are restricted by the energy δ-function to their on-shell values,

$$E'_{\beta m} = q'^2_\beta/2M_\beta + \hat{E}_{\beta m} = q^2_\alpha/2M_\alpha + \hat{E}_{\alpha n} = E_{\alpha n} \qquad (1.5.27)$$

5.3. Some More Details

Defining the AGS operators by (1.5.22), i.e., by the analog of (1.4.16), makes it particularly clear that they represent the natural generalization of the two-body T-operator. Moreover, this definition proves to be the most convenient starting point for the following algebraic manipulations. For a detailed investigation of the analytic properties of $U_{\beta\alpha}$, on the other hand, we need the three-body analog of the more explicit representation (1.4.20) of $T(z)$. Instead of the single resolvent equation (1.4.9) employed in deriving (1.4.20), we must deal with three resolvent equations in the present case,

$$G(z) = G_\gamma(z) + G_\gamma(z)\bar{V}_\gamma G(z) \qquad (1.5.28)$$

or

$$G(z) = G_\gamma(z) + G(z)\bar{V}_\gamma(z)G_\gamma(z) \qquad (1.5.29)$$

where $\gamma = 1, 2,$ or 3. Inserting these equations into one another, with γ being chosen either as α or β, we obtain a relation of the form (1.5.22) from which we read off

$$U_{\beta\alpha}(z) = \bar{\delta}_{\beta\alpha} G_0^{-1}(z) + V - V_\alpha - V_\beta + \delta_{\beta\alpha} V_\alpha + \bar{V}_\beta G(z)\bar{V}_\alpha \qquad (1.5.30)$$

From this representation we infer that the resolvent $G(z)$ occurs in

$$\langle \Phi_{\beta m, \mathbf{q}'_\beta} | U_{\beta\alpha}(z) | \Phi_{\alpha n, \mathbf{q}_\alpha} \rangle$$

sandwiched between states of the type $\bar{V}_\alpha | \Phi_{\alpha n, \mathbf{q}_\alpha} \rangle$. According to (1.5.6), these consist of a normalizable bound state $|\psi_{\alpha n}\rangle$ and a non-normalizable plane wave $|\mathbf{q}_\alpha\rangle$, with the latter multiplied by the short-range channel potential $\bar{V}_\alpha$. Between such Hilbert space vectors the resolvent displays only the dynamical poles and cuts associated with the spectrum of H. In other words, by writing $\langle \Phi_{\beta m, \mathbf{q}'_\beta} | G(z) | \Phi_{\alpha n, \mathbf{q}_\alpha} \rangle$ in the form of (1.5.24), it is separated into a well-behaved amplitude and additional kinematical singularities [compare the corresponding discussion of the representation (1.4.17)]. The steps leading from (1.5.21) via (1.5.24) to the final result (1.5.25) are therefore fully justified.

For completeness we recall that in the literature also the transition operators[62]

$$U_{\beta\alpha}^{(+)} = \bar{V}_\beta + \bar{V}_\beta G \bar{V}_\alpha \qquad (1.5.31)$$

and

$$U_{\beta\alpha}^{(-)} = \bar{V}_\alpha + \bar{V}_\beta G \bar{V}_\alpha \qquad (1.5.32)$$

are utilized. They differ from $U_{\beta\alpha}$ by terms which vanish on the energy shell when sandwiched between channel states, and hence do not change the S-matrix elements in (1.5.25):

$$U_{\beta\alpha} = \bar{\delta}_{\beta\alpha} G_\alpha^{-1} + U_{\beta\alpha}^{(+)} = \bar{\delta}_{\beta\alpha} G_\beta^{-1} + U_{\beta\alpha}^{(-)} \qquad (1.5.33)$$

However, the lack of symmetry of these operators is the source of much inconvenience in performing algebraic manipulations with them.

The AGS operators can therefore be understood as symmetrized versions of the unsymmetric operators $U_{\beta\alpha}^{(\pm)}$, and the integral equations for the $U_{\beta\alpha}$ may be obtained by rewriting the integral equations proposed by Lovelace[62] for the $U_{\beta\alpha}^{(\pm)}$. Proceeding in this way is not only technically clumsy, but also hides rather than exhibits the basic idea of the AGS-approach, which consists of translating the two-body theory into a structurally equivalent three-body theory in the most symmetric way. A good example is the derivation of the integral equations for the $U_{\beta\alpha}$ given in the following subsection. The advantage of using definitions of the type (1.5.22), however, will become particularly evident when going over to the Coulomb-modified theory.

With this definition, the relationship between the wave-function and the S-matrix approach is immediately established.[63] Inserting (1.5.22) in (1.5.20) we obtain the three relations (for α fixed and $\beta = 1, 2, 3$)

$$|\Psi_{\alpha n, \mathbf{q}_\alpha}^{(\pm)}\rangle = \delta_{\beta\alpha} |\Phi_{\alpha n, \mathbf{q}_\alpha}\rangle + G_\beta(E_{\alpha n} + i0) U_{\beta\alpha}(E_{\alpha n} + i0) |\Phi_{\alpha n, \mathbf{q}_\alpha}\rangle \qquad (1.5.34)$$

If we make use of the spectral decomposition of G_β, the standard argumentation[7,10] known from the two-body case provides us with the position–space asymptotics,

$$\Psi_{\alpha n,\mathbf{q}_\alpha}^{(\pm)}(\boldsymbol{\rho}_\alpha, \mathbf{r}_\alpha) \underset{\rho_\beta \to \infty}{\sim} \frac{1}{(2\pi)^{3/2}} \left[\psi_{\alpha n}(\mathbf{r}_\alpha) e^{i\mathbf{q}_\alpha \cdot \boldsymbol{\rho}_\alpha} \delta_{\beta\alpha} \right.$$
$$\left. + \left(\frac{M_\beta}{M_\alpha}\right)^{1/2} \sum_m f_{\beta m, \alpha n}(\mathbf{q}'_\beta, \mathbf{q}_\alpha) \psi_{\beta m}(\mathbf{r}_\beta) \frac{e^{iq'_\beta \rho_\beta}}{\rho_\beta} + \text{breakup terms} \right] \qquad (1.5.35)$$

with $\mathbf{q}'_\beta = q'_\beta \mathbf{\rho}_\beta / \rho_\beta$ and q'_β fixed by (1.5.27), i.e., $q'_\beta = \sqrt{2M_\beta(E_{\alpha n} - \hat{E}_{\beta m})}$. The amplitudes multiplying the outgoing two-fragment spherical waves are given by the on-shell matrix elements (1.5.26):

$$f_{\beta m,\alpha n}(\mathbf{q}'_\beta, \mathbf{q}_\alpha) = -(2\pi)^2 (M_\beta M_\alpha)^{1/2} T_{\beta m,\alpha n}(\mathbf{q}'_\beta, \mathbf{q}_\alpha) \qquad (1.5.36)$$

In other words, apart from a constant factor the $f_{\beta m,\alpha n}(\mathbf{q}'_\beta, \mathbf{q}_\alpha)$ coincide with the transition amplitudes of the S-matrix representation (1.5.25). The differential cross section therefore reads

$$\frac{d\sigma_{\beta m,\alpha n}}{d\Omega} = (2\pi)^4 M_\beta M_\alpha \frac{q'_\beta}{q_\alpha} |T_{\beta m,\alpha n}(\mathbf{q}'_\beta, \mathbf{q}_\alpha)|^2 \qquad (1.5.37)$$

a result which can also be verified in a more sophisticated manner within the framework of time-dependent theory (see, e.g., Amrein et al.[1] and Dollard[64]).

5.4. The AGS Equations

In the two-body case, the LS equation for $T(z)$ was derived by inserting the definition (1.4.16) into the resolvent equation and dropping the free resolvents. Analogously, we insert the defining equation (1.5.22) for $U_{\beta\alpha}$ into the three-body resolvent equation (1.5.28), written in the form

$$G = G_\beta + G_\beta \sum_\gamma \bar{\delta}_{\beta\gamma} V_\gamma G \qquad (1.5.38)$$

Cancelling the factors G_β and G_α, we obtain

$$U_{\beta\alpha} = \bar{\delta}_{\beta\alpha} G_0^{-1} + \sum_\gamma \bar{\delta}_{\beta\gamma} V_\gamma G_\gamma U_{\gamma\alpha} \qquad (1.5.39)$$

or, on making use of the two-body relation (1.4.25), the more convenient form

$$U_{\beta\alpha} = \bar{\delta}_{\beta\alpha} G_0^{-1} + \sum_\gamma \bar{\delta}_{\beta\gamma} T_\gamma G_0 U_{\gamma\alpha} \qquad (1.5.40)$$

Similarly one obtains

$$U_{\beta\alpha} = \bar{\delta}_{\beta\alpha} G_0^{-1} + \sum_\gamma U_{\beta\gamma} G_0 T_\gamma \bar{\delta}_{\gamma\alpha} \qquad (1.5.41)$$

These are the coupled sets of Faddeev-type[65] integral equations for the operators $U_{\beta\alpha}$ introduced by Alt, Grassberger, and Sandhas (AGS).[29] The similarity of our derivation with the one leading in the two-body case to the LS equation (1.4.23) indicates that the present result should have an analogous algebraic structure. In fact, introducing the matrix operators[30]

$$\mathbf{T}_{\beta\alpha} = U_{\beta\alpha} \tag{1.5.42}$$

$$\mathbf{V}_{\beta\alpha} = \bar{\delta}_{\beta\alpha} G_0^{-1} \tag{1.5.43}$$

$$\mathbf{G}_{0,\beta\alpha} = \delta_{\beta\alpha} G_0 T_\alpha G_0 \tag{1.5.44}$$

Eqs. (1.5.40) and (1.5.41) show the expected structure of the two versions of the LS equation (1.4.23),

$$\mathbf{T}_{\beta\alpha} = \mathbf{V}_{\beta\alpha} + \sum_{\gamma,\delta} \mathbf{V}_{\beta\gamma} \mathbf{G}_{0,\gamma\delta} \mathbf{T}_{\delta\alpha} \tag{1.5.45}$$

and

$$\mathbf{T}_{\beta\alpha} = \mathbf{V}_{\beta\alpha} + \sum_{\gamma,\delta} \mathbf{T}_{\beta\gamma} \mathbf{G}_{0,\gamma\delta} \mathbf{V}_{\delta\alpha} \tag{1.5.46}$$

The correspondence of the AGS approach with the two-body formalism, emphasized repeatedly and exhibited most clearly by (1.5.45) and (1.5.46), will turn out to be of considerable technical advantage in all subsequent considerations.

5.5. Breakup Reactions

With slight modifications the above formalism also covers the breakup into three asymptotically free particles. The relevant channel states are products of the relative momentum eigenstates for the three particles,

$$|\Phi_{0;\mathbf{p}_\beta,\mathbf{q}_\beta}\rangle = |\mathbf{p}_\beta\rangle|\mathbf{q}_\beta\rangle \tag{1.5.47}$$

with $\beta = 1, 2$, or 3. The Møller operators are given by

$$\Omega_0^{(\pm)} = \underset{t \to \mp\infty}{\text{s-lim}}\ e^{iHt}e^{-iH_0 t} \tag{1.5.48}$$

and the definition (1.5.12) of the scattering states goes over to

$$|\Psi_{0;\mathbf{p}_\beta,\mathbf{q}_\beta}^{(\pm)}\rangle = \Omega_0^{(\pm)}|\Phi_{0;\mathbf{p}_\beta,\mathbf{q}_\beta}\rangle \tag{1.5.49}$$

The S-matrix element for the transition from a two-fragment state $|\Phi_{\alpha n, \mathbf{q}_\alpha}\rangle$ to a three-particle final state therefore is

$$S_{0,\alpha n}(\mathbf{q}'_\beta, \mathbf{p}'_\beta; \mathbf{q}_\alpha) = \langle \Psi^{(-)}_{0;\mathbf{p}'_\beta,\mathbf{q}'_\beta} | \Psi^{(+)}_{\alpha n, \mathbf{q}_\alpha} \rangle$$
$$= \langle \Phi_{0;\mathbf{p}'_\beta,\mathbf{q}'_\beta} | \Omega_0^{(-)\dagger} \Omega_\alpha^{(+)} | \Phi_{\alpha n, \mathbf{q}_\alpha} \rangle$$
$$= \langle \mathbf{q}'_\beta | \langle \mathbf{p}'_\beta | S_{0\alpha} | \psi_{\alpha n} \rangle | \mathbf{q}_\alpha \rangle \quad (1.5.50)$$

Proceeding as in Section 5.2, instead of (1.5.21) we obtain

$$S_{0,\alpha n}(\mathbf{q}'_\beta, \mathbf{p}'_\beta; \mathbf{q}_\alpha) = \lim_{t \to \infty} \lim_{\varepsilon \to 0} e^{i(E' - E_{\alpha n})t} i\varepsilon \langle \mathbf{q}'_\beta | \langle \mathbf{p}'_\beta | G(E_{\alpha n} + i\varepsilon) | \Phi_{\alpha n, \mathbf{q}_\alpha} \rangle \quad (1.5.51)$$

where $E' = q'^2_\beta/2M_\beta + p'^2_\beta/2\mu_\beta$. Defining the breakup operators $U_{0\alpha}$ by (1.5.22), without the first term and with G_β being replaced by the channel resolvent G_0 for three free particles, i.e., by

$$G(z) = G_0(z) U_{0\alpha} G_\alpha(z) \quad (1.5.52)$$

the steps which led from (1.5.21) to (1.5.25) provide us now with

$$S_{0,\alpha n}(\mathbf{q}'_\beta, \mathbf{p}'_\beta; \mathbf{q}_\alpha) = -2\pi i \delta(E' - E_{\alpha n}) \langle \mathbf{q}'_\beta | \langle \mathbf{p}'_\beta | U_{0\alpha}(E_{\alpha n} + i0) | \psi_{\alpha n} \rangle | \mathbf{q}_\alpha \rangle \quad (1.5.53)$$

Comparing the definitions (1.5.22) and (1.5.52) of the rearrangement and breakup operators, respectively, shows that the latter are obtained from the former by putting $\beta = 0$, where

$$V_0 \equiv 0 \quad (1.5.54)$$

This, in particular, means that the AGS equations (1.5.40) go over to

$$U_{0\alpha} = G_0^{-1} + \sum_\gamma T_\gamma G_0 U_{\gamma\alpha} \quad (1.5.55)$$

It should be noted that (1.5.55) is no longer an integral equation, but expresses the breakup operators $U_{0\alpha}$ algebraically by means of quadratures on the rearrangement operators $U_{\gamma\alpha}, \gamma \neq 0$. However, by putting $\beta = 0$ in (1.5.41) we obtain an integral equation for $U_{0\alpha}$,

$$U_{0\alpha} = G_0^{-1} + \sum_\gamma U_{0\gamma} G_0 T_\gamma \bar{\delta}_{\gamma\alpha} \quad (1.5.56)$$

The fact that $U_{0\alpha}$ is determined both by this integral equation and by the algebraic relation (1.5.55) reflects a general aspect of multichannel scattering theory. Slightly reordered, Eq. (1.5.55) can be written in the form

$$U_{0\alpha} = \delta_{\beta\alpha} G_0^{-1} + (1 + T_\beta G_0) U_{\beta\alpha}, \quad \text{with } \beta = 1, 2, \text{ or } 3 \quad (1.5.57)$$

which, when inserted in (1.5.53), yields

$$S_{0,\alpha n}(\mathbf{q}'_\beta, \mathbf{p}'_\beta; \mathbf{q}_\alpha) = -2\pi i \delta(E' - E_{\alpha n}) \langle \mathbf{q}'_\beta | \langle \mathbf{p}'^{(-)}_\beta | U_{\beta\alpha}(E_{\alpha n} + i0) | \Phi_{\alpha n, \mathbf{q}_\alpha} \rangle \quad (1.5.58)$$

In other words, the on-shell breakup amplitudes are given either by means of the corresponding breakup operators $U_{0\alpha}$ multiplied from the left with plane wave states, or by the rearrangement operators $U_{\beta\alpha}$ multiplied from the left with

$$|\mathbf{p}^{(-)}_\beta\rangle |\mathbf{q}_\beta\rangle = [1 + G_0(E_\beta - i0) T_\beta(E_\beta - i0)] |\mathbf{p}_\beta\rangle |\mathbf{q}_\beta\rangle \quad (1.5.59)$$

The representation (1.5.58) is thus of the form (1.5.25) for the rearrangement amplitudes, with the outgoing subsystem bound state $|\psi_{\beta m}\rangle$ being replaced by the subsystem scattering state $|\mathbf{p}^{(-)}_\beta\rangle$.

Note that (1.5.58) can also be derived directly by inserting (1.5.22), instead of (1.5.52), into (1.5.51). From a more general point of view, this result appears as a consequence of the fact that the Møller operators (1.5.48) can be written in the form[66]

$$\Omega_0^{(\pm)} = \underset{t \to \mp \infty}{\text{s-lim}} \underset{\tau \to \mp \infty}{\text{s-lim}} e^{iHt} e^{-iH_\beta t} e^{iH_\beta \tau} e^{-iH_0 \tau}$$

$$= \Omega_\beta^{(\pm)} \hat{\Omega}_\beta^{(\pm)} \quad (1.5.60)$$

where the two-body Møller operators $\hat{\Omega}_\beta^{(\pm)}$ map the plane wave state $|\mathbf{p}_\beta\rangle$ onto the corresponding scattering states $|\mathbf{p}_\beta^{(\pm)}\rangle$.

We note here that some of these as well as other aspects of the quantum mechanical few-body scattering theory for short-range potentials are the content of several monographs.[58,67-69] For the generalization to N-body scattering within the present framework see Sandhas,[66] Grassberger and Sandhas,[70] and Alt et al.[71,72]

6. THREE-BODY COLLISION THEORY FOR CHARGED PARTICLES

Equipped with the formalism presented in the preceding section, it is comparatively straightforward to generalize the two-body screening technique developed in Section 4 to three particles.

Taking over the methods from two-body scattering literally, it seems natural at first sight to employ in the three-body case an analogous splitting of the channel interaction into its short-range and full Coulomb parts. However, such a procedure[74,75] does not sufficiently take into account that the essential Coulomb complications in composite-particle collisions concern only the relative motion of the centers of mass of the colliding fragments. This fact is more appropriately accounted for by employing a splitting of the channel interaction into a center-of-mass Coulomb potential and a term consisting of the short-range interactions and the Coulombic remainder.

The Alt–Sandhas–Ziegelmann (ASZ) approach[28,31-33] is based on this concept, and has led to a clear prescription of how to define and in particular also calculate the charged-particle amplitudes as zero-screening limits. The main part of this section is therefore devoted to this approach. Our presentation follows most closely Alt and Sandhas.[31] In addition, we show that the ASZ construction of the transition amplitudes coincides with the definitions following from Dollard's time-dependent Coulomb scattering theory.[23]

6.1. Potentials, Hamiltonians, and Resolvents

Let the full interaction

$$V^{(R)} = V^S + V^R = \sum_\gamma V^S_\gamma + \sum_\gamma V^R_\gamma \qquad (1.6.1)$$

consist of short-range potentials $V^S_\alpha(\mathbf{r}'_\alpha, \mathbf{r}_\alpha)$ and exponentially screened Coulomb potentials

$$V^R_\alpha(\mathbf{r}_\alpha) = \frac{e_\beta e_\gamma}{r_\alpha} e^{-r_\alpha/R} \qquad (1.6.2)$$

Here e_β and e_γ denote the charges of the particles in the $(\beta\gamma)$-subsystem. According to (1.5.11) the channel potential $\bar{V}^{(R)}_\alpha$, representing the interaction of particle α with each of the particles of the bound pair, is composed of a short-range part

$$\bar{V}^S_\alpha = \sum_{\gamma \neq \alpha} V^S_\gamma = \sum_\gamma \bar{\delta}_{\alpha\gamma} V^S_\gamma \qquad (1.6.3)$$

and a Coulomb part, given in coordinate space as

$$\bar{V}^R_\alpha(\mathbf{r}_\alpha, \mathbf{\rho}_\alpha) = \frac{e_\alpha e_\beta}{r_\gamma} e^{-r_\gamma/R} + \frac{e_\alpha e_\gamma}{r_\beta} e^{-r_\beta/R} \qquad (1.6.4)$$

($\mathbf{r}_\beta$ and $\mathbf{r}_\gamma$ are to be expressed as linear combinations of $\mathbf{r}_\alpha, \boldsymbol{\rho}_\alpha$). For the following considerations it is essential to introduce besides the actual Coulomb channel potential (1.6.4) the corresponding center-of-mass (CM) potential

$$v_\alpha^R(\boldsymbol{\rho}_\alpha) = \frac{e_\alpha(e_\beta + e_\gamma)}{\rho_\alpha} e^{-\rho_\alpha/R} \tag{1.6.5}$$

i.e., the Coulomb interaction of particle α with the total charge of the $(\beta\gamma)$-subsystem concentrated in its center of mass.

With these definitions the full, the channel, and the Coulomb-distorted channel Hamiltonians are, respectively,

$$H^{(R)} = H_0 + V^{(R)} = H_0 + V^S + V^R \tag{1.6.6}$$

$$H_\alpha^{(R)} = H_0 + V_\alpha^{(R)} = H_0 + V_\alpha^S + V_\alpha^R \tag{1.6.7}$$

$$h_\alpha^R = H_\alpha^{(R)} + v_\alpha^R \tag{1.6.8}$$

The corresponding resolvents are given by

$$G^{(R)}(z) = (z - H^{(R)})^{-1} \tag{1.6.9}$$

$$G_\alpha^{(R)}(z) = (z - H_\alpha^{(R)})^{-1} \tag{1.6.10}$$

$$g_\alpha^R(z) = (z - h_\alpha^R)^{-1} \tag{1.6.11}$$

Recall that in the two-body case we introduced the Coulomb resolvent (1.4.28) by adding in the free resolvent (1.4.29) the Coulomb potential V^R to H_0. The straightforward generalization of this definition seems to be to add in the channel resolvent (1.6.10) the Coulomb channel interaction $\bar{V}_\alpha^R$ to $H_\alpha^{(R)}$. Instead we have chosen in (1.6.8), and hence in (1.6.11), the CM Coulomb potential v_α^R. This apparently slight modification will turn out to be crucial in the following considerations.

As pointed out repeatedly, the most appropriate way of introducing transition operators consists of extracting from the full resolvents the free or channel resolvents and thus the corresponding singularities [compare (1.5.22)]. In the present case this suggests defining the full transition operator $U_{\beta\alpha}^{(R)}$, the CM Coulomb transition operator t_α^R, and the Coulomb-modified short-range operator $U_{\beta\alpha}^{SR}$ by

$$G^{(R)} = \delta_{\beta\alpha} G_\alpha^{(R)} + G_\beta^{(R)} U_{\beta\alpha}^{(R)} G_\alpha^{(R)} \tag{1.6.12}$$

$$g_\alpha^R = G_\alpha^{(R)} + G_\alpha^{(R)} t_\alpha^R G_\alpha^{(R)} \tag{1.6.13}$$

and

$$G^{(R)} = \delta_{\beta\alpha} g_\alpha^R + g_\beta^R U_{\beta\alpha}^{SR} g_\alpha^R \qquad (1.6.14)$$

Combining these equations, we obtain immediately the fundamental relation

$$U_{\beta\alpha}^{(R)} = \delta_{\beta\alpha} t_\alpha^R + (1 + t_\beta^R G_\beta^{(R)}) U_{\beta\alpha}^{SR} (1 + G_\alpha^{(R)} t_\alpha^R) \qquad (1.6.15)$$

Note that the derivation of (1.6.15) follows exactly the steps leading from (1.4.30)–(1.4.32) to (1.4.39).

6.2. The Two-Potential Formula

For two-body scattering Eq. (1.4.39) implied the two-potential formula (1.4.41). The three-body analog of (1.4.39), the fundamental relation (1.6.15), is therefore expected to provide a suitable generalization of (1.4.41). In fact, this generalization will turn out to represent the appropriate starting point for the ASZ screening approach.

When applying (1.6.15) onto $|\Phi_{\alpha n, \mathbf{q}_\alpha}\rangle = |\psi_{\alpha n}\rangle |\mathbf{q}_\alpha\rangle$, we have to study the effect of the operator $1 + G_\alpha^{(R)} t_\alpha^R$ on these channel states. The resolvent and the transition operator in this expression depend on the channel Hamiltonian (1.6.7), whose action on $|\Phi_{\alpha n, \mathbf{q}_\alpha}\rangle$ yields

$$\begin{aligned} H_\alpha^{(R)} |\Phi_{\alpha n, \mathbf{q}_\alpha}\rangle &= (Q_\alpha^2/2M_\alpha + P_\alpha^2/2\mu_\alpha + V_\alpha^S + V_\alpha^R) |\psi_{\alpha n}\rangle |\mathbf{q}_\alpha\rangle \\ &= (Q_\alpha^2/2M_\alpha + \hat{E}_{\alpha n}) |\psi_{\alpha n}\rangle |\mathbf{q}_\alpha\rangle \end{aligned} \qquad (1.6.16)$$

Apart from a shift of the energy parameter z to $z - \hat{E}_{\alpha n}$, the operator

$$1 + G_\alpha^{(R)}(z) t_\alpha^R(z)$$

therefore acts exclusively on $|\mathbf{q}_\alpha\rangle$. Comparison with (1.4.21) shows that for $z = E_{\alpha n} \pm i0$ it plays the role of a two-body Møller operator, mapping the plane wave state $|\mathbf{q}_\alpha\rangle$ onto the corresponding scattering states $|\mathbf{q}_{\alpha,R}^{(\pm)}\rangle$:

$$\lim_{\varepsilon \to 0} [1 + G_\alpha^{(R)}(E_{\alpha n} \pm i\varepsilon) t_\alpha^R(E_{\alpha n} \pm i\varepsilon)] |\Phi_{\alpha n, \mathbf{q}_\alpha}\rangle = |\psi_{\alpha n}\rangle |\mathbf{q}_{\alpha,R}^{(\pm)}\rangle \qquad (1.6.17)$$

Similar considerations yield

$$\langle \mathbf{q}'_\alpha | \langle \psi_{\alpha m} | t_\alpha^R(E_{\alpha n} + i0) | \psi_{\alpha n}\rangle | \mathbf{q}_\alpha\rangle = \delta_{mn} t_\alpha^R(\mathbf{q}'_\alpha, \mathbf{q}_\alpha) \qquad (1.6.18)$$

the r.h.s. being the two-body CM Coulomb scattering amplitude associated with the CM potential v_α^R.

Taken between channel states the basic relation (1.6.15) therefore goes over into

$$\langle \mathbf{q}'_\beta | \langle \psi_{\beta m} | \langle U^{(R)}_{\beta \alpha}(E_{an} + i0) | \psi_{an} \rangle | \mathbf{q}_\alpha \rangle$$
$$= \delta_{\beta \alpha} \delta_{mn} t^R_\alpha(\mathbf{q}'_\alpha, \mathbf{q}_\alpha) + \langle \mathbf{q}'^{(-)}_{\beta,R} | \langle \psi_{\beta m} | U^{SR}_{\beta \alpha}(E_{an} + i0) | \psi_{an} \rangle | \mathbf{q}^{(+)}_{\alpha,R} \rangle \quad (1.6.19)$$

This two-potential formula clearly represents the three-body generalization of (1.4.41) corresponding to the splitting

$$\bar{V}^S_\alpha + \bar{V}^R_\alpha = v^R_\alpha + (\bar{V}^S_\alpha + \bar{V}^R_\alpha - v^R_\alpha) \quad (1.6.20)$$

The advantage of choosing v^R_α in (1.6.8) and hence in (1.6.11), or equivalently, of employing the splitting (1.6.20), is that the amplitude t^R_α and the states $|\mathbf{q}^{(+)}_{\alpha,R}\rangle$ and $\langle \mathbf{q}'^{(-)}_{\beta,R}|$ in (1.6.19) concern only the relative motion of the two fragments without any reference to their internal structure. In other words, they are pure two-body Coulomb quantities whose properties are well-known from elementary two-body Coulomb theory. This is of particular relevance when studying the zero-screening limit of (1.6.19). All effects arising from the compositeness of the bound two-body subsystems are contained in $\langle \psi_{\beta m} | U^{SR}_{\beta \alpha}(E_{an} + i0) | \psi_{an} \rangle$.

In contrast, the splitting of the channel interaction into its full Coulomb and short-range parts leads to an analogous two-potential formula, introduced earlier by Noble[74] and Bencze,[75] in which the first term is the full three-body Coulomb amplitude, and the second one a short-range transition operator sandwiched between the full three-body Coulomb scattering states. Since neither the three-body Coulomb amplitude nor the scattering states are known analytically, this particular splitting does not lead to a calculable expression and therefore requires approximations, e.g., as proposed by Bencze et al.[75,76]

6.3. Details on the ε-Limit

In the above derivation, the two-potential formula followed as an immediate consequence of the algebraic relation (1.6.15) in the limit $\varepsilon \to 0$. However, what has tacitly been assumed in the transition from (1.6.15) to (1.6.19) is that the ε-limit can be performed independently in the states $(1 + G^{(R)}_\alpha t^R_\alpha) | \psi_{an} \rangle | \mathbf{q}_\alpha \rangle$ and the operators $U^{SR}_{\beta \alpha}$. In order to show that the originally occurring limit of a product of three terms really equals the product of their limits, we employ, instead of the definition (1.6.14), the more explicit representation

$$U^{SR}_{\beta \alpha} = \bar{\delta}_{\beta \alpha}(g^R_\alpha)^{-1} + \bar{V}^S_\beta + \bar{V}^R_\beta - v^R_\beta + (\bar{V}^S_\beta + \bar{V}^R_\beta - v^R_\beta) G^{(R)}(z)(\bar{V}^S_\alpha + \bar{V}^R_\alpha - v^R_\alpha)$$
$$(1.6.21)$$

obtained by means of the same steps which led from (1.5.22) via (1.5.28) and (1.5.29) to (1.5.30). In (1.6.21) the resolvent $G^{(R)}$ occurs multiplied on each side by short-range operators $\bar{V}_\alpha^S + \bar{V}_\alpha^R - v_\alpha^R$, and hence will be sandwiched between proper Hilbert-space vectors when going over from (1.6.15) to (1.6.19). The simultaneous or independent transition to the real axis can therefore be performed without any problem in this part of (1.6.21).

It remains to prove that the same holds true with respect to the term $\bar{\delta}_{\beta\alpha}[g_\alpha^R(E_{\alpha n} + i\varepsilon)]^{-1}$. Since it gives zero contribution for $\varepsilon = 0$ when being inserted into the final expression (1.6.19), it has to be shown that it also vanishes in the transition from (1.6.15) to (1.6.19). According to (1.6.13) we have

$$(g_\alpha^R)^{-1}(1 + G_\alpha^{(R)} t_\alpha^R) = G_\alpha^{(R)-1} \qquad (1.6.22)$$

and by comparing (1.6.13) with the resolvent equation for g_α^R and $G_\alpha^{(R)}$ we infer $t_\beta^R G_\beta^{(R)} = v_\beta^R g_\beta^R$. The contribution of $(g_\alpha^{(R)})^{-1}$ to (1.6.15), sandwiched between the channel states, is therefore given by

$$i\varepsilon\langle\Phi_{\beta m,\mathbf{q}_\beta'}|[1 + v_\beta^R g_\beta^R(E_{\alpha n} + i\varepsilon)]|\Phi_{\alpha n,\mathbf{q}_\alpha}\rangle$$
$$= i\varepsilon\langle\Phi_{\beta m,\mathbf{q}_\beta'}|[1 + v_\beta^R(q_\beta'^2/2M_\beta + i\varepsilon - Q_\beta^2/2M_\beta - v_\beta^R)^{-1}]|\Phi_{\alpha n,\mathbf{q}_\alpha}\rangle \qquad (1.6.23)$$

where in the second step Eq. (1.5.9) and $E_{\beta m} = E_{\alpha n}$ have been used. The matrix element on the right-hand side of (1.6.23) is evidently nonsingular for $\beta \neq \alpha$. Being multiplied by a factor ε this term thus vanishes in the limit $\varepsilon \to 0$.

Let us recall in this context that terms such as $\bar{\delta}_{\beta\alpha}(g_\alpha^R)^{-1}$ are typical for AGS-type operators. They guarantee the high symmetry of these operators, an aspect which is most convenient in algebraic manipulations such as the ones leading to the basic relation (1.6.15). However, this property is of real advantage only if these additional terms vanish on the energy shell. According to the above discussion, this is true in the present case, which means that the algebraic relation (1.6.15), characterized by a purely diagonal first term on the right-hand side, induces the same feature for the on-shell result (1.6.19).

Of course, the above definitions and algebraic manipulations can be taken over to any splitting of the underlying channel interaction into two terms such that one ends up with an operator relation of the form (1.6.15). However, it is clear a priori that in general this structure does not persist when going over to the corresponding on-shell relation. That is, in contrast to statements to the contrary, made on the basis of purely algebraic manipulations,[77,78] the symmetric form (1.6.19) of the two-potential formula is not generally valid. Otherwise, by choosing a splitting into the whole interaction and a zero remainder, one could even draw the conclusion that any rearrangement amplitude is diagonal, which is trivially wrong.

The question whether the limit of a product of quantities equals the product of their limits is of relevance not only in the present context. Ignoring this crucial point has led in the past to erroneous conclusions or apparent contradictions in various other derivations of formal scattering theory, e.g., concerning the validity of homogeneous vs. inhomogeneous LS equations. We refer in this context to Faddeev and Yakubovsky[79] and Levin and Sandhas.[80]

6.4. Zero-Screening Limit

The characteristic feature of the two-potential formula (1.6.19) is that the amplitude $t_\alpha^R(\mathbf{q}'_\alpha, \mathbf{q}_\alpha)$ and the states $|\mathbf{q}_{\alpha,R}^{(+)}\rangle$ and $|\mathbf{q}_{\beta,R}^{(-)}\rangle$ are two-body quantities representing the motion of the centers of mass of the two (charged) fragments under the influence of the (screened) Coulomb interaction v_α^R. Their zero-screening behavior can therefore be taken over immediately from the genuine two-body case. According to (1.4.44) and (1.4.45) we have

$$t_\alpha^R(\mathbf{q}'_\alpha, \mathbf{q}_\alpha) Z_{\alpha,R}^{-1}(q_\alpha) \xrightarrow[R \to \infty]{} t_\alpha^C(\mathbf{q}'_\alpha, \mathbf{q}_\alpha) \qquad (1.6.24)$$

$$|\mathbf{q}_{\alpha,R}^{(\pm)}\rangle Z_{\alpha,R}^{\mp 1/2}(q_\alpha) \xrightarrow[R \to \infty]{} |\mathbf{q}_{\alpha,C}^{(\pm)}\rangle \qquad (1.6.25)$$

where now [compare (1.3.4) and (1.3.6)]

$$Z_{\alpha,R}(q_\alpha) \underset{R \to \infty}{\sim} \exp\{-2ie_\alpha(e_\beta + e_\gamma)M_\alpha q_\alpha^{-1}[\ln(2q_\alpha R) - C]\} \qquad (1.6.26)$$

What remains to be shown is that the limit $R \to \infty$ can be performed without any problems in $U_{\beta\alpha}^{SR}$. According to the explicit definition (1.6.21) of this operator, the resolvent $G^{(R)}$ occurs in the second term of (1.6.19) between states $(\bar{V}_\alpha^S + \bar{V}_\alpha^R - v_\alpha^R)|\psi_{\alpha n}\rangle|\mathbf{q}_{\alpha,R}^{(\pm)}\rangle$ which are normalizable not only for finite R, but also in the limit $R \to \infty$. This is obvious for the term containing $\bar{V}_\alpha^S$, but also holds true for the Coulombic part

$$\langle \boldsymbol{\rho}_\alpha, \mathbf{r}_\alpha | \bar{V}_\alpha^C - v_\alpha^C | \psi_{\alpha n} \rangle | \mathbf{q}_{\alpha,C}^{(\pm)} \rangle$$
$$= [e_\alpha e_\beta(r_\gamma^{-1} - \rho_\alpha^{-1}) + e_\alpha e_\gamma(r_\beta^{-1} - \rho_\alpha^{-1})]\psi_{\alpha n}(\mathbf{r}_\alpha)\langle \boldsymbol{\rho}_\alpha | \mathbf{q}_{\alpha,C}^{(+)} \rangle \qquad (1.6.27)$$

This assertion may be proved by expressing $\mathbf{r}_\gamma$ and $\mathbf{r}_\beta$ in terms of the independent pair of Jacobi coordinates $\mathbf{r}_\alpha, \boldsymbol{\rho}_\alpha$ and then making a multipole expansion of r_γ^{-1} and r_β^{-1} in powers of ρ_α^{-1}. The spatial localization of

$\psi_{\alpha n}(\mathbf{r}_\alpha)$ implies that the r.h.s. of (1.6.27) decreases for large ρ_α at least as fast as ρ_α^{-2}, which is sufficient to guarantee normalizability. Because of this normalizability, in the second term on the r.h.s. of (1.6.19) the transition from the full screened resolvent $G^{(R)}$ to its unscreened counterpart G, and hence the transition from $U_{\beta\alpha}^{SR}$ to $U_{\beta\alpha}^{SC}$, is justified.

When going over to the zero-screening limit of (1.6.19), renormalization factors are needed only for the CM Coulomb amplitudes and states in this expression, in order to guarantee the existence of

$$\lim_{R\to\infty} Z_{\beta,R}^{-1/2}(q'_\beta)\langle\mathbf{q}'_\beta|\langle\psi_{\beta m}|U_{\beta\alpha}^{(R)}(E_{\alpha n}+i0)|\psi_{\alpha n}\rangle|\mathbf{q}_\alpha\rangle Z_{\alpha,R}^{-1/2}(q_\alpha)$$
$$= \delta_{\beta\alpha}\delta_{mn}t_\alpha^C(\mathbf{q}'_\alpha,\mathbf{q}_\alpha) + \langle\mathbf{q}'^{(-)}_{\beta,C}|\langle\psi_{\beta m}|U_{\beta\alpha}^{SC}(E_{\alpha n}+i0)|\psi_{\alpha n}\rangle|\mathbf{q}^{(+)}_{\alpha,C}\rangle \quad (1.6.28)$$

Note that this result followed by a direct generalization of the two-body screening technique. In view of the existence of the limit it makes sense to define the full charged-particle arrangement amplitudes $T_{\beta m,\alpha n}(\mathbf{q}'_\beta,\mathbf{q}_\alpha)$ by the left-hand side of (1.6.28), i.e., as the zero-screening limit of the renormalized full screened amplitudes. This definition is not only mathematically correct but is also physically convincing. For the amplitudes defined in this way we read off the representation [compare Eq. (1.4.47)]

$$T_{\beta m,\alpha n}(\mathbf{q}'_\beta,\mathbf{q}_\alpha) = \delta_{\beta\alpha}\delta_{mn}t_\alpha^C(\mathbf{q}'_\alpha,\mathbf{q}_\alpha) + \langle\mathbf{q}'^{(-)}_{\beta,C}|\langle\psi_{\beta m}|U_{\beta\alpha}^{SC}(E_{\alpha n}+i0)|\psi_{\alpha n}\rangle|\mathbf{q}^{(+)}_{\alpha,C}\rangle$$
$$= \delta_{\beta\alpha}\delta_{mn}t_\alpha^C(\mathbf{q}'_\alpha,\mathbf{q}_\alpha) + \tilde{T}_{\beta m,\alpha n}^{SC}(\mathbf{q}'_\beta,\mathbf{q}_\alpha) \quad (1.6.29)$$

For the sake of completeness we note the obvious fact that, if one of the colliding (composite) particles in the initial or final state is neutral, then the corresponding CM Coulomb scattering state is to be replaced by a plane wave, and the Coulomb amplitude t_α^C vanishes. Finally we note that an expression of the form (1.6.29) can also be derived for two-cluster reactions of an arbitrary number of particles.[31,81]

6.5. Practical Construction of $\tilde{T}_{\beta m,\alpha n}^{SC}$ as Zero-Screening Limit

The above screening approach not only represents a means for defining the relevant transition amplitudes but, equally important, constitutes a practical procedure for their determination. In view of the structural similarity with the two-body case, it suffices for this purpose simply to adapt the steps given in Section 4.8:

1. Calculate the full (screened) amplitudes $\langle\mathbf{q}'_\beta|\langle\psi_{\beta m}|U_{\beta\alpha}^{(R)}|\psi_{\alpha n}\rangle|\mathbf{q}_\alpha\rangle$ by any method of short-range theory. It seems natural though not mandatory to use the AGS equations (1.5.40) for the operators $U_{\beta\alpha}^{(R)}$.

2. Before going over to the zero-screening limit, subtract from these amplitudes the CM Coulomb amplitude $\delta_{\beta\alpha}\delta_{mn}t_\alpha^R(\mathbf{q}'_\alpha, \mathbf{q}_\alpha)$. This is absolutely necessary in all approaches based on partial-wave decompositions since, as pointed out at length in the corresponding two-body context, neither this part nor the full amplitude can be approximated by a finite number of partial-wave contributions in the limit $R \to \infty$.

3. Perform the zero-screening limit for this difference, after renormalization, in order to obtain the Coulomb-modified short-range amplitude,

$$Z_{\beta,R}^{-1/2}(q'_\beta)[\langle \mathbf{q}'_\beta | \langle \psi_{\beta m} | U_{\beta\alpha}^{(R)} | \psi_{\alpha n} \rangle | \mathbf{q}_\alpha \rangle - \delta_{\beta\alpha}\delta_{mn}t_\alpha^R(\mathbf{q}'_\alpha, \mathbf{q}_\alpha)]Z_{\alpha,R}^{-1/2}(q_\alpha)$$

$$\xrightarrow[R\to\infty]{} \langle \mathbf{q}'^{(-)}_{\beta,C} | \langle \psi_{\beta m} | U_{\beta\alpha}^{SC} | \psi_{\alpha n} \rangle | \mathbf{q}^{(+)}_{\alpha,C} \rangle \qquad (1.6.30)$$

4. Add to the numerically constructed amplitude (1.6.30) the analytically known two-body Coulomb amplitude $\delta_{\beta\alpha}\delta_{mn}t_\alpha^C(\mathbf{q}'_\alpha, \mathbf{q}_\alpha)$, thereby obtaining the full amplitude (1.6.29).

Although structurally equivalent, the two- and the three-body cases show a characteristic difference. In the former the amplitude T^{SC} is governed by the short-range potential V^S. A limited, usually small, number of partial waves is therefore sufficient for its construction. On the other hand, from (1.6.21) we know that $U_{\beta\alpha}^{SC}|\psi_{\alpha n}\rangle$ depends in an essential manner on the fairly long-range contribution $(\bar{V}_\alpha^C - v_\alpha^C)|\psi_{\alpha n}\rangle$. This is a consequence of the compositeness, and hence polarizability, of one of the two colliding fragments. A rather large number of nonnegligible partial-wave contributions must therefore be taken into account when constructing (1.6.30), and the zero-screening limit is only reached for fairly large values of the screening radius R. In any concrete situation this assertion must be checked numerically.

An important advantage of the above procedure is that any two- and three-body computer program which works for short-range potentials can immediately be employed to calculate the full (screened) transition amplitudes $t_\alpha^R(\mathbf{q}'_\alpha, \mathbf{q}_\alpha)$ and $\langle \Phi_{\beta m, \mathbf{q}'_\beta}|U_{\beta\alpha}^{(R)}(E_{\alpha n}+i0)|\Phi_{\alpha n, \mathbf{q}_\alpha}\rangle$ and thus, via (1.6.30), the Coulomb-modified short-range amplitude. The disadvantage is that it represents a purely numerical and rather time-consuming process, based on repeated calculations of the relevant amplitudes for increasing R until the limit (1.6.30) is reached.

6.6. Integral Equation for $U_{\beta\alpha}^{SC}$

In view of the preceding comment, the direct treatment of the unscreened operators $U_{\beta\alpha}^{SC}$ by means of AGS-type integral equations may appear as a promising alternative (Alt and Sandhas, unpublished; see Alt[36]). In other words, we are looking for a three-body generalization of the two-body integral equation (1.4.56) for T^{SC}. Our starting point is the resolvent equation for the unscreened full and CM Coulomb resolvents (1.6.9) and (1.6.11),

$$G = g_\beta^C + g_\beta^C(\bar{V}_\beta^S + \bar{V}_\beta^C - v_\beta^C)G \tag{1.6.31}$$

Making use of the fact that, according to (1.6.4) and (1.6.5), the Coulomb part of the channel interaction and the CM Coulomb potential are given in coordinate space by

$$\bar{V}_\beta^C = \frac{e_\beta e_\alpha}{r_\gamma} + \frac{e_\beta e_\gamma}{r_\alpha} \tag{1.6.32}$$

and

$$v_\beta^C = \frac{e_\beta e_\alpha}{\rho_\beta} + \frac{e_\beta e_\gamma}{\rho_\beta} \tag{1.6.33}$$

respectively, we see that the Coulomb contribution $\bar{V}_\beta^C - v_\beta^C$ in (1.6.31) can be written as a sum of two subsystem potentials

$$\bar{V}_\beta^C - v_\beta^C = \sum_\gamma \bar{\delta}_{\beta\gamma} \hat{V}_\gamma^C \tag{1.6.34}$$

with $\hat{V}_\gamma^C = e_\beta e_\alpha/r_\gamma + e_\beta e_\alpha/\rho_\beta$. Therefore, the resolvent equation (1.6.31) takes on the form

$$G = g_\beta^C + g_\beta^C \sum_\gamma \bar{\delta}_{\beta\gamma}(V_\gamma^S + \hat{V}_\gamma^C)G \tag{1.6.35}$$

If we insert here on both sides the expression

$$G = \delta_{\beta\alpha} g_\alpha^C + g_\beta^C U_{\beta\alpha}^{SC} g_\alpha^C \tag{1.6.36}$$

which, in analogy to (1.6.14), defines the unscreened Coulomb-modified short-range operators $U_{\beta\alpha}^{SC}$, the same steps which led to the AGS equations (1.5.39) now yield

$$U_{\beta\alpha}^{SC} = \bar{\delta}_{\beta\alpha}(g_\alpha^{C^{-1}} + V_\alpha^S + \hat{V}_\alpha^C) + \sum_\gamma \bar{\delta}_{\beta\gamma}(V_\gamma^S + \hat{V}_\gamma^C)g_\gamma^C U_{\gamma\alpha}^{SC} \tag{1.6.37}$$

This result represents the three-body analog of the two-body equation (1.4.56). To see this in more detail we make use of the spectral decomposition of the Coulomb-distorted channel resolvent (again for repulsive CM

Coulomb potentials only) in the space spanned by the relative momentum states,

$$g_\gamma^C(z) = \int d^3 q_\gamma |\mathbf{q}_{\gamma,C}^{(-)}\rangle \hat{G}_\gamma(z - q_\gamma^2/2M_\gamma)\langle \mathbf{q}_{\gamma,C}^{(-)}| \qquad (1.6.38)$$

where

$$\hat{G}_\gamma(z) = (z - P_\gamma^2/2\mu_\gamma - V_\gamma^S - V_\gamma^C)^{-1} \qquad (1.6.39)$$

denotes the channel resolvent restricted to the γ-subsystem [compare (1.6.11) and the discussion following (1.6.16)]. Inserting this representation into (1.6.37) we obtain

$$\langle \mathbf{q}_{\beta,C}^{\prime(-)}|U_{\beta\alpha}^{SC}(z)|\mathbf{q}_{\alpha,C}^{(+)}\rangle = \bar{\delta}_{\beta\alpha}\langle \mathbf{q}_{\beta,C}^{\prime(-)}|g_\alpha^{C^{-1}}(z) + V_\alpha^S + \hat{V}_\alpha^C|\mathbf{q}_{\alpha,C}^{(+)}\rangle$$
$$+ \sum_\gamma \bar{\delta}_{\beta\gamma} \int d^3 q_\gamma'' \langle \mathbf{q}_{\beta,C}^{\prime(-)}|V_\gamma^S + \hat{V}_\gamma^C|\mathbf{q}_{\gamma,C}^{\prime\prime(-)}\rangle \hat{G}_\gamma(z - q_\gamma^{\prime\prime 2}/2M_\gamma)$$
$$\times \langle \mathbf{q}_{\gamma,C}^{\prime\prime\prime(-)}|U_{\gamma\alpha}^{SC}(z)|\mathbf{q}_{\alpha,C}^{(+)}\rangle \qquad (1.6.40)$$

The similarity of this result to the two-body integral equation (1.4.57) is evident.

Sandwiching both sides of Eq. (1.6.40) between $\langle \mathbf{p}_\beta'|$ and $|\psi_{\alpha n}\rangle$ and going over to the momentum representation of $\hat{G}_\gamma$, this last relation becomes a six-dimensional (after partial-wave decomposition two-dimensional) integral equation for the amplitudes $\langle \mathbf{q}_{\beta,C}^{\prime(-)}|\langle \mathbf{p}_\beta'|U_{\beta\alpha}^{SC}|\psi_{\alpha n}\rangle |\mathbf{q}_{\alpha,C}^{(+)}\rangle$. With some care it may be treated by numerical methods known from the short-range case. Moreover, separable expansions of $\hat{G}_\gamma$, or other standard techniques such as the quasi-particle approach,[29] can be employed in order to reduce (1.6.40) to (one-dimensional) effective two-body equations.

Whatever methods or approximations are used with respect to $\hat{G}_\gamma$, the relative motion of the fragments under the influence of the CM Coulomb potential is fully taken into account in (1.6.40). Note that if V_γ^S is given in coordinate space, the inhomogeneity and the potential term in the kernel of this integral equation can be computed directly by employing the analytically known Coulomb functions $\langle \mathbf{\rho}_\alpha|\mathbf{q}_{\alpha,C}^{(\pm)}\rangle$.

As discussed after (1.6.30) in the context of the screening method, the Coulomb-modified short-range operators depend on $(\bar{V}_\alpha^R - v_\alpha^R)|\psi_{\alpha n}\rangle$, which becomes fairly long-range for $R \to \infty$. This requires large values of the screening radius and careful stability investigations when going over to the zero-screening limit. The counterpart of this complication is the occurrence of the potentials $\hat{V}_\gamma^C$ in Eq. (1.6.40). From their definition (1.6.34) we infer that, when applied onto the normalizable subsystem bound states, they decrease in general only as ρ_γ^{-2} [compare the arguments given in the context of (1.6.27)], and behave even like Coulomb potentials when acting on nonnormalizable subsystem states, e.g., two-body continuum states which necessarily occur if the spectral decomposition of $\hat{G}_\gamma$ is used. Correspondingly, the kernel of (1.6.40) is quite long-range in the coordinate representation, making the numerical treatment of this integral equation a nontrivial task. On the other hand, the occurrence of $\hat{V}_\gamma^C$ is a consequence of the γ-subsystem being composite. Neglect of these terms, as is done in many approximate treatments, therefore ignores the most characteristic feature of Coulomb-distorted composite-particle collisions.

Let us note that Eq. (1.6.40), or alternative integral equations[31,36,82–84] based on the same idea of splitting off the CM Coulomb singularities, have hardly been employed in practical scattering calculations. The screening procedure of Section 6.5, however, has been used successfully and led to the first reliable results for the Coulomb effects in proton–deuteron collisions (see Section 8).

6.7. Three-Body Breakup with Two Charged Outgoing Particles

The above formalism can be extended to the description of breakup processes. Let us consider the reaction where the charged particle α impinges on a $(\beta\gamma)$ bound state with γ charged and β neutral. From (1.5.58) we then infer the representation

$$T^{(R)}_{0,\alpha n}(\mathbf{q}'_\beta, \mathbf{p}'_\beta; \mathbf{q}_\alpha) = \langle \mathbf{q}'_\beta | \langle \mathbf{p}'^{(-)}_{\beta(R)} | U^{(R)}_{\beta\alpha}(E_{\alpha n} + i0) | \psi_{\alpha n} \rangle | \mathbf{q}_\alpha \rangle \qquad (1.6.41)$$

of the breakup amplitude, with $|\mathbf{p}^{(-)}_{\beta(R)}\rangle$ being the two-charged-particle scattering state generated by the full interaction $V^{(R)}_\beta = V^S_\beta + V^R_\beta$. Because $e_\beta = 0$, which implies $v^R_\beta = 0$ and hence $t^R_\beta = 0$, Eq. (1.6.15) simplifies to $(\beta \neq \alpha)$

$$U^{(R)}_{\beta\alpha} = U^{SR}_{\beta\alpha}(1 + G^{(R)}_\alpha t^R_\alpha) \qquad (1.6.42)$$

Inserting this relation in (1.6.41) and using (1.6.17) we obtain

$$T^{(R)}_{0,\alpha n}(\mathbf{q}'_\beta, \mathbf{p}'_\beta; \mathbf{q}_\alpha) = \langle \mathbf{q}'_\beta | \langle \mathbf{p}'^{(-)}_{\beta(R)} | U^{SR}_{\beta\alpha}(E_{\alpha n} + i0) | \psi_{\alpha n} \rangle | \mathbf{q}^{(+)}_{\alpha,R} \rangle \qquad (1.6.43)$$

As in (1.6.28) we see that after multiplication with the appropriate screening factors, the zero-screening limit of this equation exists, viz,

$$\lim_{R \to \infty} Z_R^{-1/2}(p'_\beta) T^{(R)}_{0,\alpha n}(\mathbf{q}'_\beta, \mathbf{p}'_\beta; \mathbf{q}_\alpha) Z_{\alpha,R}^{-1/2}(q_\alpha) = \langle \mathbf{q}'_\beta | \langle \mathbf{p}'^{(-)}_\beta | U^{SC}_{\beta\alpha}(E_{\alpha n} + i0) | \psi_{\alpha n} \rangle | \mathbf{q}^{(+)}_{\alpha,C} \rangle$$

$$(1.6.44)$$

Here,

$$Z_R(p_\beta) = \exp\{-2ie_\alpha e_\gamma \mu_\beta p_\beta^{-1}[\ln(2p_\beta R) - C]\} \qquad (1.6.45)$$

is the renormalization factor in the $(\alpha\gamma)$-subsystem [compare Eqs. (1.3.4) (1.3.6)], and the $|\mathbf{p}_\beta^{(\pm)}\rangle$ are the two-body scattering states corresponding to $V_\beta^S + V_\beta^C$.

The existence of the limit (1.6.44) justifies our defining the breakup amplitude $T_{0,\alpha n}(\mathbf{q}_\beta', \mathbf{p}_\beta'; \mathbf{q}_\alpha)$ as the left-hand side of it. As in the rearrangement case it can also be used for practical calculations of the breakup amplitude in the zero-screening limit. The amplitude introduced in this way is

$$T_{0,\alpha n}(\mathbf{q}_\beta', \mathbf{p}_\beta'; \mathbf{q}_\alpha) = \langle \mathbf{q}_\beta' | \langle \mathbf{p}_\beta'^{(-)} | U_{\beta\alpha}^{SC}(E_{\alpha n} + i0) | \psi_{\alpha n} \rangle | \mathbf{q}_{\alpha,C}^{(+)} \rangle \qquad (1.6.46)$$

Of course, the scattering state $|\mathbf{q}_{\alpha,C}^{(+)}\rangle$ is to be replaced by the plane wave $|\mathbf{q}_\alpha\rangle$ if one of the incident fragments is neutral.

6.8. Equivalence with the Dollard Amplitudes for Rearrangement Scattering

The Møller operators (1.5.13) or (1.5.14) do not exist if the channel interaction between the colliding fragments contains Coulomb contributions. In the above treatment convergence has been enforced by using screened Coulomb potentials, deferring the transition to the zero-screening limit to the renormalized screened amplitudes. As emphasized in the last section this is not only an intuitively convincing way of defining the correct amplitudes, but also provides a practical means for their numerical determination.

An alternative, physically equally well-motivated approach to the charged-particle scattering problem has been established by Dollard.[23] (See also Sakhnovich[85]; other approaches have been developed, e.g., by Zorbas,[21,88] Rosenberg,[86] and Chandler and Gibson[87] and are reviewed in Chandler [34] and Alt[36]). Here, convergence is achieved by introducing additional time-dependent factors in a modified definition of the Møller operators. It is instructive to show that this technically and conceptually quite different approach leads to the same transition amplitudes as the screening procedure.[32]

The total Hamiltonian H, the channel Hamiltonian H_α and the Coulomb-distorted channel Hamiltonian h_α^C considered in the following are given by (1.6.6), (1.6.7), and (1.6.8) without the screening factors. Introducing the kinetic energy operator for the relative motion of the fragments

$$h_\alpha^0 = Q_\alpha^2/2M_\alpha \qquad (1.6.47)$$

and the channel Hamiltonian restricted to the α-subsystem

$$\hat{H}_\alpha = P_\alpha^2/2\mu_\alpha + V_\alpha^S + V_\alpha^C \tag{1.6.48}$$

the operators H_α and h_α^C can be written as

$$H_\alpha = h_\alpha^0 + \hat{H}_\alpha \tag{1.6.49}$$

and

$$h_\alpha^C = H_\alpha + v_\alpha^C = (h_\alpha^0 + v_\alpha^C) + \hat{H}_\alpha \tag{1.6.50}$$

We also introduce the operator

$$\hat{h}_\alpha^0(t) = \varepsilon(t) e_\alpha (e_\beta + e_\gamma) \sqrt{M_\alpha/2h_\alpha^0} \ln(4h_\alpha^0|t|) \tag{1.6.51}$$

where $\varepsilon(t) = \pm 1$ for $t \gtrless 0$ is the sign function [for a transparent introduction of $\hat{h}_\alpha^0(t)$ see Kulish and Faddeev[89]].

According to Dollard, the strong limits

$$\Omega_\alpha^{(\pm)} = \text{s-lim}_{t \to \mp\infty} e^{iHt} e^{-iH_\alpha t} e^{-i\hat{h}_\alpha^0(t)} \mathbb{P}_\alpha \tag{1.6.52}$$

representing the channel Møller operators, exist. The same holds true a fortiori for the

$$\omega_{\alpha,C}^{(\pm)} = \text{s-lim}_{t \to \mp\infty} e^{i(h_\alpha^0 + v_\alpha^C)t} e^{-ih_\alpha^0 t} e^{-i\hat{h}_\alpha^0(t)} \tag{1.6.53}$$

which act as two-particle Møller operators on the relative momentum state $|\mathbf{q}_\alpha\rangle$, mapping it onto the Coulomb scattering states $|\mathbf{q}_{\alpha,C}^{(\pm)}\rangle$. Writing the product of exponentials in (1.6.52) as

$$e^{iHt} e^{-iH_\alpha t} e^{-i\hat{h}_\alpha^0(t)} = e^{iHt} e^{-ih_\alpha^C t} e^{ih_\alpha^C t} e^{-iH_\alpha t} e^{-i\hat{h}_\alpha^0(t)}$$

$$= e^{iHt} e^{-ih_\alpha^C t} e^{i(h_\alpha^0 + v_\alpha^C)t} e^{-ih_\alpha^0 t} e^{-i\hat{h}_\alpha^0(t)} \tag{1.6.54}$$

we see that the channel Møller operators can be represented in the form

$$\Omega_\alpha^{(\pm)} = \Omega_{\alpha,SC}^{(\pm)} \omega_{\alpha,C}^{(\pm)} \mathbb{P}_\alpha \tag{1.6.55}$$

with the Coulomb-modified short-range Møller operators being given by

$$\Omega_{\alpha,SC}^{(\pm)} = \text{s-lim}_{t \to \mp\infty} e^{iHt} e^{-ih_\alpha^C t} \tag{1.6.56}$$

Since the difference $\bar{V}_\alpha^C - v_\alpha^C$, occurring in the difference of H and h_α^C, decreases as ρ_α^{-2} when acting on bound states $|\psi_{\alpha n}\rangle$, the existence of (1.6.56) is at least plausible. There is, however, no need to go through a detailed direct proof based on this property. For, from the existence of $\Omega_\alpha^{(\pm)}$ and $\omega_{\alpha,C}^{(\pm)}$, guaranteed by Dollard's proof, the existence of $\Omega_{\alpha,SC}^{(\pm)}$ on states $\omega_{\alpha,C}^{(\pm)}|\Phi_{\alpha n}\rangle$ is rigorously inferred (compare the analogous proof given for short-range operators in Sandhas[66]).

Making use of (1.6.55), the S-matrix, introduced as in (1.5.19) but now with the present definition (1.6.52) of the Møller operators, takes the form

$$S_{\beta m,\alpha n}(\mathbf{q}_\beta', \mathbf{q}_\alpha) = \langle \mathbf{q}_\beta' | \langle \psi_{\beta m} | \omega_{\beta,C}^{(-)\dagger} \Omega_{\beta,SC}^{(-)\dagger} \Omega_{\alpha,SC}^{(+)} \omega_{\alpha,C}^{(+)} | \psi_{\alpha n}\rangle | \mathbf{q}_\alpha\rangle$$
$$= \langle \mathbf{q}_{\beta,C}^{\prime(-)} | \langle \psi_{\beta mn} | S_{\beta\alpha}^{SC} | \psi_{\alpha n}\rangle | \mathbf{q}_{\alpha,C}^{(+)}\rangle \tag{1.6.57}$$

A procedure analogous to that in Section 5.2 then leads to[32]

$$T_{\beta m,\alpha n}(\mathbf{q}_\beta', \mathbf{q}_\alpha) = \delta_{\beta\alpha}\delta_{mn} t_\alpha^C(\mathbf{q}_\alpha', \mathbf{q}_\alpha) + \langle \mathbf{q}_{\beta,C}^{\prime(-)} | \langle \psi_{\beta m} | U_{\beta\alpha}^{SC}(E_{\alpha n} + i0) | \psi_{\alpha n}\rangle | \mathbf{q}_{\alpha,C}^{(+)}\rangle \tag{1.6.58}$$

with $U_{\beta\alpha}^{SC}$ being defined by (1.6.36).

It should be emphasized that this last step requires some care. A detailed investigation of the analyticity structure of the matrix elements of $U_{\beta\alpha}^{SC}$, similar to the one of Section 6.3, is necessary. The result is that for the splitting of the channel interaction underlying the present case the representation (1.6.58) is correct; but it is not valid for any arbitrary splitting of the channel interaction. In other words the representation (1.6.58), with its diagonal first term on the right-hand side, is typical for a splitting in which the corresponding part of the potential does not induce rearrangement.

Comparing (1.6.29) with (1.6.58) we see that their right-hand sides coincide. The amplitudes $T_{\beta m,\alpha n}$ defined in the screening method as zero-screening limits, and in the Dollard approach as time limits, are therefore identical.

6.9. Equivalence with the Dollard Amplitudes for Breakup Processes

A similar argument can be applied to breakup processes in a system of one neutral and two charged particles. We choose the same configuration as in Section 6.7, i.e., $e_\beta = 0$. In analogy to (1.6.51), the operator

$$\hat{H}_\beta^0(t) = \varepsilon(t) e_\alpha e_\gamma \sqrt{\mu_\beta/2\hat{H}_\beta^0} \ln(4\hat{H}_\beta^0 |t|) \tag{1.6.59}$$

is introduced, where $\hat{H}_\beta^0 = P_\beta^2/2\mu_\beta$. Then, according to Dollard, the strong limits

$$\Omega_0^{(\pm)} = \underset{t\to\mp\infty}{\text{s-lim}}\ e^{iHt}e^{-iH_0t}e^{-i\hat{H}_\beta^0(t)} \tag{1.6.60}$$

exist, defining the Møller operators for the three-free-particle channel considered. The same holds true, a fortiori, for the two-particle Møller operators

$$\begin{aligned}\hat{\Omega}_\beta^{(\pm)} &= \underset{t\to\mp\infty}{\text{s-lim}}\ e^{iH_\beta t}e^{-iH_0 t}e^{-i\hat{H}_\beta^0(t)} \\ &= \underset{t\to\mp\infty}{\text{s-lim}}\ e^{i\hat{H}_\beta t}e^{-i\hat{H}_\beta^0 t}e^{-i\hat{H}_\beta^0(t)}\end{aligned} \tag{1.6.61}$$

which map the plane wave $|\mathbf{p}_\beta\rangle$ onto the full two-charged-particle scattering states $|\mathbf{p}_\beta^{(\pm)}\rangle$. Similarly to (1.6.55), we thus infer the representation

$$\Omega_0^{(\pm)} = \tilde{\Omega}_{\beta,SC}^{(\pm)}\hat{\Omega}_\beta^{(\pm)} \tag{1.6.62}$$

with the Coulomb-modified short-range Møller operators given as

$$\tilde{\Omega}_{\beta,SC}^{(\pm)} = \underset{t\to\mp\infty}{\text{s-lim}}\ e^{iHt}e^{-iH_\beta t} \tag{1.6.63}$$

Their existence follows rigorously, as described after (1.6.56), but is immediately plausible since the difference of the potentials occurring in H and H_β is of short range only (recall $\bar{V}_\beta = \bar{V}_\beta^s$ for $e_\beta = 0$). If we make use of the decompositions (1.6.55) and (1.6.62), the breakup S-matrix element, introduced as in (1.5.50) but now with the present definitions (1.6.52) and (1.6.60) of the Møller operators, takes the form

$$\begin{aligned}S_{0,\alpha n}(\mathbf{q}_\beta', \mathbf{p}_\beta'; \mathbf{q}_\alpha) &= \langle\mathbf{q}_\beta'|\langle\mathbf{p}_\beta'|\Omega_0^{(-)\dagger}\Omega_\alpha^{(+)}|\psi_{\alpha n}\rangle|\mathbf{q}_\alpha\rangle \\ &= \langle\mathbf{q}_\beta'|\langle\mathbf{p}_\beta'^{(-)}|\tilde{\Omega}_{\beta,SC}^{(-)\dagger}\Omega_{\alpha,SC}^{(+)}|\psi_{\alpha n}\rangle|\mathbf{q}_{\alpha,C}^{(+)}\rangle\end{aligned} \tag{1.6.64}$$

Because $e_\beta = 0$ we also have $v_\beta = 0$, and hence $g_\beta^C = G_\beta$ [compare (1.6.10) and (1.6.11)]. Thus, the defining equation (1.6.36) of the Coulomb-modified short-range operators specializes to

$$G = \delta_{\beta\alpha}g_\alpha^C + G_\beta U_{\beta\alpha}^{SC}g_\alpha^C \tag{1.6.65}$$

Proceeding as in Section 5.2 [note that the first term on the r.h.s. of (1.6.65) does not contribute because of the orthogonality of bound and scattering states in subsystem α, $\langle \mathbf{p}_\alpha^{\prime(-)}|\psi_{\alpha n}\rangle = 0$], we finally obtain

$$T_{0,\alpha n}(\mathbf{q}'_\beta, \mathbf{p}'_\beta; \mathbf{q}_\alpha) = \langle \mathbf{q}'_\beta|\langle \mathbf{p}_\beta^{\prime(-)}|U_{\beta\alpha}^{SC}(E_{\alpha n} + i0)|\psi_{\alpha n}\rangle|\mathbf{q}_{\alpha,C}^{(+)}\rangle \qquad (1.6.66)$$

which coincides with the result (1.6.46) for the breakup amplitude derived in the screening and renormalization approach.

Although in principle equivalent, the two procedures are quite different in spirit and in detail. The limits in (1.6.28), or better in (1.6.30), and in (1.6.44) are just ordinary limits of functions of a real variable R, and are therefore well suited for numerical applications. The limits in Dollard's approach are strong limits in Hilbert space, i.e., they have to be performed after integration over momentum distribution functions. Such definitions serve, of course, primarily to provide a mathematically rigorous formulation of the problem. Nevertheless they may also be used as a means for numerically constructing the corresponding amplitudes.[90]

We finally note that the transition from Dollard's Møller operators to scattering amplitudes has also been investigated for two particles by Grosse et al.[91] and Bajzer,[92] for three particles by Veselova[93] and Bajzer,[94] and for two-cluster reactions of an arbitrary number of particles by Alt and Sandhas.[32]

7. EFFECTIVE TWO-BODY FORMALISM

The screening and renormalization approach described above represents a means not only for defining but also for calculating the amplitudes for elastic, rearrangement, and breakup collisions under the action of short-range and Coulomb forces. In such practical calculations *any* method of short-range theory can be used to compute first the amplitudes for screened Coulomb potentials and then, via (1.6.28) and (1.6.44), the amplitudes of the original unscreened problem.

In this context the AGS equations represent a particularly convenient tool. As is known from the pure short-range case, they provide a suitable starting point for momentum-space calculations of the desired amplitudes. Moreover, they can be reduced to equivalent effective two-body equations in which the Coulomb corrections occur essentially as additive contributions to the effective interactions. By proceeding in this manner one arrives at a formalism which is both easily interpretable and technically manageable, and which allows one to incorporate the Coulomb effects in an exact manner, and which can be used as an alternative proof of the screening procedure developed in the preceding section. The first justification of this method was given, in fact, along this line.[28] Approximations, still unavoidable till now, have been suggested and can be studied within this framework.

The content of the present section is based on this effective two-body approach.[28,33]

7.1. Rearrangement Collisions

According to (1.5.26) the screened arrangement amplitudes are given by

$$T^{(R)}_{\beta m, \alpha n}(\mathbf{q}'_\beta, \mathbf{q}_\alpha) = \langle \mathbf{q}'_\beta | \langle \psi_{\beta m} | U^{(R)}_{\beta \alpha}(E_{\alpha n} + i0) | \psi_{\alpha n} \rangle | \mathbf{q}_\alpha \rangle \qquad (1.7.1)$$

with $U^{(R)}_{\beta\alpha}$ being defined either by (1.5.22), by (1.5.30) or by means of the AGS equations (1.5.40),

$$U^{(R)}_{\beta\alpha} = \bar{\delta}_{\beta\alpha} G_0^{-1} + \sum_\gamma \bar{\delta}_{\beta\gamma} T^{(R)}_\gamma G_0 U^{(R)}_{\gamma\alpha} \qquad (1.7.2)$$

The two-body T-operators in the kernel of these equations correspond to the full potentials $V^{(R)}_\gamma = V^S_\gamma + V^R_\gamma$, and thus satisfy the LS equations

$$T^{(R)}_\gamma = V^{(R)}_\gamma + V^{(R)}_\gamma G_0 T^{(R)}_\gamma \qquad (1.7.3)$$

Applied onto $|\Phi_{\alpha n, \mathbf{q}_\alpha}\rangle = |\psi_{\alpha n}\rangle |\mathbf{q}_\alpha\rangle$, Eq. (1.7.2) represents a well-defined (after partial-wave decomposition two-dimensional) integral equation for the functions $\langle \mathbf{q}'_\beta | \langle \mathbf{p}'_\beta | U^{(R)}_{\beta\alpha} | \psi_{\alpha n} \rangle | \mathbf{q}_\alpha \rangle$ which, after integrating over the two-body wave functions $\psi^*_{\beta m}(\mathbf{p}'_\beta)$, yield the amplitudes (1.7.1).

For purely short-range interactions, a direct solution of the two-dimensional AGS equations in terms of Padé approximants has become possible (the literature can be traced, e.g., from Glöckle,[68] Witala et al.,[95] and Glöckle et al.[96]). In principle one may therefore try to do the same in the (screened) Coulomb case. For sufficiently small screening radii this is certainly possible. Whether it remains practicable when going over to large R, as required by the limit (1.6.28), is an open question.

Given the foregoing remarks, the effective two-body formalism developed by Alt et al.[29] represents a promising practical alternative. It is based on a splitting of the amplitude $T^{(R)}_\gamma$ into a finite number of separable terms and a nonseparable remainder. Inserted in (1.7.2) such a splitting leads to an exact reduction of the original three-body equations to effective two-body equations. Depending on the technique employed in constructing these separable terms (separable potential expansions,[97–99] spectral decompositions of subsystem resolvents,[63] expansion into Sturmian functions,[100–102] etc.), structurally similar but in detail rather different final equations are obtained.

We restrict ourselves to the standard procedure based on approximating the originally given (local or nonlocal) short-range potential V^S_γ by a series of separable terms,

$$V_\gamma^S \approx \sum_r |\chi_{\gamma r}\rangle \lambda_{\gamma r} \langle \chi_{\gamma r}| \qquad (1.7.4)$$

characterized by "form factors" $|\chi_{\gamma r}\rangle$ and strength parameters $\lambda_{\gamma r}$. In other words, when considered, e.g., in momentum space this potential is approximated by a sum of terms which are separable in the variables $\mathbf{p}'$ and $\mathbf{p}$,

$$V_\gamma^S(\mathbf{p}', \mathbf{p}) \approx \sum_r \chi_{\gamma r}(\mathbf{p}') \lambda_{\gamma r} \chi_{\gamma r}^*(\mathbf{p}) \qquad (1.7.5)$$

In principle the difference between the original interaction and the approximation (1.7.4) could be taken into account within the exact AGS reduction scheme[29] as a (nonseparable) remainder. However, this leads to fairly clumsy final relations, which hide the simple idea underlying the following considerations. We therefore consider this remainder as being negligible. In fact, in numerous three-nucleon calculations it has been demonstrated that expanding V_γ^S into a few (or even only one) appropriately chosen separable terms, yields an excellent approximation to the nuclear (short-range) two-body input (see, e.g., Bartnik et al.,[103] Plessas and Heidenbauer,[104] Koike et al.,[105] and Rescigno and Orel[106]). Although for small R we could also expand V_γ^R in this manner, such an expansion would become more and more problematic as $R \to \infty$, and is fortunately not necessary. It is one of the virtues of the AGS reduction technique that V_γ^R, as any nonseparable potential part, can be taken into account without approximation.

For a short-range potential given exactly by the r.h.s. of (1.7.4), and a screened Coulomb potential V_γ^R, the subsystem transition operator $T_\gamma^{(R)}$ entering the kernel of (1.7.2) is represented by

$$T_\gamma^{(R)}(z) = \sum_{r,s} |\varphi_{\gamma r}^{(R)}(z)\rangle \Delta_{\gamma, rs}^{(R)}(z) \langle \varphi_{\gamma s}^{(R)}(z^*)| + T_\gamma^R \qquad (1.7.6)$$

Here T_γ^R denotes the pure (screened) Coulomb amplitude,

$$T_\gamma^R(z) = V_\gamma^R + V_\gamma^R G_0(z) T_\gamma^R(z) \qquad (1.7.7)$$

The Coulomb-distorted form factors and the propagators are given by, respectively,

$$|\varphi_{\gamma r}^{(R)}(z)\rangle = [1 + T_\gamma^R(z) G_0(z)] |\chi_{\gamma r}\rangle \qquad (1.7.8)$$

and

$$[\Delta_\gamma^{(R)^{-1}}(z)]_{rs} = \delta_{rs} \lambda_{\gamma r}^{-1} - \langle \chi_{\gamma r} | G_0(z) | \varphi_{\gamma s}^{(R)}(z)\rangle \qquad (1.7.9)$$

A splitting of the form (1.7.6) reduces the AGS equations to effective two-body LS equations which in matrix notation are[29]

$$\mathcal{T}^{(R)} = \mathcal{V}^{(R)} + \mathcal{V}^{(R)}\mathcal{G}_0^{(R)}\mathcal{T}^{(R)} \tag{1.7.10}$$

Here, the amplitude, potential, and free Green's function matrices are defined by

$$\mathcal{T}_{\beta m,\alpha n}^{(R)} = \langle \varphi_{\beta m}^{(R)}(z^*)|G_0(z)U_{\beta\alpha}^{(R)}(z)G_0(z)|\varphi_{\alpha n}^{(R)}(z)\rangle \tag{1.7.11}$$

$$\mathcal{V}_{\beta m,\alpha n}^{(R)} = \langle \varphi_{\beta m}^{(R)}(z^*)|G_0(z)U_{\beta\alpha}^{R}(z)G_0(z)|\varphi_{\alpha n}^{(R)}(z)\rangle \tag{1.7.12}$$

$$\mathcal{G}_{0;\beta m,\alpha n}^{(R)} = \delta_{\beta\alpha}\Delta_{\alpha,mn}^{(R)}(z) \tag{1.7.13}$$

The auxiliary operators $U_{\beta\alpha}^R$ occurring in (1.7.12) satisfy equations of the form (1.7.2), but with the full amplitude $T_\gamma^{(R)}$ being replaced by the pure Coulomb amplitude T_γ^R, i.e.,

$$U_{\beta\alpha}^R = \bar{\delta}_{\beta\alpha}G_0^{-1} + \sum_\gamma \bar{\delta}_{\beta\gamma}T_\gamma^R G_0 U_{\gamma\alpha}^R \tag{1.7.14}$$

In (1.7.11)–(1.7.13) the three-body operators are sandwiched between two-body states. The resulting quantities are therefore two-body operators acting in the space spanned by the relative momentum states of the colliding clusters. With suitable restrictions[29] on $|\chi_{\gamma r}\rangle$ and thus also on $|\varphi_{\gamma r}^{(R)}(z)\rangle$, the effective T-matrix defined by (1.7.11) goes over on the energy shell into the desired transition amplitude,

$$T_{\beta m,\alpha n}^{(R)}(\mathbf{q}'_\beta, \mathbf{q}_\alpha) = \langle \mathbf{q}'_\beta|\mathcal{T}_{\beta m,\alpha n}^{(R)}(E_{\alpha n} + i0)|\mathbf{q}_\alpha\rangle \tag{1.7.15}$$

The integral equation (1.7.14) can, e.g., be treated iteratively. If only two of the three particles are charged, and consequently only one of the three operators T_γ^R differs from zero, only one iteration occurs, yielding the exact result

$$U_{\beta\alpha}^R = \bar{\delta}_{\beta\alpha}G_0^{-1} + \sum_\gamma \bar{\delta}_{\beta\gamma}T_\gamma^R \bar{\delta}_{\gamma\alpha} \tag{1.7.16}$$

Thus, by solving (1.7.10) with the effective potential (1.7.12) determined via (1.7.16), we get a full screened amplitude which is exact with respect to the Coulomb contributions. The only approximation concerns the separable representation (1.7.4) of the short-range potential, an approximation which, as noted above, has been successfully used in many calculations with purely short-range interactions. Accordingly, the Coulomb-distorted amplitudes

(1.7.1) are obtained to the same level of sophistication as achieved in conventional treatments of short-range scattering problems. In particular, the error can be similarly minimized by increasing the number of separable terms.

The present formalism is thus a practical means for determining the screened amplitudes which, when subjected to the unscreening procedure (1.6.28), provide us with the correct Coulomb-distorted final amplitudes. Up to now the effective potential (1.7.12) could be determined only with the additional approximation

$$T_\gamma^R \to V_\gamma^R \tag{1.7.17}$$

which, however, appears to be reasonably well justified (cf. the discussion of this point in Section 8.1.2)

7.2. Breakup Reactions

Effective two-body equations of the form (1.7.10) can also be derived for the breakup amplitudes. Applying the above reduction technique to the operator relation (1.5.55), we obtain the matrix equation

$$\mathscr{T}^{bu(R)} = \mathscr{V}^{bu(R)} + \mathscr{V}^{bu(R)} \mathscr{G}_0^{(R)} \mathscr{T}^{(R)} \tag{1.7.18}$$

Here, the elements of the breakup amplitude and potential matrices are given by

$$\mathscr{T}^{bu(R)}_{\mathbf{p}'_\beta, \alpha n}(z) = \langle \mathbf{p}'_\beta | U_{0\alpha}^{(R)}(z) G_0(z) | \varphi_{\alpha n}^{(R)}(z) \rangle \tag{1.7.19}$$

and

$$\mathscr{V}^{bu(R)}_{\mathbf{p}'_\beta, \alpha n}(z) = \langle \mathbf{p}'_\beta | U_{0\alpha}^R(z) G_0(z) | \varphi_{\alpha n}^{(R)}(z) \rangle \tag{1.7.20}$$

As anticipated by the structure of (1.5.55), Eq. (1.7.18) determines the breakup amplitude by quadrature from the rearrangement amplitudes $\mathscr{T}^{(R)}$, i.e., by means of the solutions of (1.7.10). However, starting from (1.5.56) instead of (1.5.55), an integral equation rather than an algebraic relation is obtained also for $\mathscr{T}^{bu(R)}$,

$$\mathscr{T}^{bu(R)} = \mathscr{V}^{bu(R)} + \mathscr{T}^{bu(R)} \mathscr{G}_0^{(R)} \mathscr{V}^{(R)} \tag{1.7.21}$$

which has essentially the same kernel as (1.7.10) but a different inhomogeneity. Under the restrictions on $|\chi_{\gamma r}\rangle$ leading to (1.7.15), the plane wave matrix elements of the effective amplitude (1.7.19) go over on the energy shell to the breakup amplitudes

$$T^{(R)}_{0,\alpha n}(\mathbf{q}'_\beta, \mathbf{p}'_\beta; \mathbf{q}_\alpha) = \langle \mathbf{q}'_\beta | \mathcal{T}^{bu(R)}_{\mathbf{p}'_\beta,\alpha n}(E_{\alpha n} + i0) | \mathbf{q}_\alpha \rangle \tag{1.7.22}$$

After renormalization it yields in the zero-screening limit (1.6.44) the physical breakup amplitude (1.6.46).

Since only two of the three particles are assumed to be charged, the operator $U_{0\alpha}$ which determines the breakup potential (1.7.20) is again given exactly by (1.7.16) but with $\beta = 0$.

7.3. Specialization to Proton–Deuteron Scattering

It is instructive to write down the above relations in some detail for proton–deuteron scattering. For simplicity we choose separable nuclear interactions of rank one

$$V^S_\gamma = |\chi_\gamma\rangle \lambda_\gamma \langle \chi_\gamma| \tag{1.7.23}$$

The form factors $|\chi_\gamma\rangle$ and strength parameters λ_γ are chosen to reproduce the low-energy nucleon–nucleon data. Denoting the protons as particles 1 and 2, the Coulomb contribution occurs only in the subsystem 3, i.e., we have

$$V^R_\gamma(r_\gamma) = \delta_{\gamma 3} V^R_3(r_3) = \delta_{\gamma 3} \frac{e^2}{r_3} e^{-r_3/R} \tag{1.7.24}$$

Similarly we may write for the corresponding Coulomb amplitude

$$T^R_\gamma = \delta_{\gamma 3} T^R_3 \tag{1.7.25}$$

The effective LS equation (1.7.10) is then a 3 × 3 matrix relation

$$\mathcal{T}^{(R)}_{\beta\alpha} = \mathcal{V}^{(R)}_{\beta\alpha} + \sum_\gamma \mathcal{V}^{(R)}_{\beta\gamma} \mathcal{G}^{(R)}_{0;\gamma} \mathcal{T}^{(R)}_{\gamma\alpha} \tag{1.7.26}$$

the potential matrix (1.7.12) and the effective free Green function (1.7.13) being given by

$$\mathcal{V}^{(R)}_{\beta\alpha} = \bar{\delta}_{\beta\alpha} \langle \chi_\beta | G_0 + (\delta_{\alpha 3} + \delta_{\beta 3} + \bar{\delta}_{\beta 3} \bar{\delta}_{\alpha 3}) G_0 T^R_3 G_0 | \chi_\alpha \rangle + \delta_{\beta\alpha} \bar{\delta}_{\alpha 3} \langle \chi_\alpha | G_0 T^R_3 G_0 | \chi_\alpha \rangle \tag{1.7.27}$$

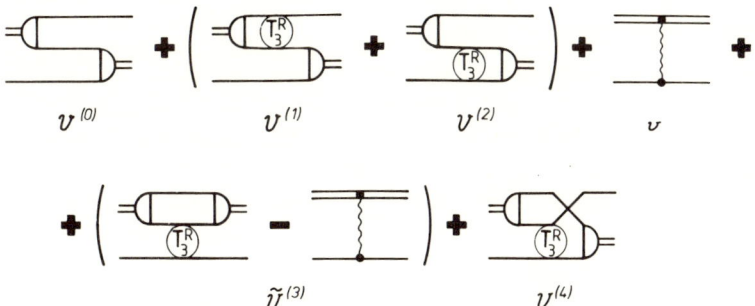

FIGURE 1. Graphical representation of the different contributions to the effective potential (1.7.27). Semicircles denote the form factors. $\mathscr{V}^{(0)}$ is the driving term (one-nucleon exchange) relevant for neutron-deuteron scattering, $\mathscr{V}^{(1)}$ and $\mathscr{V}^{(2)}$ contain the Coulomb modifications of the form factors, v is the center-of-mass Coulomb potential extracted as in (1.7.35), $\tilde{\mathscr{V}}^{(3)}$ is the remainder of the diagonal part, and $\mathscr{V}^{(4)}$ is the Coulomb exchange part.

and

$$\mathscr{G}^{(R)}_{0;\alpha} = (\lambda_\alpha^{-1} - \langle \chi_\alpha | G_0 + \delta_{\alpha 3} G_0 T_3^R G_0 | \chi_\alpha \rangle)^{-1} \quad (1.7.28)$$

In deriving these expressions we have made use of the explicit representations (1.7.8) and (1.7.16) and, in particular, of the property (1.7.25). The graphical representation of the various terms in (1.7.27) is given in Fig. 1.

We emphasize once more that, via the effective potential (1.7.27) and the Green function (1.7.28), the Coulomb contribution to the interaction is taken into account exactly. The only approximation lies in the simple choice (1.7.23) for the nuclear interaction which, however, can be optimized by choosing suitable form factors, or improved upon by taking into account a higher number of separable terms. In the latter case we end up, of course, with a higher dimension for the matrices (1.7.11)–(1.7.13) and thus of the number of coupled equations (1.7.10).

According to (1.7.18) the half-off-shell solutions of (1.7.26) can also be used to calculate the deuteron breakup by proton impact. The only additional ingredient is the effective breakup potential (1.7.20), which for nuclear forces of the type (1.7.23) takes the simple exact form (with $\beta = 3$)

$$\mathscr{V}^{bu(R)}_{\mathbf{p}'_3,\alpha} = \langle \mathbf{p}'_3 | [1 + T_3^R(z) G_0(z)] | \chi_\alpha \rangle \quad (1.7.29)$$

Solving Eq. (1.7.26) in momentum space, i.e., solving for the off-shell T-matrices $\langle \mathbf{q}'_\beta | \mathscr{T}^{(R)}_{\beta\alpha}(z) | \mathbf{q}_\alpha \rangle$, we get the rearrangement and breakup amplitudes of the screened problem and thus, via the zero-screening procedures (1.6.28) and (1.6.44), the corresponding unscreened amplitudes.

7.4. Justification of the Screening and Renormalization Method within the Effective Two-Body Formalism

As in the previous section we restrict ourselves to the case that only particles 1 and 2 are charged, particle 3 being neutral. The general case of three charged particles is treated in Alt.[33] Inspection of (1.7.27) shows that of the diagonal elements only $\mathscr{V}_{11}^{(R)}$ and $\mathscr{V}_{22}^{(R)}$ differ from zero; they are essentially determined by the Coulomb interaction between the two protons:

$$\mathscr{V}_{\alpha\alpha}^{(R)}(\mathbf{q}'_\alpha, \mathbf{q}_\alpha) = \bar{\delta}_{\alpha 3} \langle \mathbf{q}'_\alpha | \langle \chi_\alpha | G_0 T_3^R G_0 | \chi_\alpha \rangle | \mathbf{q}_\alpha \rangle$$
$$= \bar{\delta}_{\alpha 3} \langle \mathbf{q}'_\alpha | \langle \chi_\alpha | G_0 V_3^R G_0 | \chi_\alpha \rangle | \mathbf{q}_\alpha \rangle + \tilde{\mathscr{V}}_{\alpha\alpha}^{(R)}(\mathbf{q}'_\alpha, \mathbf{q}_\alpha) \quad (1.7.30)$$

To arrive at the second line we have split off from T_3^R its Born part V_3^R.
Let us introduce the function

$$F_\alpha(\mathbf{q}'_\alpha, \mathbf{q}_\alpha; z) = \int \frac{d^3k \, \chi_\alpha^*(\mathbf{k}) \chi_\alpha(\mathbf{k} - \mathbf{D}_\alpha)}{[k^2/2\mu_\alpha - (z - q'^2_\alpha/2M_\alpha)][(\mathbf{k} - \mathbf{D}_\alpha)^2/2\mu_\alpha - (z - q_\alpha^2/2M_\alpha)]} \quad (1.7.31)$$

with $\mathbf{D}_\alpha = \varepsilon_{\alpha\gamma}(\mathbf{q}'_\alpha - \mathbf{q}_\alpha) m_\gamma/(m_\beta + m_\gamma)$, where $\varepsilon_{\alpha\gamma} = -\varepsilon_{\gamma\alpha} = 1$ if α, γ is a cyclic ordering of the indices 1, 2, 3. For the deuteron channel $F_\alpha(\mathbf{q}'_\alpha, \mathbf{q}_\alpha; z)$ can easily be recognized as an off-shell extension of the deuteron body form factor,

$$\tilde{F}_\alpha(\mathbf{q}'_\alpha, \mathbf{q}_\alpha) = \int d^3k \, \psi_\alpha^*(\mathbf{k}) \psi_\alpha(\mathbf{k} - \mathbf{D}_\alpha)$$
$$= \int d^3r \, e^{i\mathbf{D}_\alpha \cdot \mathbf{r}} |\tilde{\psi}_\alpha(\mathbf{r})|^2 \quad (1.7.32)$$

i.e., of the Fourier transform (with respect to the momentum transfer) of the bound-state density. Hence, the function (1.7.31) is normalized to unity on the energy shell and for zero-momentum transfer $\mathbf{q}'_\alpha = \mathbf{q}_\alpha$. The first term in the second line of (1.7.30) is then found to be (we again omit the energy parameter z)

$$\bar{\delta}_{\alpha 3} \langle \mathbf{q}'_\alpha | \langle \chi_\alpha | G_0 V_3^R G_0 | \chi_\alpha \rangle | \mathbf{q}_\alpha \rangle = \bar{\delta}_{\alpha 3} v_\alpha^R(\mathbf{q}'_\alpha, \mathbf{q}_\alpha) F_\alpha(\mathbf{q}'_\alpha, \mathbf{q}_\alpha) \quad (1.7.33)$$

where

$$v_\alpha^R(\mathbf{q}'_\alpha, \mathbf{q}_\alpha) = \frac{e^2/4\pi}{(\mathbf{q}'_\alpha - \mathbf{q}_\alpha)^2 + R^{-2}} \quad (1.7.34)$$

is just the Fourier transform of the CM Coulomb potential (1.6.5) (recall that $e_\beta = 0$). Thus we can write (1.7.27) as

$$\mathcal{V}_{\beta\alpha}^{(R)}(\mathbf{q}'_\beta, \mathbf{q}_\alpha) = \delta_{\beta\alpha}\bar{\delta}_{\alpha 3}v_\alpha^R(\mathbf{q}'_\alpha, \mathbf{q}_\alpha) + \mathcal{V}_{\beta\alpha}^{SR}(\mathbf{q}'_\beta, \mathbf{q}_\alpha) \tag{1.7.35}$$

with

$$\mathcal{V}_{\beta\alpha}^{SR}(\mathbf{q}'_\beta, \mathbf{q}_\alpha) = \delta_{\beta\alpha}\bar{\delta}_{\alpha 3}v_\alpha^R(\mathbf{q}'_\alpha, \mathbf{q}_\alpha)[F_\alpha(\mathbf{q}'_\alpha, \mathbf{q}_\alpha) - 1] + \tilde{\mathcal{V}}_{\beta\alpha}^{SR}(\mathbf{q}'_\beta, \mathbf{q}_\alpha) \tag{1.7.36}$$

where $\tilde{\mathcal{V}}^{SR}$ combines $\tilde{\mathcal{V}}^{(R)}$ from (1.7.30) and all nondiagonal terms of (1.7.27).

Inserting the splitting (1.7.35) into (1.7.26) we obtain

$$\mathcal{T}^{(R)} = (v^R + \mathcal{V}^{SR}) + (v^R + \mathcal{V}^{SR})\mathcal{G}_0^{(R)}\mathcal{T}^{(R)} \tag{1.7.37}$$

which leads, as in the genuine two-body case, to a representation of the form (1.4.39), viz,

$$\mathcal{T}^{(R)} = t^R + (1 + t^R \mathcal{G}_0^{(R)})\mathcal{T}^{SR}(1 + \mathcal{G}_0^{(R)} t^R) \tag{1.7.38}$$

Here, t^R is an off-shell continuation[28] of the CM Coulomb T-matrix corresponding to v^R. Explicitly written, the on-shell restriction of (1.7.38) reads

$$\mathcal{T}_{\beta\alpha}^{(R)}(\mathbf{q}'_\beta, \mathbf{q}_\alpha) = \delta_{\beta\alpha}\bar{\delta}_{\alpha 3}t_\alpha^R(\mathbf{q}'_\alpha, \mathbf{q}_\alpha) + \langle \mathbf{q}'^{(-)}_{\beta, R} | \mathcal{T}_{\beta\alpha}^{SR}(E_{\alpha n} + i0) | \mathbf{q}^{(+)}_{\alpha, R} \rangle \tag{1.7.39}$$

This is a two-potential formula of the type (1.6.19) on which the ASZ screening technique is based, although it arises here within the framework of the effective two-body AGS theory. Proceeding in this way allows us to work from the very beginning with integral equations that are particularly well suited for practical applications. In contrast, the presentation of the method given in Section 6, which followed essentially Alt and Sandhas,[31] is based on resolvent and transition operator identities and therefore exhibits the underlying algebraic structure most directly.

The similarity of the above considerations with the ones of Section 4 is not surprising. For, it is a characteristic feature of the AGS approach that its basic definitions and relations show a close structural analogy to the genuine two-body formalism, both on the three-body operator level and, a fortiori, after reduction to effective two-body equations.

As emphasized repeatedly, all practical applications of this method have been made by calculating $\mathcal{T}^{(R)}$ and t^R for increasing R and then going over to the zero-screening limit via (1.6.28). This is done by determining the phase shifts $\delta_L^{(R)}$ and σ_L^R corresponding to $\mathcal{T}^{(R)}$ and t^R, respectively. The existence of the limit (1.6.28) also guarantees the existence of

$$\delta_L^{(R)} - \sigma_L^R = \delta_L^{SR} \xrightarrow[R \to \infty]{} \delta_L^{SC} \qquad (1.7.40)$$

which, by means of formulas such as (1.4.55), provides us with the unscreened Coulomb-modified short-range amplitude $\widetilde{\mathscr{T}}_{\beta m, \alpha n}^{SC}$.

8. NUMERICAL RESULTS FOR NUCLEAR THREE-BODY SYSTEMS

8.1. General Remarks

In contrast to the large number of theoretical investigations into various aspects of charged-composite-particle reaction theory, their practical application to nuclear processes has lagged behind.* Certainly one reason for this is the considerable increase in numerical complexity of the equations to be solved, as compared to that for the analogous equations without Coulomb forces. Moreover, this complexity entails approximations whose consequences are difficult to assess in many cases. Another reason is that not all theoretical problems arising in this context have yet found both a satisfactory and a practicable solution.

We will restrict ourselves to results for nuclear three-body reactions obtained by solving three-body equations, either for wave functions in coordinate space or for transition matrices in momentum space. Though the main emphasis is on the latter approach, it is nevertheless instructive to compare their respective advantages and disadvantages.

8.1.1. The Wave Function Approach in Coordinate Space. Biased by the experience with two elementary charged particles, at first sight it appears evident that an approach based on the solution of (integro-)differential equations for wave functions in coordinate space, which contain only the simple Coulomb potentials V_α^C (in addition to any short-range potential) and not the highly singular (off-shell) amplitudes T_α^C which typically occur in momentum-space integral equations, should be adequate. However, the infinite range of V_α^C is the cause of several major difficulties.

The first is connected with the fact that, at least up to now, the high dimensionality of the equations has necessitated a partial-wave decomposition of the three-body wave function, both within each given two-particle

---

*This applies even more to applications of the three-body theory to atomic charged-particle reactions. For an early review see Chen.[102]

subsystem and with respect to the relative coordinate of the third particle (see, e.g., Merkuriev[25] and Chen et al.[107]). The corresponding angular momenta are then combined to give the total orbital angular momentum. The total number of partial-wave states taken into account determines the dimensionality of the coupled system of equations to be solved.

For particles interacting via short-range forces V_α^S only, such a procedure converges relatively quickly because the influence of the V_α^S decreases rapidly with increasing subsystem angular momentum, thereby limiting, for a given value of the total orbital angular momentum, the number of partial-wave states for the third particle that need to be included. In the presence of Coulomb potentials, the situation must be expected to deteriorate considerably because the two-body Coulomb partial-wave series which are used in such an approach tend to diverge (compare the corresponding discussion in Section 2). It therefore becomes a delicate question at which (high) value of the subsystem angular momentum to truncate.

Secondly, long-range induced interactions such as polarization potentials will contribute to an increase in the required number of angular momentum states for the relative motion of the two fragments.

Finally, it is even unclear at present whether in a three-charged-particle system, for energies beyond the bound-state dissociation threshold, a truncated partial-wave formalism represents a well-defined approximation to the original problem at all (see in this context the recent investigation[108] of the difference between the asymptotic behavior of the truncated and of the original, infinite-dimensional system of coupled equations, in that part of configuration space which corresponds to three asymptotically free and approximately equally separated particles).

It is therefore mandatory, albeit probably very difficult, to ascertain whether the as-yet-unavoidable restriction in the number of subsystem Coulomb partial-wave contributions has led to a result which is significant and, if yes, of acceptable accuracy. For instance, restricting the Coulomb interaction to act in two-body S-wave states only, as has been done in the majority of calculations, eliminates polarization forces completely[109] and, depending on the system considered, sometimes even all Coulomb effects (cf. the discussion in Alt et al.[110]).

On a more practical level, ignoring the problems just mentioned, it appears to be numerically quite difficult to accurately integrate the resulting (integro-)differential equations out to such large distances that their solution can be matched reliably to some asymptotic boundary condition, a procedure which is necessary for the extraction of phase shifts or partial-wave amplitudes. The main source of difficulty resides in the long range of the induced particle/composite-particle interactions (unless they are either very weak because the composite particle is tightly bound, or are eliminated, e.g., by drastic approximations).

Hence, because of the lack of controllability of some of the as-yet-unavoidable approximations, numerical results obtained in this approach should still be looked at with some caution. Improved asymptotic expansions and improved numerical techniques may help to overcome these difficulties (cf. the Chapter 2 in this volume).

8.1.2. The ASZ Transition-Matrix Approach in Momentum Space.
Nearly all momentum-space scattering calculations performed up to now have made use of the ASZ screening and renormalization approach described in Sections 6 and 7. Since here all potentials either are of short range or are made short-range (via screening), any standard solution method of nuclear physics can be used. In practice, the effective two-body AGS equations have been employed. Their main advantage is that the Coulomb interaction is taken into account without need of a partial-wave decomposition in the two-body subsystems. The price to be paid is that the evaluation of the effective potentials, and hence of the kernels, becomes rather difficult. Thus, in numerical calculations of arrangement-scattering amplitudes, the (screened) two-body Coulomb T-matrices T_α^R have up to now always been approximated by their Born terms V_α^R [cf. Eq. (1.7.17)].*
A three-body binding energy calculation[111] with the unscreened version of Eq. (1.7.26), and systematic numerical investigations[112,113] of the effective potential (1.7.27) (with $R = \infty$) below the breakup threshold, indicated that such an approximation should be rather accurate for three-body reactions such as deuteron impact on protons, but may start failing if the proton target is replaced by a nucleus with higher charge [note that corrections are expected to be of the order $(V_\alpha^R)^2$]. At higher energies, where no such tests are available, use of this approximation introduces some uncertainty as to the accuracy of the calculated Coulomb corrections. For an additional aspect see Section 9.2.

The decisive question from the practical point of view is how fast the unscreening limit is reached, say, in the Coulomb-modified short-range phase shift δ_L^{SR}, as defined in (1.7.40). If too large values of R would be needed the results would become numerically unreliable. For, as R increases the screened Coulomb potentials, and hence also the corresponding T-matrices, approach their unscreened counterparts; that is, in momentum space they develop a quasi-singular behavior which becomes more and more pronounced, thereby rendering a reliable solution of the integral equations increasingly more difficult. A multitude of numerical tests[110,114] for the proton–deuteron system indicated that the smaller the energy and/or the larger the angular momentum value L between the colliding (composite)

---

*Nevertheless, when comparing results obtained on this basis with those of conventional model calculations, they will be called "exact," if this is the sole approximation that has been made.

particles, the larger the values of R must be in order for δ_L^{SR} to reach its asymptotic value δ_L^{SC}. In both situations this is a consequence of the increasing relative importance, either as L becomes larger or the energy smaller, of the long-range (e.g., polarization) part as compared to the shorter-range contributions in the effective potential. This entails a considerable increase in numerical effort.

Summarizing, the ASZ screening and renormalization approach provides a means for calculating the effects arising from the inclusion of the Coulomb force which is free from theoretical difficulties. It is particularly efficient if the approximation $T_\alpha^R \to V_\alpha^R$ can already be expected to yield reliable results. Indeed, this appears to be the case for nuclear reactions involving very light nuclei, such as proton–deuteron scattering to be discussed below. Of course, that approximation can be avoided altogether, albeit at the expense of a drastic increase in numerical complexity, in which case the calculated Coulomb effects would become exact.

8.2. Results for the Proton–Deuteron System

The scattering of protons off deuterons has been the primary field of application for the Coulomb-adapted three-body theories. Although the first results for phase parameters and differential cross sections were published quite some time ago,[115,116] the numerical as well as the theoretical complexities of the problem have conspired to make further progress fairly slow. Nevertheless, several interesting insights have been gained which will be addressed in the following. It is obvious that by setting all Coulomb potentials equal to zero in the three-body equations, the latter describe the corresponding neutral system, in our case neutron–deuteron (nd) scattering, with which we will compare.

8.2.1. Proton–Deuteron Elastic Scattering Phase Shifts. Calculations which cover the energy regions both below and above the deuteron breakup threshold have been performed up to now only for simple nuclear potentials: separable S-wave potentials of rank one if the work is performed in momentum space,[110,115–118] S-wave projected Yukawa-type potentials if coordinate-space equations are solved[25] (below the breakup threshold also more realistic nuclear potentials have been used[119,120]). Under such simplifying assumptions the total spin, which can take on the values $S = \frac{3}{2}$ (quartet channel) and $S = \frac{1}{2}$ (doublet channel) as well as the total orbital angular momentum of the three-nucleon system are conserved separately. As examples we show in Figs. 2 to 5 the real parts of the proton–deuteron phase shifts $^{(2S+1)}\delta_L^{SC}(pd)$ obtained in the screening approach, as compared to the corresponding neutron–deuteron phase shifts $^{(2S+1)}\delta_L(nd)$, in the quartet as well as in the doublet state, for nucleon–deuteron relative orbital

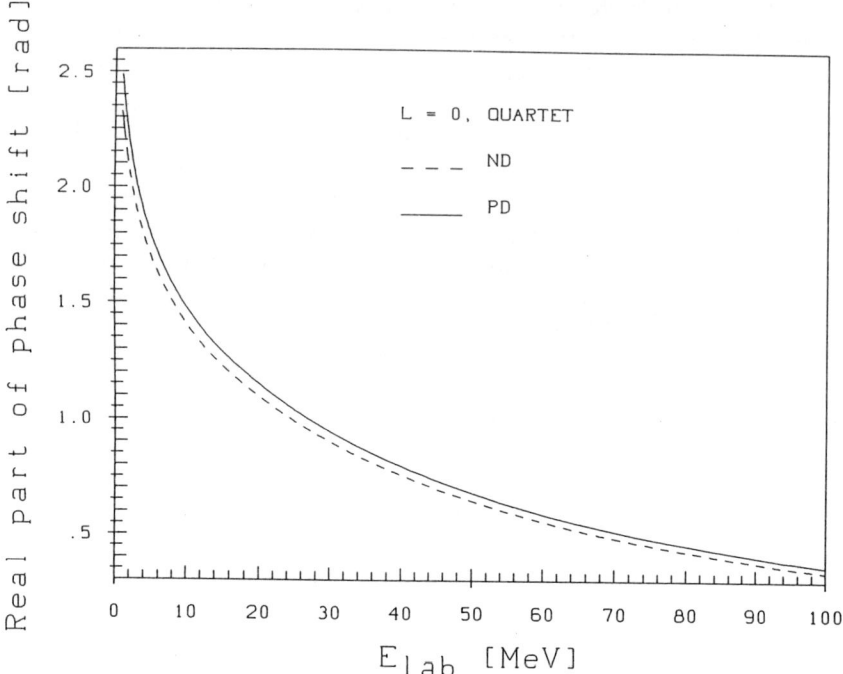

FIGURE 2. Real parts of the pd and nd S-wave phase shifts (in radians), as functions of the proton and neutron laboratory kinetic energy, for total spin $\frac{3}{2}$.

angular momenta* $L = 0$ and 1. For other partial waves and for the corresponding imaginary parts we refer to Alt et al.[110]

Typically, the real parts of the Coulomb corrections

$$\mathrm{Re}[^{(2S+1)}\delta_L^{SC}(\mathrm{pd}) - {}^{(2S+1)}\delta_L(\mathrm{nd})]$$

are of the order of 5–15%, with effects of similar size in the imaginary parts as well. It is remarkable that they apparently remain constant or decrease only slowly in percentage, either as the energy increases (as can be inferred from the figures) or as L becomes larger.[110] Such a behavior differs from what is known in two-particle scattering, but is easily understandable. Even for large L, or equivalently large impact parameters, scattering by the long-range induced interactions is still appreciable. And with certain probability, the (possibly even high) total three-particle energy is distributed among the three particles in the intermediate state in such a way that the

---

*The S-wave phase shifts are chosen to start at the value π, not only for the doublet channel where one bound state, namely ^{3}He or ^{3}H, exists (as suggested by Levinson's theorem[6,7,10,121]) but also in the quartet channel.[122]

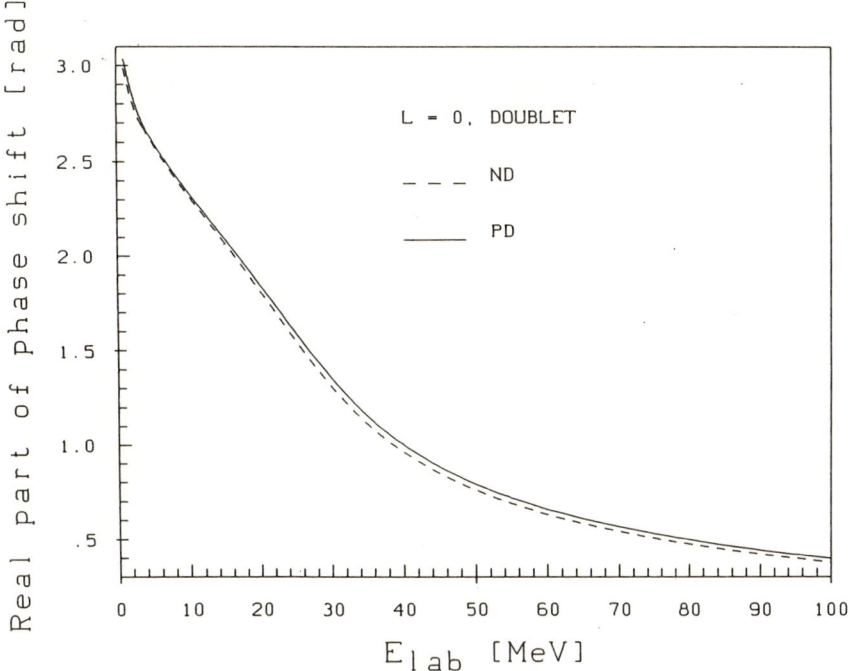

FIGURE 3. Same as in Fig. 2, but for total spin equal to $\frac{1}{2}$.

proton–proton subsystem energy is small enough to favor large two-body Coulomb effects.

One characteristic feature, common to the pd and nd phase shifts, is their behavior as functions of L for fixed energy.[118] This is shown in Fig. 6 for a proton laboratory energy of 35.0 MeV. The oscillatory nature, which is particularly pronounced in the quartet channel (as noted above, a factor π has been added to the S-wave phase shift) is typical[67] for an underlying exchange interaction. Recall that if the two-particle interactions allow for a separable representation of the form (1.7.4) with only a relatively small number of terms, the resulting effective two-body formulation of the three-body theory leads to one-particle exchange as the dominant reaction mechanism (cf. the first diagram in Fig. 1). It is therefore not surprising that this formalism is particularly well suited for the description of processes such as pd scattering which proceed mainly via one-nucleon exchange. Vice versa, for reactions which are dominated by direct-channel interactions such as polarization forces, a few-term separable approximation of the subsystem potentials, which emphasizes such a one-particle-exchange dynamics, cannot be expected to work so efficiently; see, e.g., Chen et al.[125–127] and Bürger et al.[128,129]

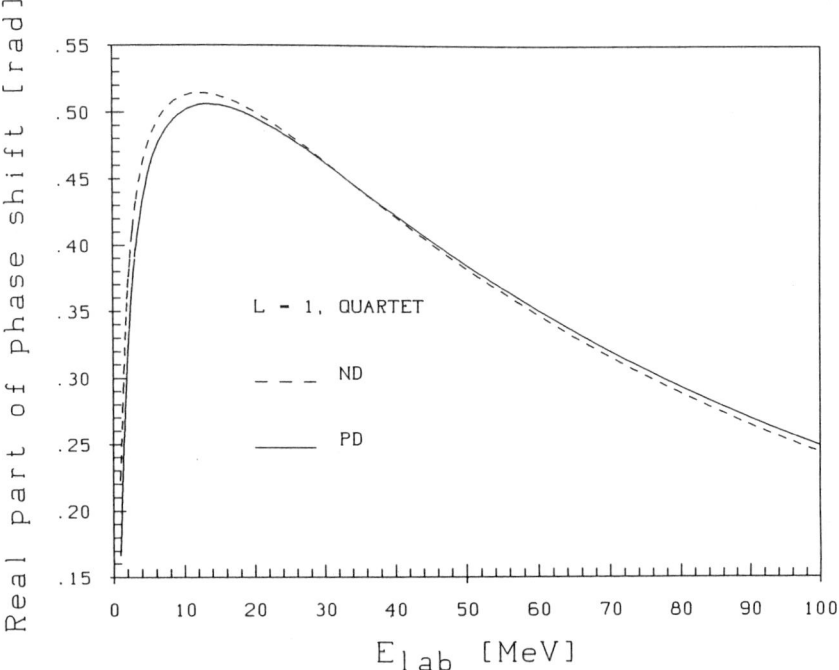

FIGURE 4. Same as in Fig. 2, but for P waves. The strong rise at low energies is due to a resonance.[110]

Another interesting qualitative result concerns the model dependence of the Coulomb corrections $^{(2S+1)}\delta_L^{SC}(pd) - {}^{(2S+1)}\delta_L(nd)$. Charge-independent and charge-symmetric nuclear forces, fitted to the same low-energy nucleon–nucleon parameters except for the $^1S_0$ state, give quite different pd and nd phase shifts but practically identical differences.[110]

We finally point to a typical feature of the Coulomb effects as calculated by means of the integral equations (1.7.26). The various Coulomb contributions to the effective potential occur additively in (1.7.27) and tend to strongly interfere with each other. This is corroborated by displaying in Table I the contributions to the real parts of the pd doublet and quartet phase shifts for $L = 0$ and 1, obtained by adding consecutively the Coulomb terms both in the effective potential (1.7.27), as represented diagrammatically in Fig. 1, and in the effective Green function (1.7.28). As is apparent, the individual contributions are often considerably larger than their sum, which is the total Coulomb correction. Furthermore, each of them varies in magnitude and often in sign for different orbital angular momenta L, and sometimes even for fixed L as a function of the energy. A similar situation prevails for the imaginary parts of the phase shifts, and for higher angular

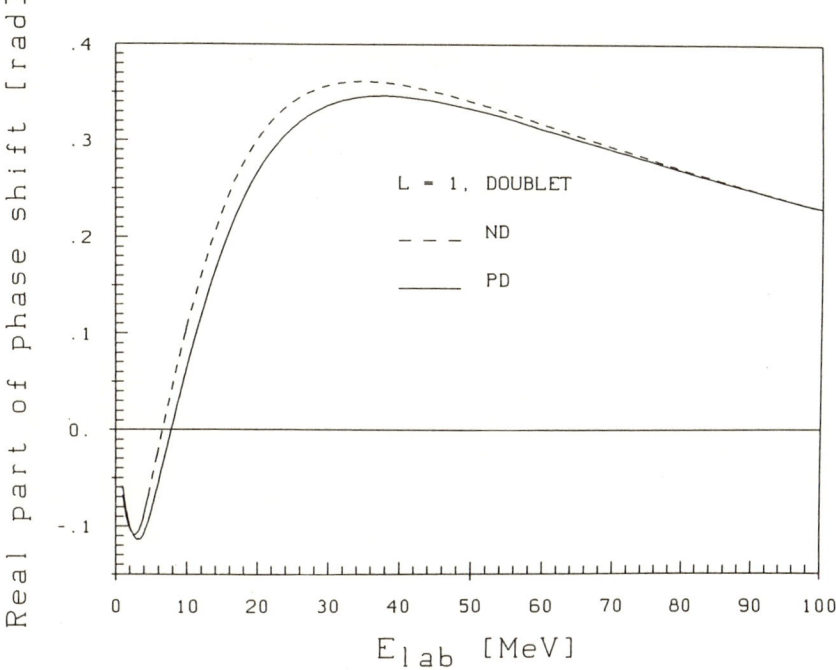

FIGURE 5. Same as in Fig. 3, but for P waves.

momenta L. Consequently, including only parts of these Coulomb correction terms will, in general, be rather misleading, not only with respect to magnitudes, but occasionally even with respect to signs. [110,130]

8.2.2. Proton–Deuteron Scattering Lengths. The quartet and doublet scattering lengths are fundamental quantities. Experimentally they are obtained by extrapolating phase shifts, extracted from analyses of cross sections measured at some low but nonzero energy, down to the elastic threshold. At present there is no agreement among different experimental groups[131–133] as to the doublet scattering length $^2a_{pd}$ (values quoted range from 1.3 to 4.0 fm), whereas the extracted values for $^4a_{pd}$ cluster around 11.4 fm. This is to be contrasted with the generally accepted nd values $^2a_{nd} = 0.65 \pm 0.04$ fm and $^4a_{nd} = 6.35 \pm 0.02$ fm. Notwithstanding this uncertainty, it is apparent that the inclusion of the Coulomb repulsion gives rise to appreciable modifications of these quantities. In other words, the scattering lengths act as a kind of magnifying glass for Coulomb effects.

Theoretically the situation is even more confused, for several reasons. From the principal (but not necessarily quantitatively most important) point of view, the most bewildering one is that different groups actually

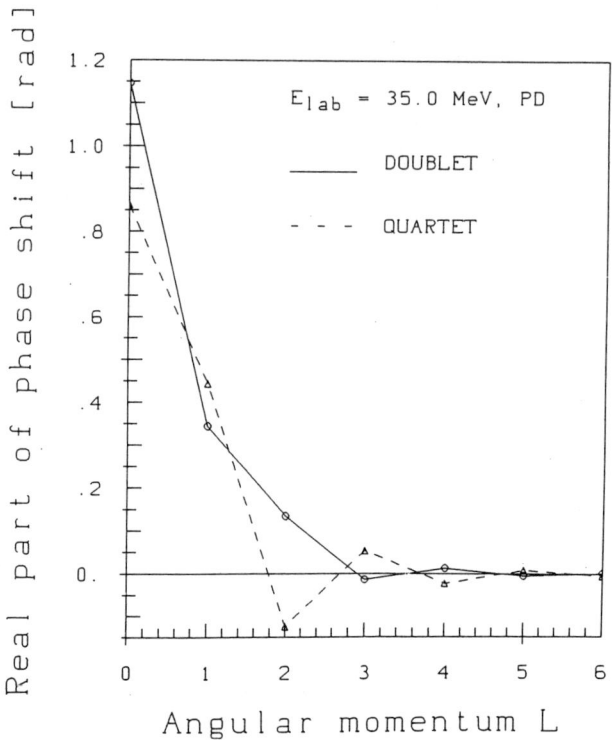

FIGURE 6. Real parts of the pd phase shifts (in radians) for total spin $S = \frac{1}{2}$ and $\frac{3}{2}$, as functions of the relative orbital angular momentum L. The points are connected by lines to guide the eye.

TABLE I

Contributions of the various terms of the effective potential (1.7.27), graphically represented in Fig. 1, and of the Coulomb correction in the effective propagator (1.7.28), to the real part of the doublet and quartet pd phase shift.[a]

	L	$\mathscr{V}^{(0)}$ +	v +	$\tilde{\mathscr{V}}^{(3)}$	$\mathscr{V}^{(4)}$	$\mathscr{V}^{(1)} + \mathscr{V}^{(2)}$	"$\mathscr{G}_0$"	$(2S+1)\delta_L^{SC}$
Doublet	0	103.5 +	1.9 −	3.1 −	0.5 +	4.3	− 0.6	105.5
	1	17.2 +	1.7 +	0.3 −	0.9 −	3.3	+ 0.3	15.3
Quartet	0	62.7 +	3.7 +	6.1 −	6.6			65.9
	1	28.7 −	0.6 +	1.4 −	1.1			28.4

[a] The proton laboratory energy is 20 MeV, and the pd relative orbital angular momenta are $L = 0$ and 1. Starting from the nd phase shift (only $\mathscr{V}^{(0)}$), one Coulomb correction term after the other is added until the final pd phase shift is obtained. In the quartet channel, the form factor and propagator modifications are absent.

calculate different quantities which they call pd scattering lengths. Namely, some perform their calculations directly at the elastic threshold. Others extrapolate effective-range functions down to threshold from higher energies; sometimes for this purpose $^{(2S+1)}\delta_0^{SC} = {}^{(2S+1)}\delta_0 - \sigma_0$ is taken, with $^{(2S+1)}\delta_0$ being the total and σ_0 the pure Coulomb S-wave phase shift, while in other cases σ_0 is replaced by some other phase shift.

Behind this confusion lies, of course, a physical reason: it is well-known that the polarizability of the deuteron is the origin of a long-range interaction with the incoming proton, which decreases asymptotically only as ρ^{-4}, ρ being the distance between the proton and the deuteron. Although rather weak it nevertheless strongly influences the very-low-energy behavior of $^{(2S+1)}\delta_0^{SC}$. For, it makes this phase shift go to zero at threshold not exponentially, as would be the case for a (e.g., exponentially bounded) short-range potential, but only with the fifth power of the pd relative momentum.[134] Consequently, the usual (Coulomb-modified) effective-range function with $^{(2S+1)}\delta_0^{SC}$ vanishes there, leading to infinite values of the scattering lengths extracted from it.

An old proposal[134] to circumvent this problem consists of introducing a phase shift

$$^{(2S+1)}\delta_0^{S,CP} = {}^{(2S+1)}\delta_0 - {}^{(2S+1)}\delta_0^{CP}$$

with $^{(2S+1)}\delta_0^{CP}$ being the phase shift due to the combined action of the CM Coulomb and polarization potentials. The correspondingly modified effective-range function has been proven to approach a finite value at threshold, and hence the scattering lengths extracted from it are finite. Unfortunately, however, these quantities cannot be determined purely experimentally. At present there appear to exist only two attempts[135,136] to calculate $^2 a_{pd}$ along this line, though both contain some arbitrariness regarding what is used as $^{(2S+1)}\delta_0^{CP}$ there.

Another way out is to follow as closely as possible the experimental procedure: extrapolate the usual (i.e., only Coulomb-corrected) effective-range function with the phase shift $^{(2S+1)}\delta_0^{SC}$ from energies where the polarization interaction is negligible (the value of that energy has been estimated in a two-body[137] and a three-body model[109]). However, this approach also has a serious disadvantage: because of the strong curvature of the effective-range function at very low energies,* reliable extrapolations are the more difficult to perform the higher the energies are from which this is carried out, a fact which had already hampered similar attempts in the nd case in the past.[139]

---

*This is due to the existence of a nearby pole on the unphysical sheet in the doublet channel, which will even move to the physical region for too attractive nuclear forces (see, e.g., Alt and Sandhas[138]).

Finally, different nuclear potentials give rather different nd and, hence, also different pd doublet scattering lengths. This impedes comparison of the calculated values, but opens up the possibility of testing the nuclear interaction. In contrast, $^4a_{nd}$ and, similarly, $^4a_{pd}$ appear to be rather insensitive. The first calculation[115] of $^4a_{pd}$, performed within the ASZ screening procedure, already gave satisfactory agreement with the data.

Concluding, there is still considerable work to be done before a theoretically acceptable and at the same time experimentally accessible pd doublet scattering length is agreed upon, which would justify renewed experimental effort on this problem. An interesting step in this direction has been made by Bencze et al.,[140] where, in a two-body model, a different kind of effective-range function (but with the same phase shift $^{(2S+1)}\delta_0^{S,CP}$) has been proposed as an alternative, physically acceptable definition of a scattering length in the presence of Coulomb and polarization interactions. Not only has it the virtue of allowing a well-defined extrapolation to zero energy, but it is also not much more difficult to implement in numerical extrapolation procedures than the standard Coulomb-modified effective-range formalism (see also Chapter 2 in this volume).

8.2.3. Elastic Proton–Deuteron Differential Cross Sections.

Differential cross sections for elastic pd scattering have been calculated both below and above the deuteron breakup threshold. Below relatively refined ansaetze for the nuclear force have been employed,[141,142] leading to satisfactory overall agreement with available experimental data. However, the enormous increase in numerical complexity when going beyond this threshold has precluded the use of other than very simple nuclear potentials (separable potentials of rank one acting in S-waves only for momentum-space calculations,[116-118] and S-wave projected Yukawa-type potentials in the coordinate-space approach[143,144]). This fact should be kept in mind when comparing these theoretical with experimental results.

Figure 7 shows elastic differential cross sections calculated,[116-118] after determination of $^{(2S+1)}\delta_L^{SC}$ via the zero-screening limit (1.7.40), from the on-shell amplitude

$$^{(2S+1)}\mathcal{T}(\mathbf{q}',\mathbf{q}) = t^C(\mathbf{q}',\mathbf{q}) + \sum_L \frac{(2L+1)P_L(\cos\Theta)e^{2i\sigma_L^C}(e^{2i^{(2S+1)}\delta_L^{SC}}-1)}{2iq} \quad (1.8.1)$$

[This is the three-body analog of the two-body formula (1.2.39) with (1.2.40).]

Despite the rather unrealistic nuclear input, the experimental data[145,146] are fairly well reproduced. The discrepancies, in particular in

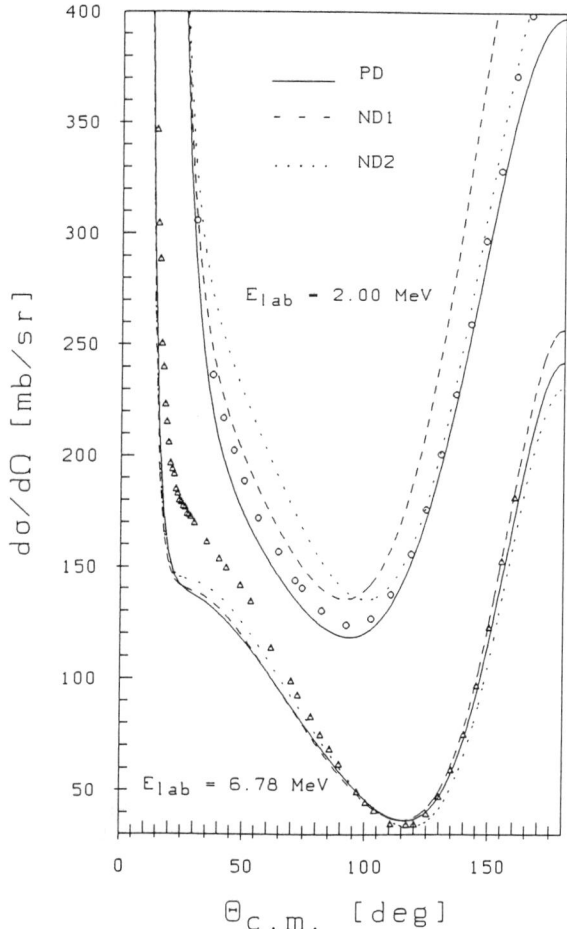

FIGURE 7. Proton–deuteron differential cross sections at 2.0 and 6.78 MeV proton bombarding energy. Experimental data are from Kocher and Clegg,[145] and Grötzschel et al.[146] Also shown are the results obtained by using the approximations (1.8.2) (ND1) and (1.8.3) (ND2).

the nuclear-Coulomb interference region, where the two terms on the r.h.s. of (1.8.1) become comparable in magnitude and thus can interfere strongly, are an artifact of the simple ansatz used for the nuclear force, which is already known to be inadequate for the description of nd scattering. In Figs. 8 and 9 the quartet and doublet contributions are displayed separately. Inspection reveals that, whereas for low energies the former completely dominates, they become comparable at higher energies, in accord with the situation in the nd case.

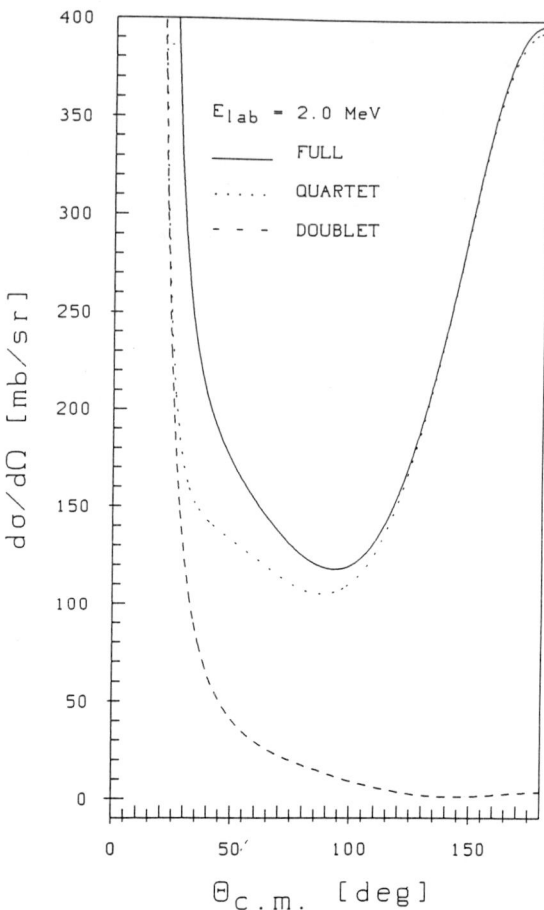

FIGURE 8. Proton–deuteron differential cross sections at 2 MeV proton bombarding energy, and the individual doublet and quartet channel contributions.

From (1.8.1) it is clear that only knowledge of the Coulomb-modified short-range phase shifts $^{(2S+1)}\delta_L^{SC}$ enables one to calculate the pd differential cross section correctly. However, from the practical point of view it is of considerable relevance to investigate whether the related huge effort is really indispensable.[116–118] Conventionally, two approximate methods are in use for obtaining differential cross sections for charged-particle scattering without solving the charged-particle equations. They are based on:

$$^{(2S+1)}\mathcal{T}(\mathbf{q}',\mathbf{q}) \approx t^C(\mathbf{q}',\mathbf{q}) + \sum_L \frac{(2L+1)P_L(\cos\Theta)(e^{2i\,{}^{(2S+1)}\delta_L^S} - 1)}{2iq} \quad (1.8.2)$$

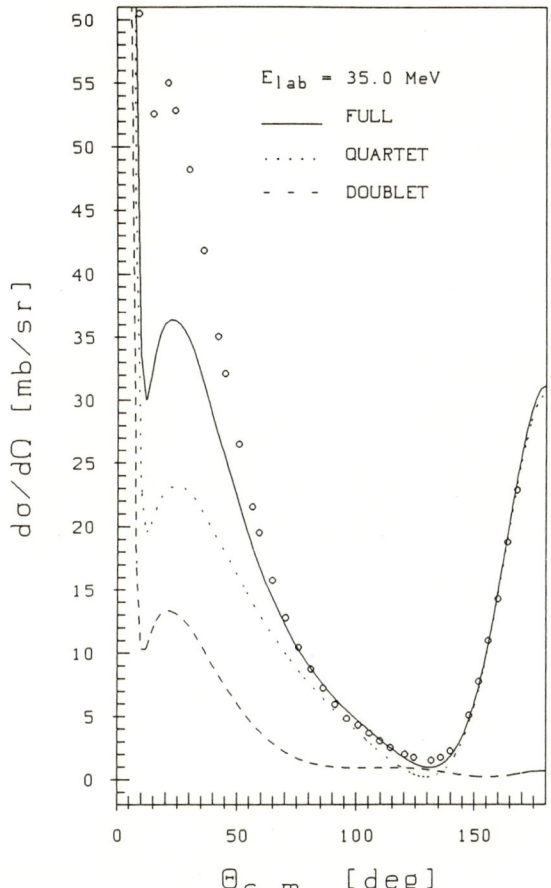

FIGURE 9. Same as in Fig. 8, but at 35 MeV proton kinetic energy. Experimental data are from Bunker et al.[147]

or

$$^{(2S+1)}\mathcal{T}(\mathbf{q}',\mathbf{q}) \approx t^C(\mathbf{q}',\mathbf{q}) + \sum_L \frac{(2L+1)P_L(\cos\Theta)e^{2i\sigma_L}(e^{2i^{(2S+1)}\delta_L^S}-1)}{2iq} \quad (1.8.3)$$

To arrive at (1.8.2), the second term on the r.h.s. of the exact formula (1.8.1), which is the Coulomb-modified short-range amplitude $^{(2S+1)}\tilde{\mathcal{T}}^{SC}$ [cf. Eq. (1.6.29)], is replaced by the pure nuclear amplitude. This is tantamount to neglecting all Coulomb effects except those coming from the additive Rutherford amplitude $t^C(\mathbf{q}',\mathbf{q})$. In the approximation (1.8.3) the Coulomb-

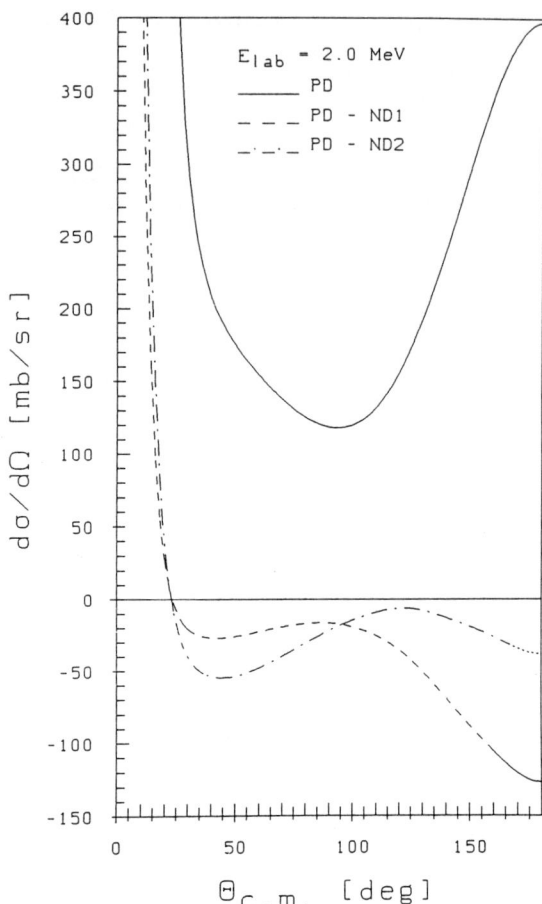

FIGURE 10. Proton–deuteron differential cross sections at 2 MeV from the exact expression (1.8.1) (PD), and the differences (PD-ND1) and (PD-ND2) between the exact cross section and the cross sections calculated from the approximate amplitudes (1.8.2) and (1.8.3), respectively.

modified short-range phase shifts $^{(2S+1)}\delta_L^{SC}$ are again replaced by the purely nuclear phase shifts $^{(2S+1)}\delta_L^{S}$, but the initial and final state Coulomb distortions are taken into account by means of the Coulomb phase shifts σ_L. In both cases only the solution of the neutral-particle equations to obtain $^{(2S+1)}\delta_L^{S}$ is necessary. In Fig. 7 approximate cross sections obtained in this way are included and compared with the exact ones. A more detailed comparison is provided in Figs. 10 and 11, where, in addition to the exact pd cross section, the difference of the cross sections obtained from the exact and the approximate amplitudes (1.8.2) and (1.8.3) are presented. These differences are appreciable both at low and at higher energies.

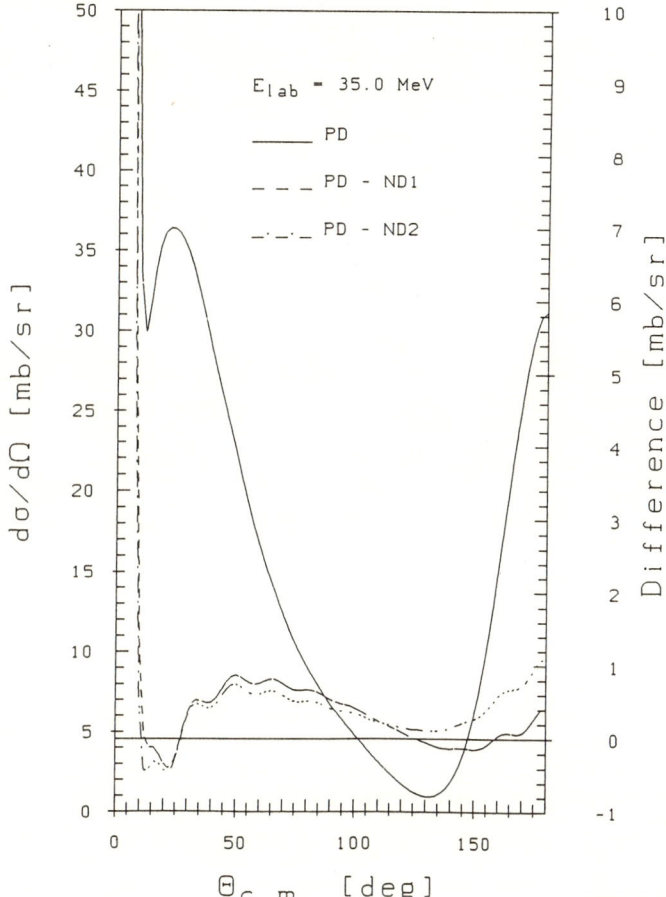

FIGURE 11. Same as in Fig. 10, but for 35 MeV.

8.2.4. The Proton-Induced Deuteron Breakup. Recently the first proton–deuteron breakup cross sections were obtained[148,149] by applying the formalism described in Section 7. Here, use was made of Eq. (1.7.18), which yields the breakup amplitudes via quadrature from the elastic amplitudes and the effective breakup potentials. The same nuclear potential as for pd elastic scattering was chosen. Whereas the elastic amplitudes were calculated with the approximation (1.7.17), the effective breakup potentials (1.7.29) have been evaluated using the full pp Coulomb amplitude.

Depending on the kinematic configuration, the effect of taking the Coulomb repulsion between the two protons into account can be sizable. Figures 12 and 13 show two examples with pronounced Coulomb effects, namely a quasi-free-scattering (QFS) and a collinear (COL) situation. [The

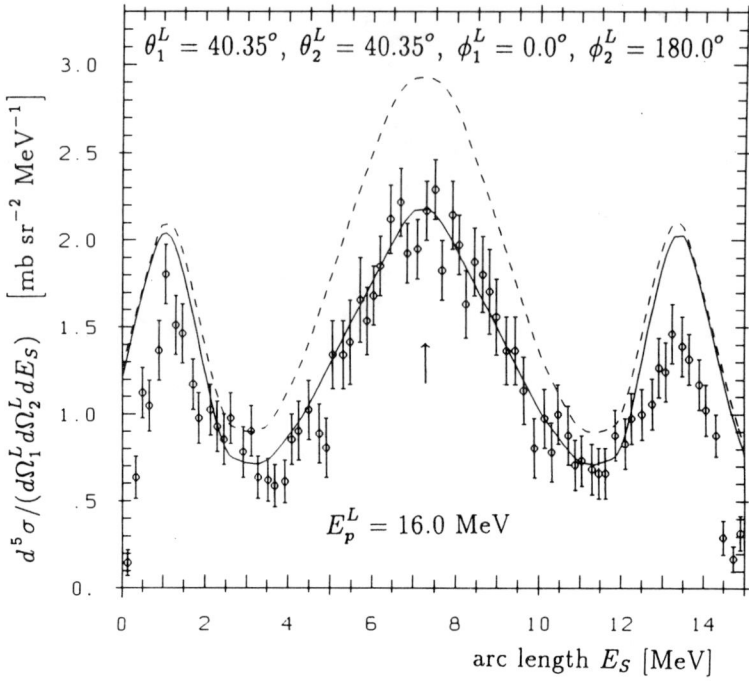

FIGURE 12. Deuteron breakup cross section for a quasi-free scattering (QFS) configuration at 16 MeV projectile energy and the indicated laboratory angles of the two detected particles, as a function of the arc length along the kinematically allowed curve: by proton impact (solid line), by neutron impact (dashed line). Experimental pd data are from Klein et al.[151] The arrow marks the point where the QFS condition is satisfied.

QFS condition is most easily visualized in the laboratory frame: the impinging proton knocks the proton out of the deuteron, leaving the neutron at rest (it acts only as a spectator). The two protons are then detected at certain angles which, in the case considered here, have been chosen identical with respect to the incident direction. A similar interpretation, but in the center-of-mass system, can be given of the COL configuration: after the deuteron breakup the neutron again remains at rest, whereas the two protons leave the reaction in opposite directions under a certain, experimentally chosen angle with respect to the incoming momentum. Interchange of protons and neutrons yields the description of the analogous configurations for the neutron-induced deuteron breakup. As is usual, breakup cross sections are presented as functions of the arc length along the kinematically allowed curve (see, e.g., Ohlsen[150]).] Evidently, the agreement with the experimental pd breakup data[151,152] is rather satisfactory. Of course, such kinematic configurations, e.g., some np final state interaction

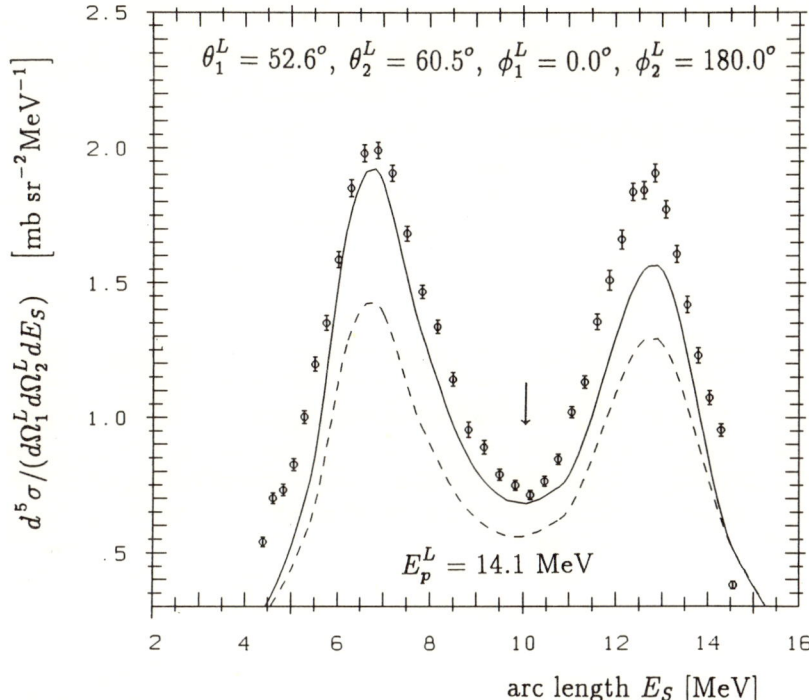

FIGURE 13. As in Fig. 12, but for a collinear (COL) configuration at 14.1 MeV projectile energy. Experimental pd data are from Karus et al.[152] The arrow indicates the point of collinearity.

regions, can also be found where the pd and nd breakup cross sections show very little difference, indicating that Coulomb corrections play a negligible role there.

8.2.5. *Approximate Calculations of Proton–Deuteron Scattering Observables.* Further investigations[153–156] have concerned different approximations for calculating the pd Coulomb-modified short-range scattering lengths and phase shifts $^{(2S+1)}\delta_L^{SC}$. For a detailed discussion of their shortcomings we refer to Alt et al.[110,117,130] It is regrettable that attempts to develop simplifying, but nevertheless quantitatively acceptable appoximation methods for calculating $^{(2S+1)}\delta_L^{SC}$ have not met with much success yet. Also, a near-threshold pd breakup calculation,[157] based on a separable expansion of the screened proton–proton T-matrix (for fixed cut-off radius R), was only partially successful due to convergence problems with respect to the number of separable terms as R was increased.

8.3. Applications to Deuteron–Nucleus Reactions and Related Processes

Applications of three-body theories with Coulomb interactions to three-body models of nuclear reactions have been sparse. Using the screening approach, the dependence of Coulomb effects on the target charge in elastic deuteron/structureless-nucleus scattering, for energies below the deuteron breakup threshold, was investigated by Aguiar et al.[158] and Brinati and Martins.[159] Both investigations are highly exploratory because, for increasing charges of the nuclei involved, the replacement $T_\gamma^R \to V_\gamma^R$ becomes more and more questionable,[112,113] and convergence in the unscreening procedure has not always been achieved. Furthermore, both coordinate-space[160] and momentum-space[161,162] methods have been applied to elastic and breakup scattering in the d^{4}He system. These investigations can, however, only be considered as first steps: in the coordinate-space calculation an S-wave projected Coulomb potential has been used [cf. the above discussion of possible effects of such a truncation], and in the momentum-space treatments only parts of the Coulomb correction terms in the effective potentials have been taken into account, which must be expected to be rather unreliable in view of the strong, largely destructive interference of the different Coulomb contributions noted earlier. A different approach was taken by Bajzer[163] to investigate peculiar Coulomb effects on the breakup amplitude in the frequently used plane-wave-impulse approximation, with applications to deuteron-induced nuclear reactions. We also draw the attention to a calculation[164] of deuteron breakup by positive muons using the driving term in the representation of the breakup amplitude derived by Alt et al.[28]

9. SOME FURTHER DEVELOPMENTS

In this section we draw attention to several other theoretical developments which are contributing to our further understanding of the intricacies of charged-composite-particle scattering.

9.1. Coulomb Effects on Analytic Properties of Particle Transfer Amplitudes

Several investigations concerned the Coulomb modifications of the leading singularity of the particle transfer amplitude (see, e.g., Mukhamedzhanov[165]) in the $\cos\Theta$-plane, where Θ is the scattering angle. For instance, by using the unscreened three-body AGS equations it was found[166–169] that taking account of the full Coulomb interaction in the intermediate state does not change the position or character of this singularity, although its strength is renormalized. This latter fact influences, e.g., the values of vertex constants extracted from extrapolations of scattering data to this singularity.[170] Another result concerned three-body Coulomb effects on the position and width of the resonance in a two-step breakup process via resonance formation in the final state.[171] Applications to low-energy nucleon transfer reactions have been reported in Blokhintsev *et al.*[172)(172]

9.2. Long-Range Behavior of the Effective Potential between a Charged Elementary and a Charged (or Neutral) Composite Particle

As a last field of application of three-body theories with Coulomb interactions, we mention the theoretical investigation of the long-range behavior of the optical potential for elastic scattering of a charged elementary particle α off a neutral or charged bound state of two other charged particles. Such has been performed in the momentum-space integral equations approach in Alt and Mukhamedzhanov,[173] where it was shown that the leading singular behavior of the optical potential as the momentum transfer Δ_α goes to zero is given by $e_\alpha(e_\beta + e_\gamma)/4\pi\Delta_\alpha^2 + C_1\Delta_\alpha + \cdots$ for a spherically symmetric target. The first term is nothing but the momentum-space representation of the CM Coulomb potential v_α^C, Eq. (1.6.5) with $R = \infty$, acting between the colliding fragments without regard to their internal structure. The "strength factor" C_1 of the nontrivial lowest-order contribution depends explicitly on the energy and the momentum. On the energy shell, however, the Fourier transform of this expression is at large intercluster separation ρ_α a local potential of the form $v_\alpha^C(\boldsymbol{\rho}_\alpha) - a_P/2\rho_\alpha^4 + \cdots$, where a_P is the polarizability of the bound state (for instance, for deuterons one finds[174,175] $a_P \approx 0.6 \text{ fm}^3$, while for hydrogen atoms[176] $a_P = 4.5 a_0^3$, with a_0 being the Bohr radius). All other contributions decrease faster asymptotically. This result, which has been derived in a nonperturbative, nonadiabatic way, is valid both below and above the bound state dissociation threshold, and corroborates the one derived in second-order perturbation theory or in adiabatic approximation.[5,6] In particular, it shows that the conventional expression for the polarizability a_P is neither modified by summing up all the higher-order perturbation-theoretical contributions, nor acquires any energy-dependence above the breakup threshold. In this context we mention similar earlier investigations performed both by means of coordinate-space differential equations[177] and by momentum-space integral equations,[178-182] all of which were unsatisfactory, however, since, apart from being restricted to energies below the breakup threshold, they contained questionable steps.

On the other hand, for screening radius R set to infinity, the effective potentials (1.7.10) appear to contain long-range components. However, at least for energies below the breakup threshold, it has been suggested[187−182] that all terms with asymptotic behavior proportional to ρ_α^{-2} and ρ_α^{-3} cancel against similar terms arising from higher iterations of the integral equations. Thus, if in the expression (1.7.12) the two-body Coulomb amplitude T_α^C is approximated by V_α^C, this cancellation would automatically be rendered incomplete. Hence, in addition to the (now only approximate) polarization term, long-range contributions proportional to ρ_α^{-2} and ρ_α^{-3} may survive. The question whether, and if yes how much, this affects three-body observables has not yet been investigated.

REFERENCES

1. W. O. AMREIN, J. M. JAUCH, AND K. B. SINHA, Scattering Theory in Quantum Mechanics, Lecture Notes and Supplements in Physics, Vol. 16 (Benjamin, Reading, 1977).
2. B. H. BRANSDEN, Atomic Collision Theory (W. A. Benjamin, New York, 1970).
3. P. G. BURKE, Potential Scattering in Atomic Physics (Plenum Press, New York, 1977).
4. C. J. JOACHAIN, Quantum Collision Theory (North Holland, Amsterdam, 1975).
5. M. R. C. MCDOWELL AND J. P. COLEMAN, Introduction to the Theory of Ion–Atom Collisions (North Holland, Amsterdam, 1970).
6. N. F. MOTT AND H. S. W. MASSEY, The Theory of Atomic Collisions, 3rd Ed. (Clarendon Press, Oxford, 1965).
7. R. G. NEWTON, Scattering Theory of Waves and Particles, 2nd Ed. (Springer, Berlin, 1982).
8. E. PRUGOVEČKI, Quantum Mechanics in Hilbert Space (Academic Press, New York, 1971).
9. M. REED AND B. SIMON, Methods of Modern Mathematical Physics: III Scattering Theory (Academic Press, New York, 1979).
10. J. R. TAYLOR, Scattering Theory (J. Wiley & Sons, New York, 1972).
11. H. VAN HAERINGEN, Charged-Particle-Interactions (Coulomb Press, Leyden, 1985).
12. J. C. Y. CHEN AND A. C. CHEN, in: Advances in Atomic and Molecular Physics (D. R. Bates and I. Esterman, eds.), Vol. 8 (Academic Press, New York, 1972), p. 71.
13. J. R. Taylor, Nuovo Cim. 23B, 313 (1974).
14. G. B. WEST, J. Math. Phys. 8, 942 (1967).
15. E. PRUGOVEČKI AND J. ZORBAS, J. Math. Phys. 14, 1398 (1973).
16. A. M. MUKHAMEDZHANOV, Theor. Math. Phys. 62, 70 (1985).
17. S. OKUBO AND D. FELDMAN, Phys. Rev. 117, 279 (1960); Phys. Rev. 117, 292 (1960).
18. J. SCHWINGER, J. Math. Phys. 5, 1606 (1964).
19. J. DUŠEK, J. Math. Phys. 24, 2471 (1983).
20. D. MASSON AND E. PRUGOVEČKI, J. Math. Phys. 17, 297 (1976).
21. J. ZORBAS, Nuovo Cim. Lett. 10, 121 (1974).
22. D. F. FREEMAN AND J. NUTTALL, J. Math. Phys. 14, 1883 (1973).
23. J. D. DOLLARD, Rocky Mount. J. Math. 1, 5 (1971); J. Math. Phys. 5, 729 (1964).
24. M. D. SEMON AND J. R. TAYLOR, Nuovo Cim. 26A, 48 (1975).
25. S. P. MERKURIEV, Ann. Phys. (NY) 130, 395 (1980); Acta Physica Austriaca, Suppl. XXIII, 65 (1981).
26. S. P. MERKURIEV AND L. D. FADDEEV, Kvantovaya Teoria Rasseyaniya Dlya Sistem Neskol'kich Tschastits (Nauka, Moscow, 1985) [in Russian].
27. H. KRÖGER, Physics Reports 210, 45 (1992).

28. E. O. Alt, W. Sandhas, and H. Ziegelmann, *Phys. Rev.* C **17**, 1981 (1978).
29. E. O. Alt, P. Grassberger, and W. Sandhas, *Nucl. Phys.* **B2**, 167 (1967).
30. W. Sandhas, in *Elementary Particle Physics* (P. Urban, ed.) [*Acta Physica Austriaca, Suppl. IX,* 33 (1972)]; in: *Few-Body Methods: Principles and Applications* (T. K. Lim, C. G. Bao, D. P. Hou, and H. S. Huber, eds.) (World Scientific, Singapore, 1986), p. 3.
31. E. O. Alt and W. Sandhas, *Phys. Rev.* C **21**, 1733 (1980).
32. E. O. Alt and W. Sandhas, in *Few Body Systems and Nuclear Forces I* (H. F. K. Zingl, M. Haftel, and H. Zankel, eds.) (Springer, Berlin, 1978), p. 373.
33. E. O. Alt, in *Few-Body Nuclear Physics* (G. Pisent, V. Vanzani, and L. Fonda, eds.), (IAEA, Vienna, 1978), p. 271.
34. C. Chandler, *Nucl. Phys.* **A353**, 129c (1981).
35. L. P. Kok, in *Few-Body Problems in Physics* (L. D. Faddeev and T. I. Kopaleishvili, eds.), (World Scientific, Singapore, 1985), p. 252.
36. E. O. Alt, in: *Few-Body Methods: Principles and Applications* (T. K. Lim, C. G. Bao, D. P. Hou, and H. S. Huber, eds.) (World Scientific, Singapore, 1986), p. 239; *Few-Body Systems, Suppl.* **1**, 79 (1986).
37. V. De Alfaro and T. Regge, *Potential Scattering* (North Holland, Amsterdam, 1965).
38. F. Calogero, *Variable Phase Approach to Potential Scattering* (Academic Press, New York, 1967).
39. H. Van Haeringen, *Nuovo Cim.* **34B**, 53 (1976).
40. S. Klarsfeld, *Nuovo Cim.* **43A**, 1077 (1966).
41. F. Gesztesy and C. B. Lang, *J. Math. Phys.* **22**, 312 (1981).
42. M. D. Semon and J. R. Taylor, *Phys. Rev. A* **16**, 33 (1977).
43. D. M. Goodmanson and J. R. Taylor, *J. Math. Phys.* **21**, 2202 (1980).
44. R. H. Dalitz, *Proc. Roy. Soc. (London) A* **206**, 509 (1951).
45. S. Weinberg, *Phys. Rev.* **140**, B516 (1965).
46. W. F. Ford, *Phys. Rev.* **133**, B1616 (1964); *J. Math. Phys.* **7**, 626 (1966).
47. N. K. Gyland and M. Kolsrud, *Phys. Norv.* **8**, 213 (1976).
48. M. Kolsrud, *J. Phys. A: Math. Gen.* **11**, 1271 (1978).
49. V. G. Gorshkov, *Sov. Phys.-JETP* **13**, 1037 (1961).
50. E. Prugovečki and J. Zorbas, *Nucl. Phys. A* **213**, 541 (1973).
51. J. Zorbas, *J. Phys. A: Math. Gen.* **7**, 1557 (1974).
52. W. Brenig and R. Haag, *Fortschr. Physik* **7**, 183 (1959) [English translation in *Quantum Scattering Theory* (M. Ross, ed.) (Indiana University Press, Bloomington, 1963), p. 13.]
53. G. Ludwig, *Foundations of Quantum Mechanics, Vol. II* (Springer, New York, 1985).
54. J. M. Jauch and F. Rohrlich, *The Theory of Photons and Electrons* (Addison-Wesley, Reading, 1955).
55. H. Rollnik, *Z. Phys.* **145**, 639 (1956).
56. M. Scadron, S. Weinberg, and J. Wright, *Phys. Rev.* **135**, B202 (1964).
57. B. Simon, *Quantum Mechanics for Hamiltonians Defined as Quadratic Forms* (Princeton University Press, Princeton, 1971).
58. K. M. Watson and J. Nuttall, *Topics in Several Particle Dynamics* (Holden-Day, San Francisco, 1967).
59. K. Dreissigacker, H. Pöpping, P. U. Sauer, and H. Walliser, *J. Phys. G: Nucl. Phys.* **5**, 1199 (1979).
60. A. C. Phillips, *Phys. Rev.* **142**, 984 (1966).
61. D. Z. Freedman, C. Lovelace, and J. M. Namyslowski, *Nuovo Cim.* **43A**, 258 (1966).
62. C. Lovelace, *Phys. Rev.* **135**, B1225 (1964).
63. P. Grassberger and W. Sandhas, *Z. Phys.* **220**, 29 (1969).
64. J. D. Dollard, *Commun. Math. Phys.* **12**, 193 (1969); *J. Math. Phys.* **14**, 708 (1973).

65. L. D. FADDEEV, *Sov. Phys.-JETP* **12**, 1014 (1961).
66. W. SANDHAS, in *Few-Body Nuclear Physics* (G. Pisent, V. Vanzani and L. Fonda, eds.), (IAEA, Vienna, 1978), p. 3.
67. E. W. SCHMID AND H. ZIEGELMANN, *The Quantum Mechanical Three-Body Problem* (Vieweg & Sons, Braunschweig, 1974).
68. W. GLÖCKLE, *The Quantum Mechanical Few-Body Problem* (Springer, Berlin, 1983).
69. V. B. BELYAEV, *Lectures on the Theory of Few-Body Systems* (Springer, Berlin, 1990).
70. P. GRASSBERGER AND W. SANDHAS, *Nucl. Phys.* **B2**, 181 (1967).
71. E. O. ALT, P. GRASSBERGER, AND W. SANDHAS, *Phys. Rev. C* **1**, 85 (1970).
72. E. O. ALT, P. GRASSBERGER, AND W. SANDHAS, JINR Report E4-6688 (1972); in *Few Particle Problems in the Nuclear Interaction* (I. Šlaus, S. A. Moszkowski, R. P. Haddock, and W. T. H. van Oers, eds.) (North Holland, Amsterdam, 1972), p. 299.
73. W. SANDHAS, in *Progress in Particle Physics* (P. Urban, ed.) [*Acta Physica Austriaca, Supp. XIII*, 679 (1974)]; *Czech. J. Phys. B* **25**, 251 (1975); in *Few Body Dynamics* (A. N. Mitra, I. Šlaus, V. S. Bhasin, and V. K. Gupta, eds.) (North Holland, Amsterdam, 1976), p. 540.
74. J. V. NOBLE, *Phys. Rev.* **161**, 945 (1967).
75. GY. BENCZE, *Nucl. Phys.* **A196**, 135 (1972).
76. GY. BENCZE, P. DOLESCHALL, C. CHANDLER, A. G. GIBSON, AND D. WALLISER, *Phys. Rev. C* **43**, 992 (1991).
77. GY. BENCZE, *Lett. Nuovo Cim.* **17**, 91 (1976).
78. GY. BENCZE AND H. ZANKEL, *Phys. Lett.* **82B**, 316 (1979).
79. L. D. FADDEEV AND O. A. YAKUBOVSKY, *Sov. J. Nucl. Phys.* **33**, 331 (1981).
80. F. S. LEVIN AND W. SANDHAS, *Phys. Rev. C* **29**, 1617 (1984).
81. E. O. ALT AND W. SANDHAS, in *Few Body Systems and Nuclear Forces I* (H. F. K. Zingl, M. Haftel, and H. Zankel, eds.) (Springer, Berlin, 1978), p. 375.
82. L. D. FADDEEV, in *Three Body Problem in Nuclear and Particle Physics* (J. S. C. McKee and P. M. Rolph, eds.) (North Holland, Amsterdam, 1970), p. 154.
83. A. M. VESELOVA, *Theor. Math. Phys.* **3**, 542 (1971); *Theor. Math. Phys.* **35**, 395 (1978).
84. V. F. KHARCHENKO AND S. A. STOROZHENKO, Preprint ITP-75-53E, Institute for Theoretical Physics, Kiev (1975).
85. L. A. SAKHNOVICH, *Theor. Math. Phys.* **13**, 1239 (1974).
86. L. ROSENBERG, *Phys. Rev. D* **8**, 1833 (1973).
87. C. CHANDLER AND A. G. GIBSON, *J. Math. Phys.* **15**, 291 (1974); A. G. GIBSON AND C. CHANDLER, *J. Math. Phys.* **15**, 1366 (1974).
88. J. ZORBAS, *J. Math. Phys.* **17**, 498 (1976); *J. Math. Phys.* **18**, 1112 (1977).
89. P. P. KULISH AND L. D. FADDEEV, *Theor. Math. Phys.* **4**, 745 (1970).
90. H. KRÖGER, *J. Math. Phys.* **24**, 1509 (1983); *J. Math. Phys.* **25**, 1875 (1984).
91. H. GROSSE, F. NARNHOFER, AND W. THIRRING, in *Particle Physics, Proc. of the Adriatic Summer Meeting* (Martini, ed.) (North Holland, Amsterdam, 1974), p. 22; H. GROSSE, H. R. GRÜMM, H. NARNHOFER, AND W. THIRRING, *Acta Physica Austriaca* **40**, 97 (1974).
92. Ž. BAJZER, *Z. Phys. A-Atomic Nuclei* **278**, 97 (1976).
93. A. M. VESELOVA, *Theor. Math. Phys.* **13**, 1200 (1974).
94. Ž. BAJZER, in *Few-Body Nuclear Physics* (G. Pisent, V. Vanzani, and L. Fonda, eds.) (IAEA, Vienna, 1978), p. 365.
95. H. WITALA, TH. CORNELIUS, AND W. GLÖCKLE, *Few-Body Systems* **3**, 123 (1988).
96. W. GLÖCKLE, H. WITALA, AND TH. CORNELIUS, *Nucl. Phys.* **A508**, 115c (1990).
97. L. S. FERREIRA, in *Models and Methods in Few-Body Physics,* Lecture Notes in Physics Vol. 273 (Springer, Berlin, 1987), p. 100.
98. A. L. ZUBAREV, *Sov. J. Part. Nucl.* **7**, 215 (1976).
99. D. J. ERNST, C. M. SHAKIN, AND R. M. THALER, *Phys. Rev. C* **8**, 46 (1973).

100. S. Weinberg, *Phys. Rev.* **131**, 440 (1963).
101. L. J. Dubé and J. T. Broad, *J. Phys. B: At. Mol. Phys.* **22**, L503 (1989); *J. Phys. B: At. Mol. Phys.* **23**, 1711 (1990).
102. J. C. Y. Chen, in *Case Studies in Atomic Physics* (E. W. McDaniel and M. R. C. McDowell, eds.) Vol. 3 (North-Holland, Amsterdam, 1973), p. 305.
103. E. A. Bartnik, H. Haberzettl, and W. Sandhas, *Phys. Rev. C* **34**, 1520 (1986).
104. W. Plessas and J. Haidenbauer, *Few-Body Systems, Suppl.* **2**, 185 (1989); W. Plessas, in *Models and Methods in Few-Body Physics*, Lecture Notes in Physics, Vol. 273, (Springer, Berlin, 1987), p. 137.
105. Y. Koike, J. Haidenbauer, and W. Plessas, *Phys. Rev. C* **35**, 396 (1987).
106. T. N. Rescigno and A. E. Orel, *Phys. Rev. A* **23**, 1134 (1981).
107. C. R. Chen, G. L. Payne, J. L. Friar, and B. F. Gibson, *Phys. Rev. C* **39**, 1261 (1989).
108. A. A. Kvitsinskii and D. M. Latypov, *Sov. J. Nucl. Phys.* **53**, 953 (1991).
109. A. A. Kvitsinskii and S. P. Merkuriev, *Sov. J. Nucl. Phys.* **41**, 412 (1985).
110. E. O. Alt, W. Sandhas, and H. Ziegelmann, *Nucl. Phys.* **A445**, 429 (1985); Erratum: *Nucl. Phys. A* **465**, 755 (1987).
111. L. P. Kok and H. van Haeringen, *Czech. J. Phys. B* **32**, 311 (1982).
112. L. P. Kok and H. van Haeringen, *Phys. Rev. C* **21**, 512 (1980).
113. L. P. Kok, D. J. Struik, and H. van Haeringen, University of Groningen, Internal Report 151 (1979); L. P. Kok, D. J. Struik, J. E. Holwerda, and H. van Haeringen, University of Groningen, Internal Report 170 (1981).
114. E. O. Alt and W. Sandhas, unpublished.
115. E. O. Alt, in *Few Body Dynamics* (A. N. Mitra, I. Šlaus, V. S. Bhasin, and V. K. Gupta, eds.) (North Holland, Amsterdam, 1976), p. 76.
116. E. O. Alt, W. Sandhas, H. Zankel, and H. Ziegelmann, *Phys. Rev. Lett.* **37**, 1537 (1976).
117. E. O. Alt, in *Few-Body Approaches to Nuclear Interactions in Tandem and Cyclotron Energy Regions* (S. Oryu and T. Sawada, eds.) (World Scientific, Singapore, 1987), p. 62.
118. E. O. Alt, in *Few Body and Quark-Hadronic Systems* (V. K. Lukyanov, ed.) (Joint Institute for Nuclear Research, Dubna, 1987), p. 25.
119. G. H. Berthold, A. Stadler, and H. Zankel, *Phys. Rev. C* **41**, 1365 (1990).
120. C. R. Chen, G. L. Payne, J. L. Friar, and B. F. Gibson, *Phys. Rev. C* **44**, 50 (1991).
121. N. Levinson, K. Danske Vid. Selsk. Mat. Fys. Medd. **25**, 9 (1949).
122. S. A. Sofianos, A. Papastylianos, H. Fiedeldey, and E. O. Alt, *Phys. Rev. C* **42**, R506 (1990).
123. R. D. Amado, *Phys. Rev.* **132**, 485 (1963).
124. R. Aaron, R. D. Amado, and Y. Y. Yam, *Phys. Rev.* **140**, B1291 (1965).
125. J. C. Y. Chen and T. Ishihara, *Phys. Rev.* **186**, 25 (1969).
126. J. C. Y. Chen and K. T. Chung, *Phys. Rev. A* **2**, 1449 (1970).
127. P. J. Kramer and J. C. Y. Chen, *Phys. Rev. A* **3**, 568 (1971).
128. H. Bürger, W. Sandhas, and E. O. Alt, *Phys. Rev. A* **30**, 2965 (1984).
129. H. Bürger and W. Sandhas, *Phys. Rev. A* **33**, 2284 (1986).
130. E. O. Alt, in *Dynamics of Few-Body Systems* (Gy. Bencze, P. Doleschall, and J. Revai, eds.) (KFKI, Budapest, 1985), p. 367.
131. W. T. H. van Oers and K. W. Brockman, *Nucl. Phys.* **A92**, 561 (1967).
132. J. Arvieux, *Nucl. Phys.* **A221**, 253 (1974).
133. E. Huttel, W. Arnold, H. Baumgart, H. Berg, and G. Clausnitzer, *Nucl. Phys.* **A406**, 443 (1983).
134. R. O. Berger and L. Spruch, *Phys. Rev.* **138**, B1106 (1965).
135. V. F. Kharchenko, M. A. Navrotski, and S. A. Shadchin, *Nucl. Phys.* **A512**, 294 (1990).
136. G. H. Berthold and H. Zankel, *Phys. Rev. C* **34**, 1203 (1986).
137. Gy. Bencze and C. Chandler, *Phys. Lett.* **163B**, 21 (1985).

138. E. O. ALT AND W. SANDHAS, in *Few Particle Problems in the Nuclear Interaction* (I. Šlaus, S. A. Moszkowski, R. P. Haddock, and W. T. H. van Oers, eds.) (North Holland, Amsterdam, 1972), p. 314.
139. W. T. H. VAN OERS AND J. D. SEAGRAVE, *Phys. Lett.* **B24**, 562 (1967).
140. GY. BENCZE, C. CHANDLER, J. L. FRIAR, A. G. GIBSON, AND G. L. PAYNE, *Phys. Rev. C* **35**, 1188 (1987).
141. G. H. BERTHOLD, A. STADLER, AND H. ZANKEL, *Phys. Rev. Lett.* **61**, 1077 (1988).
142. G. H. BERTHOLD, A. STADLER, AND H. ZANKEL, *Phys. Rev. C* **38**, 444 (1988).
143. S. POZDNEEV, *J. Phys. G: Nucl. Phys.* **8**, 1509 (1982).
144. YU. A. KUPERIN, S. P. MERKURIEV, AND A. A. KVITSINSKII, *Sov. J. Nucl. Phys.* **37**, 857 (1983).
145. D. C. KOCHER AND T. B. CLEGG, *Nucl. Phys.* **A132**, 455 (1969).
146. R. GRÖTZSCHEL, B. KÜHN, H. KUMPF, K. MÖLLER, AND J. MÖSNER, *Nucl. Phys.* **A174**, 301 (1971).
147. S. N. BUNKER, J. M. CAMERON, R. F. CARLSON, J. R. RICHARDSON, P. TOMAS, W. T. H. VAN OERS, AND J. W. VERBA, *Nucl. Phys.* **A113**, 461 (1968).
148. E. O. ALT AND M. RAUH, *Few-Body Systems*, Suppl. **7**, 160 (1994); *Phys. Rev. C* **49**, R2285 (1994).
149. E. O. ALT AND M. RAUH, *Few-Body Systems* **17**, 121 (1994).
150. G. G. OHLSEN, *Nucl. Instr. and Meth.* **37**, 240 (1965).
151. H. KLEIN, H. EICHNER, H. J. HELTEN, H. KRETZER, K. PRESCHER, H. STEHLE, AND W. W. WOHLFARTH, *Nucl. Phys.* **A199**, 169 (1973).
152. M. KARUS, M. BUBALLA, J. HELTEN, B. LAUMANN, R. MELZER, P. NIESSEN, H. OSWALD, G. RAUPRICH, J. SCHULTE-UEBBING, AND H. PAETZ GEN. SCHIECK, *Phys. Rev. C* **31**, 1112 (1985).
153. W. TIMM AND M. STINGL, *J. Phys. G: Nucl. Phys.* **2**, 551 (1976).
154. D. EYRE, A. C. PHILLIPS, AND F. ROIG, *Nucl. Phys.* **A275**, 13 (1977).
155. M. I. HAFTEL AND H. ZANKEL, *Phys. Rev. C* **24**, 1322 (1981).
156. H. ZANKEL AND G. M. HALE, *Phys. Rev. C* **24**, 1384 (1981).
157. P. DOLESCHALL, H. KRÖGER, AND R. J. SLOBODRIAN, *Phys. Rev. C* **37**, 927 (1988).
158. C. E. M. AGUIAR, J. R. BRINATI, AND M. H. P. MARTINS, *Nucl. Phys.* **A460**, 381 (1986).
159. J. R. BRINATI AND M. H. P. MARTINS, *Phys. Rev. C* **46**, 1607 (1992).
160. YU. A. KUPERIN, D. M. LATYPOV, S. P. MERKURIEV, M. BRUNO, AND F. CANNATA, *Sov. J. Nucl. Phys.* **53**, 582 (1991).
161. M. BRUNO, F. CANNATA, M. D'AGOSTINO, B. JENNY, W. GRÜEBLER, V. KÖNIG, P. A. SCHMELZBACH, AND P. DOLESCHALL, *Nucl. Phys.* **A407**, 29 (1983).
162. M. BRUNO, F. CANNATA, M. D'AGOSTINO, M. L. FIANDRI, M. FRISONI, H. OSWALD, P. NIESSEN, J. SCHULTE-UEBBING, H. PAETZ GEN. SCHIECK, P. DOLESCHALL, AND M. LOMBARDI, *Phys. Rev. C* **35**, 1563 (1987).
163. Ž. BAJZER, *Few-Body Systems* **2**, 9 (1987).
164. Ž. BAJZER, *Phys. Lett.* **84B**, 289 (1979).
165. A. M. MUKHAMEDZHANOV, *Czech. J. Phys. B* **32**, 298 (1982).
166. L. D. BLOKHINTSEV, A. M. MUKHAMEDZHANOV, AND A. N. SAFRONOV, *Sov. J. Part. Nucl.* **15**, 580 (1984).
167. G. V. AVAKOV, L. D. BLOKHINTSEV, A. M. MUKHAMEDZHANOV, AND R. YARMUKHAMEDOV, *Sov. J. Nucl. Phys.* **43**, 524 (1986).
168. A. M. MUKHAMEDZHANOV AND I. BORBÉLY, *Few-Body Systems* **5**, 21 (1988).
169. SH. S. KAJUMOV, A. M. MUKHAMEDZHANOV, R. YARMUKHAMEDOV, AND I. BORBÉLY, *Z. Phys. A – Atomic Nuclei* **336**, 297 (1990).
170. I. BORBÉLY, W. GRÜEBLER, V. KÖNIG, P. A. SCHMELZBACH, AND A. M. MUKHAMEDZHANOV, *Phys. Lett.* **160B**, 17 (1985).

171. A. R. Ashurov, D. A. Zubarev, A. M. Mukhamedzhanov, and R. Yarmukhamedov, *Sov. J. Nucl. Phys.* **53**, 97 (1991).
172. L. D. Blokhintsev, A. M. Mukhamedzhanov, and D. A. Savin, in *Dynamics of Few-Body Systems* (Gy. Bencze, P. Doleschall, and J. Revai, eds.) (KFKI, Budapest, 1985), p. 381.
173. E. O. Alt and A. M. Mukhamedzhanov, *Phys. Rev. A* **51**, 3852 (1995).
174. N. L. Rodning, L. D. Knutson, W. G. Lynch, and M. B. Tsang, *Phys. Rev. Lett.* **49**, 909 (1982).
175. J. L. Friar and S. Fallieros, *Phys. Rev. C* **29**, 232 (1984).
176. R. E. Johnson, *Introduction to Atomic and Molecular Collisions* (Plenum Press, New York, 1982).
177. A. A. Kvitsinskii and S. P. Merkuriev, *Sov. J. Nucl. Phys.* **48**, 79 (1988).
178. V. F. Kharchenko, S. A. Shadchin, and M. L. Zepalova, *J. Phys. B: At. Mol. Phys.* **18**, 949 (1985).
179. V. F. Kharchenko and S. A. Shadchin, *Sov. J. Nucl. Phys.* **45**, 210 (1987).
180. V. F. Kharchenko, S. A. Shadchin, and S. A. Permyakov, *Phys. Lett.* **199B**, 1 (1987).
181. V. F. Kharchenko and S. A. Shadchin, *Few-Body Systems* **6**, 45 (1989).
182. M. L. Zepalova, *Z. Phys. D – Atoms, Molecules and Clusters* **17**, 245 (1990).

CHAPTER 2

Proton–Deuteron Scattering and Reactions

J. L. Friar and G. L. Payne

1. INTRODUCTION

The n–p–p system, comprised of a neutron and two protons, provides an excellent example of the diversity, complexity, and sophistication which is possible in the field of nuclear physics. A distinct subdiscipline of this field has developed around the n–p–p system and the n–n–p system, comprised of two neutrons and one proton.

The diversity is reflected in the many possible states and reactions. There is a single n–p–p bound state (^{3}He), together with its isospin partner (^{3}H) formed from the n–p–p system. The angular momentum and parity J^π of these trinucleon bound states is $\frac{1}{2}^+$; they form an isodoublet ($T = \frac{1}{2}$). There are a variety of observables which reflect the internal structure of these ground states. The size is typically measured by electron scattering,[1] which determines the mean-square charge radius $\langle r^2 \rangle$, or the average distance from the nuclear center-of-mass (CM) to a proton. The nonzero spin of these systems (and the constituent nucleons as well) guarantees magnetic moments which are different and are primarily determined by the intrinsic magnetic moments of the nucleons.[2] The size of the "tails" of the wave function (the asymptotic normalization constants) can be measured using a variety of reactions. These asymptotic constants paradoxically reflect the interior dynamics.[3]

At a slightly higher internal energy, the bound states can barely dissociate into the nucleon–deuteron (N–d) system (i.e., either n–d or p–d).

J. L. Friar • Theoretical Division, Los Alamos National Laboratory, Los Alamos, New Mexico 87545.   G. L. Payne • Department of Physics and Astronomy, University of Iowa, Iowa City, Iowa 52242.

Coulomb Interactions in Nuclear and Atomic Few-Body Collisions, edited by Frank S. Levin and David A. Micha. Plenum Press, New York, 1996.

The threshold behavior of these scattering problems has been well studied.[4] Ignoring the Coulomb interaction at very low energies, the two separated clusters are primarily in a relative s-wave. The spin and parity of the free nucleon ($J^\pi = \frac{1}{2}^+$) couples with that of the deuteron ($J^\pi = 1^+$) to produce doublet ($\frac{1}{2}^+$) and quartet ($\frac{3}{2}^+$) states, which have very different properties. At these energies the scattering length is the quantity which epitomizes the scattering.

At still higher energies (but with a total energy less than zero), more partial waves become important in the N–d system, but it must remain asymptotically as two clusters if the available (relative) energy is insufficient ($< |E_d|$, the deuteron binding energy) to break up the deuteron. The various phase shifts for the partial waves exemplify the scattering.

The fourth energy region is above the deuteron-breakup threshold. In addition to elastic N–d scattering, the system can dissociate into the much more complex N–p–n configuration. Elastic phase shifts together with a set of breakup amplitudes which depend on spin, isospin, and all possible kinematic variables specify the scattering.

Thus, there are four interesting and distinct kinematic regions for the n–p–p and n–n–p systems (shown in Fig. 1) which lie within the narrow energy span of 10 MeV:

1. The bound state, with energy $E = -E_B$.
2. The threshold doublet and quartet N–d states, with energy $E = -|E_d|$.
3. Below-breakup N–d scattering, with energy $E < 0$.
4. Above-breakup N–d scattering and breakup, with energy $E > 0$.

In addition to these reactions which are largely driven by the strong interactions, transitions can be initiated between various regimes by "external" electromagnetic or weak interactions. Important examples of these

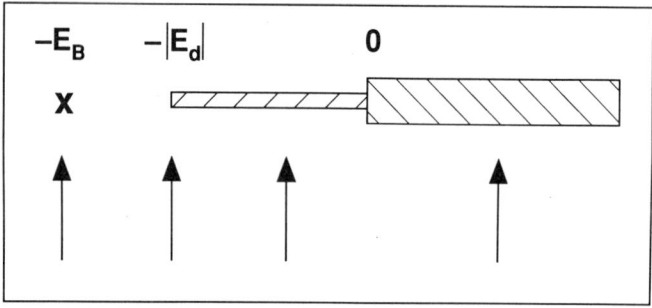

FIGURE 1. Energy regimes for low energy p–d processes, which include a bound state (cross), scattering threshold ($-|E_d|$), and breakup threshold ($E = 0$).

reactions and their initial and final regimes are:

1. The ^{3}H → ^{3}He β-decay[5]: [1 → 1].
2. Inelastic electron scattering from ^{3}He, leading to p + d or p + p + n[6]: [1 → 3,4].
3. Photoabsorption by ^{3}He, leading to p + d or p + p + n[7]: [1 → 3,4].
4. p–d radiative capture, leading to ^{3}He[8]: [3,4 → 1].
5. p–d fusion in muonic molecules, leading to ^{3}He[9]: [2 → 1].
6. μ-capture by ^{3}He[10]: [1 → 1,3,4].

We will subsequently discuss reaction (5), which is currently a "hot" problem, in some detail.

The aforementioned complexity of these systems ensues from the intrinsic complexity of the nuclear force and the presence of the Coulomb force in the n–p–p, p–d, and ^{3}He systems. The short-range nucleon–nucleon (N–N) force has a very complex structure because of the many possible spin and isospin combinations. The n–p and n–n forces are intrinsically different (even if isospin is an exact symmetry); the generalized Pauli principle allows us to separate the n–p interaction into an isotriplet ($T = 1$) part corresponding to the n–n (or p–p) system, and an isosinglet component ($T = 0$) which is unique to n–p. This difference is exemplified by the fact that only the n–p system has a bound state (the isoscalar deuteron). The two spin configurations ($S = 1$ and $S = 0$) are characterized by rather different forces. In particular, the spin-triplet component contains a very strong tensor force which couples together neighboring orbital waves (e.g., $^3S_1 - {}^3D_1$). In addition, the dominant feature of the s-wave interaction is a strong short-range repulsion which produces "holes" in the wave function. These holes play an important role in reactions because they suppress sensitivity to short-range reaction-mechanism components, whch are by far the least well understood. Forces which have the full complexity are called "realistic."

Coupled with the many degrees of freedom associated with the short-range strong interaction[11] is the long-range Coulomb force between the protons in ^{3}He or the p–d system. This force has always been an interesting complication. From the earliest days of charged-particle reactions (which led to the development of Coulomb-modified phase shifts, effective-range functions, and scattering lengths[12]) to current efforts to calculate the structure of three-nucleon systems, the "weak" Coulomb force has provided a constant challenge to theorists. Several problems now considered "solved" for n–d scattering remain to be completed for the p–d case.

Finally, the aforementioned sophistication in few-nucleon problems lies in our ability to solve the Schrödinger equation for most of these cases, even when the N–N potential has the full complexity described above. A wide variety of methods have been used for solving the Schrödinger equation containing realistic potentials for the few-nucleon bound-state problems, but only the Faddeev reformulation[13] of that equation has been successfully applied to all four of the energy regimes we tabulated earlier. In this chapter we will therefore restrict the discussion to that method formulated in configuration space,[14] where it is known how to treat the Coulomb effects in a straightforward way.[15,16]

In recent years it has been possible to generate solutions to few-nucleon problems which are "complete" or "exact." That is, the numerical error in computing observables is very small, typically one percent or less. This progress has resulted from the recent availability of large, fast computers and the implementation of sophisticated numerical methods and techniques, and allows for the first time a direct comparison of experimental data with calculations based on "first principles." Such comparisons, if successful, imply an understanding of the relevant underlying physics, while failure means that our models are inadequate and need refinement. For the first time we can check details of the nuclear force and reaction mechanisms which are not accessible in the two-nucleon problem. This exciting prospect has led many groups to concentrate on "complete" solutions for "realistic" potentials,[17–23] which can include three-nucleon forces (which we describe later). Such solutions have been obtained for the ^{3}H and ^{3}He ground states (including a Coulomb interaction between the protons in the latter case,[17,18] the ground state of ^{4}He,[19] the low-lying continuum of ^{5}He,[20] threshold n–d and p–d scattering (including a Coulomb interaction in the latter case[21]), n–d scattering above breakup threshold,[22] and electromagnetic transitions between the threshold and bound states.[23] At the present time in almost every case the calculations (properly formulated and interpreted) are in agreement with the data.

One of the areas of disagreement is p–d scattering,[21,22] and it is not yet known whether these problems are experimental or the result of an inadequate treatment (or neglect) of the Coulomb interaction. Complete p–d scattering calculations remain to be done, except at threshold. In addition, ambitious programs for studying the electromagnetic interactions of few-nucleon systems have been proposed at new experimental facilities.[24] Because ^{3}H targets are radioactive, as a practical matter all of this research involves ^{3}He and ^{4}He, where the Coulomb interaction in initial and final states is unavoidable. This further motivates a detailed and comprehensive study of few-nucleon systems which contain a Coulomb force.

We adopt as our purview in this chapter the p–p–n, p–d, and ^{3}He systems, except for those instances where comparison with the non-Coulomb case is informative. Moreover, we restrict our consideration largely to complete calculations performed with realistic potentials. In

Section 2 we discuss in a pedagogical fashion those interesting and relevant aspects of the nuclear Hamiltonian which are the subject of intense current research. In Section 3 we relate a number of elementary properties of the two-body Coulomb problem to establish our notation, point out the pitfalls, and motivate our treatment of the few-nucleon problem. In Section 4 we motivate the use of variational estimates. In Section 5 we introduce the Faddeev equations for three nucleons interacting via short-range, pairwise forces, categorize the (nonunique) ways we can incorporate three-nucleon forces and, finally, incorporate the Coulomb interaction using several different schemes. In Section 6 we discuss the boundary conditions for solving the Faddeev equations, for both the Coulomb and non-Coulomb cases, for the bound states and for scattering above and below breakup threshold and at the threshold. Because the boundary conditions present the only difficulties in principle when solving differential equations, we present a rather detailed treatment. In Section 7, we discuss the polarization potential, a subject of intense recent concern, and its relevance to threshold p–d scattering. In Section 8 we illustrate the partial-wave form of the Faddeev equations with examples. In Sections 9, 10, and 11 we discuss the ^{3}He bound state, p–d scattering, and p–d reactions in muon-catalyzed fusion. Finally, we summarize the chapter in Section 12, list the outstanding unsolved problems, and give our prognosis for the future, which promises to be as interesting and successful as the recent past.

2. STRONG-INTERACTION DYNAMICS

Much of the recent progress and focus for future work in few-body nuclear physics concerns the nuclear force. Unraveling details of this force in the context of the few-nucleon problems (and especially in the scattering problem) remains the highest priority of the field. In this context we present below a schematic view of the two-body and three-body forces, with an emphasis on recent progress which highlights the fundamental aspects of the problem.

The traditional approach to calculating the N–N force is identical to that of atomic physics: exchange virtual quanta and determine the potential which produces the same effect. In atoms the quanta are photons, while in nuclear processes the quanta are mesons. In many instances (low-density systems) the long-range forces dominate. Just as van der Waals forces dominate the long-range part of the atom–atom interaction, the exchange of pions dominates that part of the N–N interaction. One manifestation of chiral symmetry in strong-interaction dynamics[25] is the existence of a boson (the pion) with a tiny mass (on the hadronic scale). Using the familiar argument that the Compton wavelength ($\hbar/m_\pi c$) of the exchanged particle determines the range of the force, a small mass generates a force with a relatively long range. This force, the one-pion-exchange potential (OPEP), which contains a strong tensor interaction, has recently been shown to be dominant in few-nucleon systems.[26] Moreover, a secondary effect of chiral symmetry is a general "weakness" of the pion–nucleon interaction, which makes the exchange of two pions between nucleons weaker than one-pion exchange.[27] Thus in spite of the notorious and historical difficulties in treating strong-interaction processes in nuclei, OPEP stands out as the "Coulomb potential" of nuclear physics, both in terms of its "simplicity" and its importance.

One older and two recent developments have led to this picture. Many years ago, it was shown[28] that a suitably defined OPEP (whose single parameter was adjusted to produce the correct deuteron binding energy) was capable of adequately reproducing most deuteron properties. In most respects this is hardly surprising since the weak binding of the deuteron is its most notable feature. Nevertheless, observables which depend in an essential way on the tensor force were also well reproduced. If the $^3S_1 - {}^3D_1$ (tensor-coupled) partial waves of a realistic potential are replaced by that "pure" OPEP force, the binding energy of the triton is changed by less than $\frac{1}{2}$MeV, which is approximately 1% of the total potential energy. These partial waves account for roughly three-quarters of the total potential energy.

More recently, the Nijmegen group (Klomp et al.[11]) have performed a new phase-shift analysis of N–N data as part of a continuing and comprehensive program. This work is unique in that they do not *fix* the pion masses in OPEP but rather *fit* them into the data analysis. They find the masses of the neutral and charged pions to be $m_{\pi^0} = 135.6(13)$ MeV and $m_{\pi^\pm} = 139.4(10)$ MeV. Their results agree with the free pion masses (which are different) and the small error bars emphasize the importance of OPEP in the data and consequently in the nuclear potential.

Finally, a number of calculations have been performed recently from which the overall importance of OPEP can be discerned. Triton calculations with the new Nijmegen potential[29] show that the expectation value of OPEP, $\langle V_\pi \rangle$, is roughly 80% of the total potential energy $\langle V \rangle$. A similar result[26] is found for the Argonne V_{14} potential for ^{3}He and ^{4}He. Roughly two-thirds of the potential energy of ^{3}H, ^{3}He, and ^{4}He is generated by the tensor force (much of it coming from OPEP).

Thus, a case can be made that OPEP with its tensor force binds the few-nucleon systems. The latter condition makes nuclear physics qualitatively different from atomic physics, where central forces predominate. Moreover, the dominance in the few-nucleon systems of OPEP, which is the best understood part of the nuclear potential, may lead to a credible description of those systems, in spite of the well-known difficulty in describing strong-interaction dynamics from first principles.

One of first results to come from complete bound-state calculations was the inability of any realistic pairwise force to produce the observed binding of either ^{3}H (^{3}He) or ^{4}He. The calculated binding energy of ^{3}H (experimentally: 8.48 MeV) is typically 0.5–1.0 MeV too low.[17] This may be an indication of systematic underbinding in the three-nucleon systems from realistic pairwise forces. At this level of precision, relatively tiny effects become important. The small difference between the n–n and (isotriplet) n–p forces can account for 0.1–0.2 MeV in the triton binding.[30] In addition, a simple estimate of the size of relativistic corrections is 1–2% of the total kinetic or potential energy, or 0.5–1.0 MeV, although most calculations find much less.[31]

Another category of modification to the nuclear Hamiltonian has occupied the attention of much of the few-nucleon community engaged in performing complete calculations. In addition to pairwise forces, the Hamiltonian can contain three-body forces, which reflect an important aspect of nuclear dynamics. The "fundamental" constituents of a nucleus are nucleons, which are not elementary since they possess a rich excitation spectrum. Anything striking them can induce virtual excitation and de-excitation. Clearly these mechanisms are a normal part of the N–N force, but they also generate forces which are not present in a two-nucleon system and only manifest themselves in nuclei with three or more nucleons. All realistic three-nucleon forces *increase* the binding in few-nucleon systems.

The archetype of such forces is classical and displays all of their generic characteristics. Consider an idealized earth and moon (in Fig. 2), which are spherical and uniform. A small satellite orbiting the earth feels the force of both, which can be simply computed as the *pairwise* sum of its separate and independent interactions with the earth and moon. If we relax the requirement that the earth be rigid and allow the earth's oceans to develop tides, the forces on the satellite can no longer be considered pairwise interactions. The earth's tidal bulge follows the course of the moon, and this matter redistribution affects the satellite's motion in a complicated way. No longer can the forces on the satellite be computed in a purely pairwise fashion; the earth's matter distribution is now correlated with the position of the moon. We can also separate the earth–satellite interaction into two parts: the (original) two-body interaction with the solid earth and the (additional) three-body interaction with the tidal bulge. Such a separation is conventional in atomic and nuclear problems and emphasizes the special role that the deformation plays in generating a qualitatively different type of force.

We can extend this argument to include the sun, which also affects the tides. The alignment of the sun, earth, and moon leads to the exceptionally strong spring tides, while a right-triangle alignment leads to weaker neap tides. This is a small perturbation on the tidal effect but illustrates a significant feature of *all* many-body forces: they have a very complex dependence on the relative (angular) orientation of the bodies. The latter

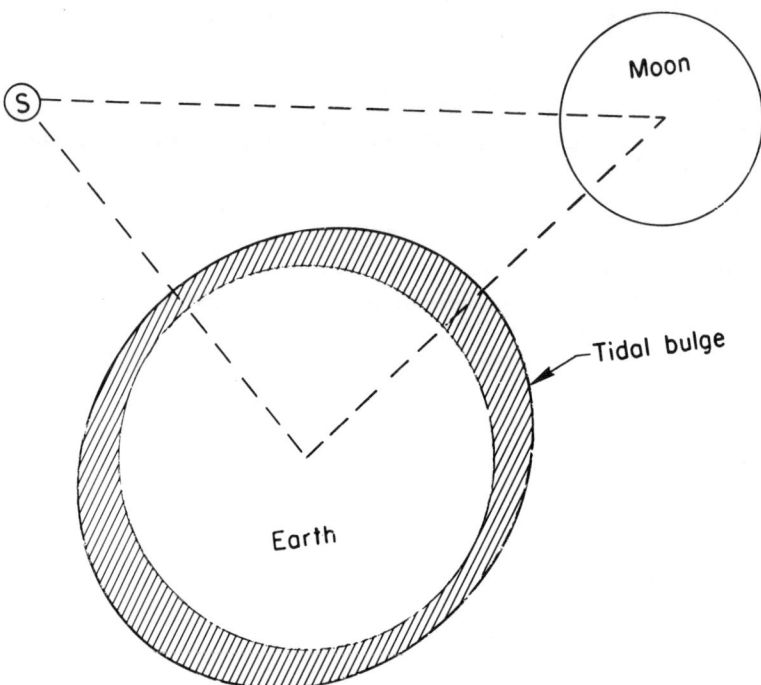

FIGURE 2. Classical three-body force in the earth–moon–satellite system, produced by the moon's deformation of the earth's oceans.

property has led to speculation that experimental exploitation of various special kinematic conditions, such as measuring and comparing the $p + d \to n + p + p$ reaction in final collinear or equilateral triangle configurations (shown in Fig. 3), could elucidate the role played by three-nucleon forces. In view of the weakness of these forces, this remains a hope.

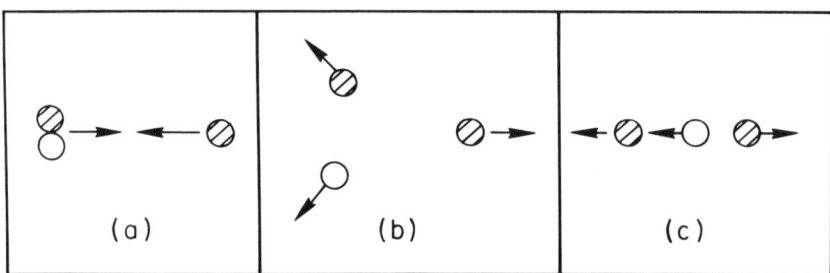

FIGURE 3. Scenario for detecting the effect of a three-nucleon force in p–d collisions. The (shaded) proton in (a) causes breakup of the deuteron into equilateral (b) and collinear (c) configurations. Detailed comparisons of (b) and (c) may show sensitivity to three-nucleon forces.

Considerable research is currently focused on this idea. We will treat three-body forces in Section 5.

3. TWO-NUCLEON PROBLEM

Much of the machinery and logic of the three-nucleon problem is an obvious extension of the two-nucleon problem[32] when one works in configuration space. We present below an elementary discussion of a few aspects of the latter which have an intimate connection to the less familiar three-nucleon problem. This allows us to establish our notation in a familiar context. Our goal will be to develop the three-nucleon problem for nonexperts in such a way that we can demystify many of its complexities. Our treatment is neither complete nor sophisticated but will be largely self-contained.

3.1. Non-Coulomb Case

We begin with the stationary-state form of the Schrödinger equation for two nucleons with equal masses M at coordinates $\mathbf{r}_1$ and $\mathbf{r}_2$, which are separated by a distance x, and interact by means of a short-range central potential $V(x)$:

$$\left[-\frac{\hbar^2}{2M}(\nabla_{r_1}^2 + \nabla_{r_2}^2) + V(x)\right]\Psi(\mathbf{r}_1,\mathbf{r}_2) = E\Psi(\mathbf{r}_1,\mathbf{r}_2) \qquad (2.3.1)$$

Transforming to the usual relative $(\mathbf{x} = \mathbf{r}_1 - \mathbf{r}_2)$ and center-of-mass $[\mathbf{R} = \frac{1}{2}(\mathbf{r}_1 + \mathbf{r}_2)]$ coordinates leads to

$$\left[-\frac{\hbar^2}{2\mu}\nabla_x^2 - \frac{\hbar^2}{2M_t}\nabla_R^2 + V(x)\right]\Psi(\mathbf{x},\mathbf{R}) = E\Psi(\mathbf{x},\mathbf{R}) \qquad (2.3.2)$$

where $\mu = M/2$ and $M_t = 2M$. Only the use of coordinates which connect the origin to the center-of-mass of the two nucleons allows the kinetic energy to be written in a form without cross terms in $\mathbf{x}$ and $\mathbf{R}$. Consequently, we can write Ψ as a product of forms describing the relative motion $\Psi(\mathbf{x})$ and the free motion of the center-of-mass $\Psi_{CM}(\mathbf{R})$, which we eliminate by working in the CM reference frame. We therefore obtain

$$\left[E + \frac{\hbar^2}{2\mu}\nabla_x^2 - V(x)\right]\Psi(\mathbf{x}) = 0 \qquad (2.3.3)$$

where E is now the CM energy. In order to solve this equation (either analytically or numerically) for an arbitrary short-range potential $V(x)$, we only need to specify two boundary conditions, one for very small x and one for x sufficiently large that $V(x) \sim 0$.

The easiest way to accomplish the former is to define a reduced wave function,

$$\Psi(\mathbf{x}) \equiv \frac{\varphi(\mathbf{x})}{x} \qquad (2.3.4)$$

and the requirement that Ψ be finite everywhere mandates that $\phi(\mathbf{x})$ vanish at the origin:

$$\phi(0) \equiv 0 \qquad (2.3.5)$$

This condition is extremely easy to implement in numerical calculations. In the rest of this chapter we will use ϕ generically for a reduced wave function and either Ψ or ψ generically for the unreduced one.

We restrict ourselves to s-waves for simplicity and deduce the (asymptotic) boundary conditions for large x by solving the potential-free version of Eq. (2.3.3),

$$\left(E + \frac{\hbar^2}{2\mu} \frac{d^2}{dx^2} \right) \phi(x) = 0 \qquad (2.3.6)$$

for those regimes we wish to treat. These regimes are: (1) a bound state, for which $E \equiv -\hbar^2\kappa^2/2\mu$; (2) a zero-energy scattering state, for which $E \equiv 0$; and (3) a positive-energy scattering (continuum) state, for which $E \equiv \hbar^2 k^2/2\mu$. These three cases will be denoted by superscripts: b, 0, c.

In the first case, we find two solutions. One is

$$\phi^b_{\text{asy}}(x) \sim e^{-\kappa x} \qquad (2.3.7)$$

while the other solution grows exponentially and must be rejected on the grounds that it violates the asymptotic boundary conditions: finiteness. The third case produces two solutions:

$$\phi^c_{\text{asy}}(x) \sim \frac{\sin(kx)}{k} \quad \text{(regular)} \qquad (2.3.8)$$

$$\sim e^{ikx}, e^{-ikx}, \cos(kx) \quad \text{(irregular)} \qquad (2.3.9)$$

We have chosen those combinations which we need in order to implement the asymptotic boundary conditions: an incoming plane wave, $\Psi_0(\mathbf{x}) = e^{i\mathbf{k}\cdot\mathbf{x}}$, plus an outgoing scattered (spherical) wave:

$$\Psi^c_{asy}(x) \sim e^{i\mathbf{k}\cdot\mathbf{x}} + f(\hat{\mathbf{k}}\cdot\hat{\mathbf{x}})\frac{e^{ikx}}{x} \qquad (2.3.10)$$

The s-wave projection of the plane wave is the regular solution of $\phi^c_{asy}(x)/x$. The complete asymptotic form must be

$$\phi^c_{asy}(x) \sim \frac{\sin(kx)}{k} + f_0 e^{ikx} \qquad (2.3.11)$$

where the s-wave part f_0 of the scattering amplitude f can be expressed in terms of the s-wave phase shift δ_0:

$$f_0 = \frac{e^{2i\delta_0} - 1}{2ik} = e^{i\delta_0}\frac{\sin\delta_0}{k} \qquad (2.3.12)$$

and leads to the alternative (but equivalent) asymptotic form:

$$\phi^c_{asy}(x) \sim \frac{e^{i\delta_0}}{k}\sin(kx + \delta_0) \qquad (2.3.13a)$$

which is the properly normalized phase-shifted incident plane wave. This equation defines both the normalization convention for our scattering wave functions and the scattering amplitude f_0. The form of the wave function in Eqs. (2.3.11) and (2.3.13a) is sometimes called the "t-matrix" form. An alternative form (which is real) results from factoring out the complex phase $e^{i\delta_0}$, and is the representation of choice for many numerical calculations. The K-matrix form is generated by factoring out an additional $\cos\delta_0$ from the wave function. The result

$$\phi^c_{asy}(x) \sim \frac{\sin(kx)}{k} + \frac{\tan\delta_0}{k}\cos(kx) \qquad (2.3.13b)$$

is real. The complete wave function for the scattering problem is always the same, and with these various forms we are just selecting which portions to treat.

The remaining (second) case is easily seen to be the $k \to 0$ limit of the previous case:

$$\phi^0_{asy}(x) \sim x \quad \text{(regular)} \qquad (2.3.14)$$

$$\sim 1 \quad \text{(irregular)} \qquad (2.3.15)$$

Using the result that for very low energies,[12] $\delta_0 \to -ka + O(k^3)$, we find that the complete asymptotic form is

$$\phi^0_{\text{asy}}(x) \sim x - a \tag{2.3.16}$$

in terms of the scattering length a. This result will be important for us later. In order to obtain the scattering length, it is conventional to form the s-wave effective-range function, $K_0(E)$, which is meromorphic in the small-k regime and, therefore, has a well-behaved power-series expansion[12]

$$K_0(E) = k \cot \delta_0 \sim -\frac{1}{a} + \tfrac{1}{2} r_0 k^2 + \cdots \tag{2.3.17}$$

where r_0 is the effective range. It is expected that in most cases of physical interest involving short-range forces, the size of r_0 will reflect the range of the force. However, an important three-nucleon problem discussed in Section 10 will contradict this expectation.

3.2. Coulomb Case

The introduction of a Coulomb force changes many of the previous ingredients in interesting (and sometimes unpleasant) ways. The difficulties are already apparent in the first Born approximation to the elastic scattering amplitude. The Coulomb potential between clusters of particles with charges Z_1 and Z_2 (in units of e, the positive fundamental charge) is given by

$$V_C(x) = \frac{Z_1 Z_2 \alpha \hbar c}{x} \tag{2.3.18}$$

where $e^2 = \alpha \hbar c$ in Gaussian units and α is the fine-structure constant. The amplitude for scattering in the first Born approximation from an incident momentum $\mathbf{k}$ to a final momentum $\mathbf{k}'$ is given by[32]

$$f_C^{(1)}(\mathbf{q}^2) = \frac{-\mu}{2\pi \hbar^2} \int d^3 x \, e^{i\mathbf{q}\cdot\mathbf{x}} V_C(x)$$

$$= -\frac{2\mu Z_1 Z_2 \alpha c}{\hbar \mathbf{q}^2} = -\frac{2\eta k}{\mathbf{q}^2} \tag{2.3.19}$$

where $\eta = \mu Z_1 Z_2 \alpha c / \hbar k$ is the usual Coulomb parameter, and $\mathbf{q} = \mathbf{k} - \mathbf{k}'$ is the momentum transfer. Because $\mathbf{q}^2 = 2k^2(1 - \cos\theta)$ in the CM frame, the amplitude is infinite for a vanishing scattering angle θ. Moreover, the partial-wave series for $f_C^{(1)}$ is very poorly behaved; for example, the s-wave projection of $\mathbf{q}^{-2}$ can easily be seen to have a logarithmic divergence. Nevertheless, because *each* term in the partial-wave series has the *same* divergence, the series can be formally summed[33] except for $\theta = 0$.

We can make this simple scattering model more realistic by introducing the finite size of one (or both) of the clusters. The Coulomb potential corresponding to a spherical charge distribution $\rho(r)$ is given by[34]

$$\bar{V}_C(x) = Z_1 Z_2 \alpha \hbar c \int d^3r \frac{\rho(r)}{|\mathbf{r} - \mathbf{x}|} \qquad (2.3.20)$$

and the corresponding scattering amplitude in first Born approximation is given by

$$\overline{f_C^{(1)}} = \frac{-2\eta k}{\mathbf{q}^2} F(\mathbf{q}^2) = -\frac{2\eta k}{\mathbf{q}^2}(1 - \mathbf{q}^2 \langle r^2 \rangle/6 + \cdots) \equiv f_C^{(1)} + \Delta f_C^{(1)} \qquad (2.3.21)$$

where

$$F(\mathbf{q}^2) = \int d^3r e^{i\mathbf{q}\cdot\mathbf{r}} \rho(r) \qquad (2.3.22)$$

is the deuteron charge form factor,[1] whose small momentum-transfer behavior is determined by the mean-square charge radius: $\langle r^2 \rangle = \int d^3r\, r^2 \rho(r)$. Note that if the point-Coulomb contribution $f_C^{(1)}$ is removed from the complete amplitude $\overline{f_C^{(1)}}$, what remains, $\Delta f_C^{(1)}$, is well-behaved for small $\mathbf{q}^2$, reflecting the "nice" properties of short-range interactions. In this and later sections we will use an overscore to distinguish between point-Coulomb properties and those modified by the short-range nuclear interaction.

This argument demonstrates that we can avoid many problems with the point-Coulomb potential by using well-known, conventional forms for the point-Coulomb scattering amplitude and by calculating only the *additional* scattering which results from the presence of the strong interactions. In the case of p–d scattering, there are several mechanisms for the latter:

1. The deuteron has a finite size and a modified Coulomb potential which reflects the static deuteron charge distribution (deuteron finite-size effect).
2. The Coulomb interaction "distorts" the deuteron (which is not rigid) and affects the asymptotic scattering (the long-range polarization potential).

3. The usual strong interaction dynamics present in n–d scattering is modified by the Coulomb interaction in p–d scattering (the Coulomb-modified strong-scattering amplitude).

The first and last mechanisms have a short range, while the second also has a long-range component. The first two have been the subject of several misconceptions and will be discussed later in some detail. The last mechanism is closely related to the two-body Coulomb problem, which we discuss next.

In order to separate the point-Coulomb interaction from the interesting strong-interaction dynamics, we need to examine the boundary conditions for Coulomb scattering. If the wave function for scattering from a point-like Coulomb potential is calculated in first-order perturbation theory, we find that a logarithmic divergence develops (from large x) which is an overall constant phase in the amplitude, plus a long-range logarithmic modification of the boundary conditions that we developed earlier. This modification is the essence of the Coulomb problem, and we discuss it next in a somewhat unconventional way.

The Schrödinger equation for a short-range potential $V(x)$ plus a point-Coulomb potential $V_C(x)$ has the form in the CM frame

$$\left[E + \frac{\hbar^2}{2\mu} \nabla_x^2 - V(x) - V_C(x) \right] \Psi_C(\mathbf{x}) = 0 \qquad (2.3.23)$$

For distances such that $V(x) \sim 0$, we can ignore the short-range potential, but not V_C. For simplicity we treat only s-waves, but this will not affect the argument below. As before, we replace $\Psi_C(\mathbf{x})$ by the reduced wave function $\phi_C(\mathbf{x})/x$ in order to simplify the Schrödinger equation and implement the boundary condition at $x = 0$. We find for $E > 0$ that

$$\left[\frac{d^2}{dx^2} + \left(k^2 - \frac{2\eta k}{x} \right) \right] \phi_C^c(x) = 0 \qquad (2.3.24)$$

Because of the perturbation theory argument given above, we might expect that the asymptotic solutions (which define the boundary conditions) are slowly varying functions which moderate the non-Coulomb solutions $e^{\pm ikx}$. Assuming solutions of the form

$$\phi_C^c(x) \sim e^{\int^x f(x')dx'} e^{\pm ikx} \qquad (2.3.25)$$

leads to a generalized Riccati[35] equation for $f(x)$, which is very convenient for determining asymptotic forms:

$$f^2(x) + f'(x) \pm 2ikf - U(x) = 0 \qquad (2.3.26)$$

where $U = 2\eta k/x$. The last two terms dominate, which leads to $f = \mp i\eta/x + \cdots$. Performing the integration in Eq. (2.3.25) generates an explicit prediction for the (irregular) asymptotic scattering functions:

$$\phi^c_{C,\text{asy}}(x) \sim e^{\pm ikx} e^{\mp i\eta \ln(x)} [1 + O(x^{-1})] \qquad (2.3.27)$$

Unfortunately, although this method gives the correct asymptotic form, it cannot predict overall factors or constant phases, which are partly dynamic (e.g., the phase shifts) and partly conventional. If we solve Eq. (2.3.24) in the usual way with the usual conventions,[32] taking the large-x limit produces instead of Eq. (2.3.27)

$$\phi^c_{C,\text{asy}}(x) \sim e^{\pm ikx} e^{\mp i\eta \ln(2kx)} \qquad (2.3.28)$$

which differs by a (constant) phase, $e^{\mp i\eta \ln(2k)}$. This phase is the standard convention although it is arbitrary. An analogous result will be obtained for three-nucleon scattering and breakup.

Everything can be combined to produce an asymptotic form similar to the non-Coulomb case if we define the argument $z_0 = kx - \eta \ln(2kx)$:

$$\phi^c_{C,\text{asy}}(x) = \frac{\sin(z_0)}{k} + f_0 e^{iz_0} = \frac{e^{i\sigma_0}}{k} \sin(z_0 + \sigma_0) \qquad (2.3.29)$$

where the s-wave Coulomb amplitude is $f_0 = (e^{2i\sigma_0} - 1)/2ik$, expressed in terms of σ_0, the s-wave Coulomb phase shift, which will be defined in Eq. (2.3.45). If we introduce a short-range modification of the Coulomb potential, there will be an additional (Coulomb-modified) nuclear phase shift, δ_0 (i.e., $\sigma_0 \to \delta_0 + \sigma_0$). Rearranging the expression which results, we find

$$\bar{\phi}^c_{C,\text{asy}}(x) \sim \phi^c_{C,\text{asy}}(x) + \bar{f}_0(e^{2i\sigma_0} e^{iz_0}) \sim \frac{e^{i(\sigma_0 + \delta_0)}}{k} \sin(z_0 + \delta_0 + \sigma_0) \qquad (2.3.30)$$

where

$$\bar{f}_0 = \frac{e^{2i\delta_0} - 1}{2ik} \qquad (2.3.31)$$

while the complete scattering amplitude is $f_0 + \bar{f}_0 e^{2i\sigma_0}$. Clearly, $\bar{f}_0$ vanishes if we eliminate the strong interactions. The *complete* Coulomb wave function replaces the Coulomb-modified plane wave and the irregular scattering solution picks up a Coulomb phase in going from Eq. (2.3.29) to Eq. (2.3.30). In numerical solutions, the complex phase in $\bar{\phi}^c_{C,\text{asy}}$ is usually ignored. We are also free to adopt any other reference potential with respect to which we define our phase shifts. However, we do not recommend this.

The Coulomb bound-state asymptotic forms can be obtained in the usual way by replacing $k \to \pm i\kappa$ in Eq. (2.3.26), leading to the asymptotic form:

$$\phi^b_{C,\text{asy}}(x) \sim e^{-\kappa x - \eta_b \ln(2\kappa x)} \tag{2.3.32}$$

where

$$\eta_b = \frac{\mu Z_1 Z_2 \alpha c}{\hbar \kappa} \tag{2.3.33}$$

Thus, if the energy κ is kept fixed, the Coulomb bound-state wave functions fall off more rapidly for large x. This is an important property.

The zero-energy case is radically different. If we set $k = 0$ in Eq. (2.3.24) and define $z = \xi x$ with

$$\xi = \frac{2Z_1 Z_2 \alpha \mu c}{\hbar} \tag{2.3.34}$$

we obtain

$$\frac{d^2 \phi^0_C(z)}{dz^2} - \frac{1}{z} \phi^0_C(z) = 0 \tag{2.3.35}$$

This equation can be solved using the representation in Eq. (2.3.25) with $U = \xi/x$ and $k = 0$. This eliminates a dominant term and completely changes the character of the solution. We easily find that $f = \pm \sqrt{(\xi/x)} + 1/4x + O(x^{-3/2})$, or

$$\phi^0_{C,\text{asy}} \sim x^{1/4} e^{\pm 2\sqrt{\xi x}} [1 + O(x^{-1/2})] \tag{2.3.36}$$

where the positive (negative) sign corresponds to the regular (irregular) solution.

The exponential behavior of these functions proves to be especially troublesome for numerical calculations. If we use properly defined regular and irregular asymptotic solutions the scattering from a short-range-plus-Coulomb potential can be determined by matching to a linear combination of these solutions at any distance x_m outside the short-range potential. However, if we choose larger and larger values for x_m, the ratio of these functions grows exponentially. Any numerical "noise" in the algorithm used to solve the differential equation will cause exponentially growing noise in the scattering length. We will see subsequently that large matching distances for the three-nucleon problem are required, and this produces a rather delicate numerical problem. Because of this peculiarity, the scattering-length regime is more difficult in many ways than the higher-energy regimes.

3.3. Exact Forms

3.3.1. Non-Coulomb Case. The previous two sections focused on asymptotic forms for s-waves, which will prove useful later. For completeness we need to define (or give proper reference to) exact regular and irregular functions for all angular momenta for the scattering problem, particularly for the Coulomb case. Many of the asymptotic expansions for these functions converge slowly, and it is always more efficient to match a numerically generated solution to its asymptotic boundary conditions at matching radii which are as small as possible. This can be done by using exact boundary functions, rather than their asymptotic forms. Writing

$$\Psi(\mathbf{x}) = \sum_{l=0}^{\infty} (2l+1) i^l P_l(\mu) \psi_l(x) \qquad (2.3.37)$$

where P_l is the lth Legendre polynomial, $\mu = \hat{\mathbf{k}} \cdot \hat{\mathbf{x}}$, and $\psi_l(x) = \phi_l(x)/x$, leads to the Schrödinger equation for the (partial-wave) reduced wave function ϕ_l,

$$\frac{d^2 \phi_l}{dx^2} + \left[k^2 - \frac{l(l+1)}{x^2} - U(x) \right] \phi_l = 0 \qquad (2.3.38)$$

in terms of $U(x) = 2\mu V(x)/\hbar^2$.

The regular and irregular solutions, ϕ_l, appropriate for the lth partial wave of the non-Coulomb case are the spherical Bessel functions, whose reduced forms are the Riccati–Bessel functions. The former[32] are $j_l(kx)$ (regular) and $n_l(kx)$ (irregular) (which are often[36] denoted by y_l) and they are real. The latter have asymptotic forms appropriate for the "K-matrix" form of the scattering wave function:

$$kx j_l(kx) \sim \sin(kx - l\pi/2) \qquad (2.3.39a)$$

$$kx n_l(kx) \sim -\cos(kx - l\pi/2) \qquad (2.3.39b)$$

Note the overall sign in the irregular solution. With these functions the complete wave function for the nuclear scattering problem has an asymptotic form expressed in terms of the phase shift δ_l,

$$\psi_l^{asy}(x) \sim e^{i\delta_l}[j_l(kx)\cos\delta_l - n_l(kx)\sin\delta_l] \tag{2.3.40}$$

The corresponding scattering amplitude is

$$f(\mu) = \frac{1}{k}\sum_{l=0}^{\infty}(2l+1)e^{i\delta_l}\sin\delta_l P_l(\mu) \tag{2.3.41}$$

and for each partial wave we can define an effective-range function

$$K_l(E) = k^{2l+1}\cot\delta_l \to -\frac{1}{a_l} + \tfrac{1}{2}r_l k^2 + \cdots \tag{2.3.42}$$

Note that the scattering length a_l has dimension: (length)$^{2l-1}$.

3.3.2. Coulomb Case. The Coulomb functions are solutions of Eq. (2.3.23) with the appropriate boundary conditions. The regular solution corresponding to $V = 0$ is

$$\Psi_C(\mathbf{x}) = C_0 e^{i\sigma_0} e^{i\mathbf{k}\cdot\mathbf{x}}{}_1F_1[-i\eta, 1; i(kx - \mathbf{k}\cdot\mathbf{x})] \tag{2.3.43}$$

where ${}_1F_1$ is the usual confluent hypergeometric function,[36] and C_0 is the $l = 0$ case of

$$C_l = 2^l e^{-\pi\eta/2}\frac{|\Gamma(l+1+i\eta)|}{(2l+1)!} \tag{2.3.44}$$

Note that there are alternative definitions of C_l, some of which are complex. The Coulomb phase shifts are defined by

$$\sigma_l = \arg[\Gamma(l+1+i\eta)] \tag{2.3.45}$$

This wave function can be expanded in partial waves,

$$\Psi_C(\mathbf{x}) = \sum_{l=0}^{\infty}(2l+1)i^l P_l(\mu)\psi_l^C(\eta, kx) \tag{2.3.46}$$

where

$$\psi_l^C(\eta, kx) = (kx)^l C_l e^{i\sigma_l}[e^{ikx}\,_1F_1(l+1+i\eta, 2l+2; -2ikx)]$$
$$\equiv \frac{F_l(\eta, kx)}{kx} e^{i\sigma_l} \qquad (2.3.47)$$

The factors in the bracket can be shown to be real using Kummer's transformation (see Abramowitz and Stegun[36]). This is the t-matrix form of the *unreduced* wave function, which defines the *reduced, real K-matrix* form F_l in the usual way. Complementary to the latter function is the (real) irregular function G_l; together these functions have the asymptotic form[35]

$$G_l + iF_l \sim e^{i(z_0 - l\pi/2 + \sigma_l)} \qquad (2.3.48)$$

where $z_0 = kx - \eta \ln(2kx)$. For $\eta = 0$ the Coulomb functions reduce to

$$F_l \to kxj_l(kx) \to \sin(kx - l\pi/2) \qquad (2.3.49a)$$
$$G_l \to -kxn_l(kx) \to \cos(kx - l\pi/2) \qquad (2.3.49b)$$

Note the sign in the latter-case limit [cf. Eq. (2.3.39b)]. The asymptotic form of the complete wave function for Coulomb-plus-nuclear scattering is given by

$$\bar{\Psi}_{C,\text{asy}}(\mathbf{x}) = \sum_{l=0}^{\infty} (2l+1)i^l P_l(\mu)$$
$$\times e^{i(\delta_l + \sigma_l)} \frac{F_l(\eta, kx)\cos\delta_l + G_l(\eta, kx)\sin\delta_l}{kx} \qquad (2.3.50)$$

where δ_l is the *additional* (Coulomb-modified) nuclear phase shift, which leads to the complete scattering amplitude

$$\bar{f}_C = f_C + \frac{1}{k}\sum_{l=0}^{\infty}(2l+1)P_l(\mu)e^{i(2\sigma_l+\delta_l)}\sin\delta_l \equiv f_C + \Delta f_C \qquad (2.3.51)$$

and the point-Coulomb scattering amplitude is given by[12,32]

$$f_C = -\frac{2\eta k}{\mathbf{q}^2} e^{-i\eta\ln(\mathbf{q}^2/4k^2) + 2i\sigma_0} \qquad (2.3.52)$$

Thus, Δf_C determines the (Coulomb-modified) short-range nuclear scattering. Our Born-approximation example in Eq. (2.3.21) corresponds to linearizing f_C and Δf_C in the fine-structure constant. We argued earlier that the partial-wave expansion of f_C is not well-behaved. There is no such problem with Δf_C, because the expansion is controlled by the strong-interaction (short-range) phase shifts δ_l.

The Coulomb-modified effective-range function is now much more complicated, and is forced to agree with Eq. (2.3.42) in the $\eta = 0$ limit[12]:

$$K_l^C(E) \equiv \left[(2l + 1)!! \frac{C_l}{C_0}\right]^2 k^{2l+1}[2\eta h(\eta) + C_0^2 \cot \delta_l]$$

$$\sim -\frac{1}{a_l^C} + \tfrac{1}{2} r_l^C k^2 + \cdots \qquad (2.3.53)$$

where

$$h(\eta) = -\ln(\eta) + \text{Re}[\psi(1 + i\eta)] \qquad (2.3.54)$$

is defined in terms of the usual ψ (digamma) function.[36] The Coulomb-modified scattering length results from taking the $k \to 0$ limit, where $\eta h(\eta)$ also vanishes.

If we take the limit of vanishing k, the wave functions appropriate for the scattering-length regime are

$$\Psi_C^0 = \lim_{k \to 0} \frac{\Psi_C(\mathbf{x})}{C_0 e^{i\sigma_0}} = I_0[\sqrt{2\xi x(1 - \mu)}]$$

$$= \sum_{l=0}^{\infty} (2l + 1)(-1)^l P_l(\mu) \mathscr{I}_l(\xi x) \qquad (2.3.55)$$

where the regular, zero-energy wave function

$$\mathscr{I}_l(\xi x) = \lim_{k \to 0} \frac{F_l}{kxC_0} = \frac{I_{2l+1}[2\sqrt{(\xi x)}]}{\sqrt{(\xi x)}} \sim \frac{(\xi x)^l}{(2l + 1)!} \qquad (2.3.56a)$$

and the corresponding irregular function

$$\mathscr{K}_l(\xi x) = \lim_{k \to 0} G_l C_0 = 2\sqrt{\xi x}\, K_{2l+1}(2\sqrt{\xi x}) \sim \frac{(2l)!}{(\xi x)^l} \qquad (2.3.56b)$$

are defined in terms of modified Bessel functions,[36] and are given together with their small-argument limits. Our choice of dimensionless zero-energy Coulomb functions has the unfortunate consequence that $x\mathscr{I}_l$ and $\mathscr{K}_l$ are the reduced wave functions, which introduces a notational asymmetry.

In order to define these zero-energy functions properly, it has been necessary to remove from the Coulomb functions a factor of $C_0 = [2\pi\eta/(e^{2\pi\eta} - 1)]^{1/2}$ (the Coulomb barrier penetration factor, which vanishes exponentially), leaving a residue which is well-behaved. We also note that a peculiarity of the Coulomb problem is the replacement of the k^l (angular-momentum-barrier-induced) factors in the non-Coulomb case by ξ^l factors. All partial waves survive at zero momentum in the Coulomb case, even those normally "closed" by the angular momentum barrier. This is not surprising, since the Coulomb force dominates the centrifugal force at long distances. Even in the non-Coulomb case, higher partial waves can survive if there is a tensor force. For example, at zero energy the tensor-coupled $^3S_1 - {}^3D_1$ reduced wave function components for n–p scattering have s- and d-wave components, $(\bar{u}, \bar{w})$, which have the asymptotic forms

$$\bar{u} \sim x - a \tag{2.3.57a}$$

$$\bar{w} \sim b/x^2 \tag{2.3.57b}$$

This follows immediately from the zero-energy Schrödinger equation, whose irregular solutions are x^{-l} and whose regular solutions are x^{l+1}.

The regular and irregular functions, $\mathscr{I}_l$ and $\mathscr{K}_l$, have asymptotic $(+)$ and $(-)$ forms determined by Eq. (2.3.36), which are independent of l. Using the zero-momentum relations, $\sigma_l \to l\pi/2 + \sigma_0$ (note that σ_0 is not well-behaved in this limit) and $\delta_l \to 0$, we can define the complete asymptotic zero-momentum wave function as

$$\bar{\Psi}^0_{C,\text{asy}}(\mathbf{x}) \equiv \lim_{k \to 0} \frac{\bar{\Psi}_{C,\text{asy}}(\mathbf{x})}{C_0 e^{i\sigma_0}} = \sum_{l=0}^{\infty} (2l + 1)(-1)^l P_l(\mu)\left(\mathscr{I}_l - \frac{\tilde{a}^C_l \mathscr{K}_l}{x}\right) \tag{2.3.58}$$

where

$$\tilde{a}^C_l = \lim_{k \to 0} (-kC_0^2 \cot \delta_l)^{-1} = \left[\left(\frac{\xi}{2}\right)^l \frac{1}{l!}\right]^2 a^C_l \tag{2.3.59}$$

The l-dependent factors in Eq. (2.3.59) follow from the definition in Eq. (2.3.53) of the Coulomb-modified scattering length a^C_l. The Coulomb-modified scattering amplitude in this limit becomes

$$\Delta f^0_C = \lim_{k \to 0} \frac{\Delta f_C}{C_0^2 e^{2i\sigma_0}} = -\sum_{l=0}^{\infty} (2l + 1)(-1)^l P_l(\mu) \left[\frac{\xi^{2l}}{4^l(l!)^2}\right] a^C_l \tag{2.3.60}$$

Normally the terms with $l > 0$ are greatly suppressed because ξ is small. The quantities a^C_l should have a "normal" size, since in the small-ξ limit they become the non-Coulomb quantities a_l. The quantities in the bracket in Eq. (2.3.59) guarantee this. Note that Δf_C vanishes in the small-k limit because it contains the Coulomb barrier factor C_0^2.

4. VARIATIONAL ESTIMATES

Although it is always desirable to solve the Schrödinger equation to the desired accuracy, sometimes this is not possible. Moreover, when we are solving any equation for the first time, errors of all kinds are possible and independent checks are highly desirable. One easy and powerful way to check the solutions is to use variational estimates. Typically, variational techniques are used to solve equations rather than to check solutions generated by other methods, but the principles are the same.

We first form a functional $F[\Psi]$ from a *normalized* bound-state wave function Ψ and a Hamiltonian H,

$$F[\Psi] = \langle \Psi | H - E_0 | \Psi \rangle \tag{2.4.1}$$

where E_0 is the lowest eigenvalue of H. Variation of the wave function ($\Psi \to \Psi_0 + \delta\Psi$) about the lowest normalized eigenfunction of H, Ψ_0, produces a variation in F which is of second order in $\delta\Psi$:

$$\delta F \sim \langle \delta\Psi | H - E_0 | \delta\Psi \rangle \tag{2.4.2}$$

Thus, a first-order error in the wave function produces a second-order error in $\langle H \rangle$, and therein lies the power of variational estimates.[32] It can easily be shown that the quantity on the right-hand side of Eq. (2.4.2) is positive and, hence, $\langle H \rangle$ bounds the lowest eigenvalue from above.

In order for $\langle H \rangle$ to be rather close to E_0, it is only necessary that Ψ be close to Ψ_0 on the average. For precision work Ψ should be close to Ψ_0 *locally*. This can be tested by performing the integral in Eq. (2.4.1) over small regions ($F[\Psi]$ is the sum over all the regions) and looking for large contributions. This knowledge can be used to improve the wave function.

A similar procedure can be applied to scattering problems. Forming the functional

$$I[\Psi] = \langle \Psi | H - E | \Psi \rangle \tag{2.4.3}$$

and varying the wave function $\Psi \to \Psi_0 + \delta\Psi$, subject to the boundary conditions, leads to

$$\delta I = \langle \Psi_0 | H - E | \delta\Psi \rangle + \langle \delta\Psi | H - E | \delta\Psi \rangle \tag{2.4.4}$$

where we have used $(H - E)\Psi_0 = 0$. If the kinetic energy were Hermitian with respect to the functions Ψ_0 and $\delta\Psi$, the first term would vanish. Unfortunately, this is not the case for scattering wave functions; there are surface terms which remain and which satisfy

$$\langle\Psi_0|T|\delta\Psi\rangle = \langle T\Psi_0|\delta\Psi\rangle + \delta S \qquad (2.4.5)$$

$$\delta S = -\frac{\hbar^2}{2\mu}\int d\mathbf{S}\cdot[\Psi_0^\dagger(\mathbf{x})\nabla\delta\Psi(\mathbf{x}) - \nabla\Psi_0^\dagger(\mathbf{x})\delta\Psi(\mathbf{x})] \qquad (2.4.6)$$

Thus, Eq. (2.4.4) leads to Kohn's second-order variational result[37]

$$\delta(I[\Psi] - S) = \langle\delta\Psi|H - E|\delta\Psi\rangle \qquad (2.4.7)$$

demonstrating that $I[\Psi] - S$ is an extremum. Note that Kohn's original derivation defined a different functional (without complex conjugation). In the example below we use a wave function which is real, so that the distinction is lost.

We first consider an s-wave example with a Coulomb-plus-nuclear potential. We write the wave function in K-matrix form, introducing the Coulomb factor C_0 for later convenience:

$$\Psi_0 \sim \frac{F_0(kx)}{kxC_0} + \beta_C\left[\frac{G_0(kx)C_0}{x}\right] \qquad (2.4.8)$$

where

$$\beta_C = \frac{\tan(\delta_0)}{kC_0^2} \qquad (2.4.9)$$

and

$$\delta\Psi \sim \delta\beta_C\left[\frac{G_0(kx)C_0}{x}\right] \qquad (2.4.10)$$

There is no "plane wave" in Eq. (2.4.10), because it is already included in Eq. (2.4.8), as dictated by the boundary conditions. Manipulation of Eq. (2.4.6) leads to

$$\delta S = \frac{\hbar^2}{2\mu}4\pi\delta\beta_C \qquad (2.4.11)$$

or

$$\delta\left(I_C - \frac{4\pi\hbar^2 \beta_C}{2\mu}\right) \sim 0 \qquad (2.4.12)$$

Furthermore, because we vary the trial solution about the exact solution (with $I_C = 0$), we can write

$$\beta_C \cong \beta_C^{tr} - \frac{2\mu I_C^{tr}}{4\pi\hbar^2} \qquad (2.4.13)$$

Given a trial wave function which generates β_C^{tr} and I_C^{tr}, we can obtain a better estimate of $\tan \delta_0$ using Eq. (2.4.13).

For the case of vanishing momentum and no Coulomb interaction, this is equivalent to $\Psi_{tr}^0 \sim 1 - a^{tr}/x$, and

$$a \cong a^{tr} + \frac{2\mu}{4\pi\hbar^2} I^{tr} \qquad (2.4.14)$$

In the Coulomb case the factors of C_0 necessary to extrapolate to zero momentum are already in place. This leads to the corresponding Coulomb-modified s-wave result:

$$a_C \cong a_C^{tr} + \frac{2\mu}{4\pi\hbar^2} I_C^{tr} \qquad (2.4.15)$$

We summarize by giving three examples of the utility of variational techniques:

1. Applying variational corrections allows us to extrapolate and obtain a more accurate estimate (or confirmation) of binding energies[17] or phase shifts.[21]
2. Making "local" variational estimates is an excellent technique[4] for determining where wave functions are least accurate.
3. For problems which are defined as a sequence of smaller problems (such as developing the nuclear or Coulomb potential in partial-wave form and solving the problem with a limited number of partial waves), variational estimates can accelerate the convergence of the sequence.[18]

Each of these examples is useful and powerful. The last case is particularly useful in the Coulomb problem if the Hamiltonian (including a

Coulomb potential) is developed in a (slowly converging) partial-wave form. Performing the variational estimate without expanding the Coulomb potential accelerates the convergence of the sequence.

5. THREE-NUCLEON PROBLEM

The previous sections detailed important aspects of the two-nucleon problem, many of which can be applied *mutatis mutandis* to the three-nucleon problem. Along the way we highlighted certain points which will allow us to resolve analogous difficulties with systems of three nucleons. In what follows, we will adopt a layered approach to resolving difficulties in handling the three-nucleon dynamics. We will present in sequence more complex forms as more sophistication is required to progress from simple two-body forces to three-body forces to the Coulomb interaction. The key to all approaches to the problem is permutation symmetry, and we will tailor our notation to this need.

5.1. Geometric Aspects

The three-nucleon system is significantly more complex than the two-nucleon one, even if we ignore spin and isospin. Specifying the positions of two bodies requires two vectors $\mathbf{r}_1$ and $\mathbf{r}_2$, one combination of which determines the CM. The remaining three coordinates determine one length, as well as two angles which merely serve to orient the two particles with respect to the CM coordinate axes. The specification of a typical three-body system requires three vectors, $\mathbf{r}_1$, $\mathbf{r}_2$, and $\mathbf{r}_3$, one combination of which simply fixes the CM. Three particles determine a plane, whose orientation with respect to the CM coordinate axes is specified by the three Euler angles. The remaining three coordinates (typically, two lengths and one angle) determine the internal configuration of a three-nucleon system and are our independent variables.

The choice of coordinates is motivated by a desire to keep the Schrödinger equation as structurally simple as possible. We defined CM and relative coordinates in the two-body problem so that they decoupled and produced separate contributions in the resultant kinetic energy [Eq. 2.3.2)]. The Jacobi coordinates for a many-body system are constructed analogously. We assume equal-mass particles. Particles 2 and 3 are selected arbitrarily and their relative coordinate denoted by $\mathbf{x}$,

$$\mathbf{x} = \mathbf{r}_2 - \mathbf{r}_3 \qquad (2.5.1a)$$

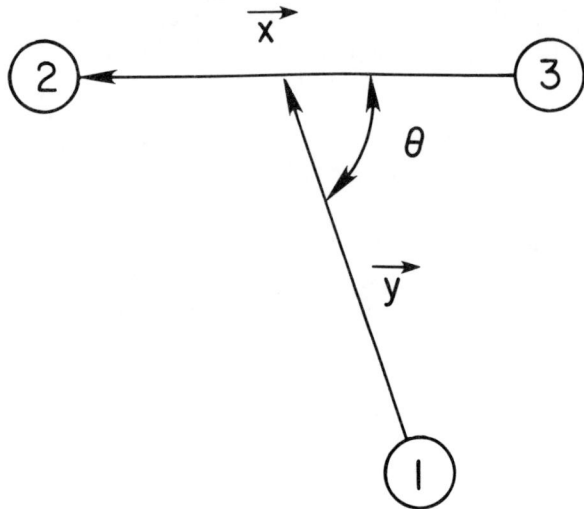

FIGURE 4. Jacobi coordinates for three particles.

while the coordinate of the third particle points to the CM of the other two:

$$\mathbf{y} = \tfrac{1}{2}(\mathbf{r}_2 + \mathbf{r}_3) - \mathbf{r}_1 \qquad (2.5.1\text{b})$$

The lengths of $\mathbf{x}$ and $\mathbf{y}$ are denoted by x and y, respectively, while the angle between them, θ, is determined from $\cos\theta \equiv \hat{\mathbf{x}} \cdot \hat{\mathbf{y}} = \mu$. These coordinates are shown in Fig. 4. Our three independent variables have now been specified, using a procedure which is inherently asymmetric. An "interacting pair" is singled out, leaving a "spectator" nucleon. This scheme seems somewhat artificial, but is well adopted to our problem, as we will see below. Note that the much more difficult alpha-particle problem requires six internal coordinates.

We can immediately write down the kinetic energy in the CM frame based on Eq. (2.3.2):

$$T = \frac{-\hbar^2 \nabla_x^2}{2\mu_x} - \frac{\hbar^2 \nabla_y^2}{2\mu_y} = T_x + T_y \qquad (2.5.2)$$

where the reduced mass for particles 2 and 3 is $\mu_x = M/2$, while that of particle 1 relative to the 2–3 system is $\mu_y = 2M/3$, for nucleons with identical masses M. Our choice of coordinates guarantees no cross terms in $\mathbf{x}$ and $\mathbf{y}$.

5.2. Boundary Conditions

The Schrödinger equation for three particles can be written in the form

$$[E - T - V(23) - V(12) - V(13)] \Psi(\mathbf{x}, \mathbf{y}) = 0 \quad (2.5.3)$$

where the total potential, $V = V(12) + V(13) + V(23)$, is the sum of two-body potentials, $V(ij)$, between nucleons i and j. The presence of three such potentials is responsible for an essential difficulty which did not exist for two potentials. Imagine that particle 1 is incident on particles 2 and 3, which are bound (i.e., a "deuteron"). Elastic scattering is clearly possible, which corresponds asymptotically to

$$1 + (23) \to 1 + (23) \quad (2.5.4a)$$

It is potential $V(23)$ which allows this configuration to exist asymptotically and correspond to small values of the variable x ($<x_d$, the "size" of the deuteron) and large values of y. Other reactions are also possible, such as the breakup reaction:

$$1 + (23) \to 1 + 2 + 3 \quad (2.5.4b)$$

where asymptotically there are no interactions in the final state. Problems arise with the rearrangement channels corresponding to

$$1 + (23) \to 2 + (13) \quad (2.5.4c)$$

or

$$1 + (23) \to 3 + (12) \quad (2.5.4d)$$

made possible by sufficiently strong potentials, $V(13)$ or $V(12)$, respectively. These asymptotic configurations correspond to $y \cong \frac{1}{2}x$ in Fig. 4, and $\theta = 0$ or π.

We might guess that the breakup reaction is nothing more than a complicated outgoing wave (and it is). The rearrangement channels, on the other hand, are fundamentally different, since those final states could also be associated with initial plane waves. This complexity is illustrated in Fig. 5 for collinear scattering, where ordinary elastic scattering (small x, large y) is depicted by the deuteron strip, and whose boundary conditions are simple extensions of the two-body case (which will be shown later). For conveni-

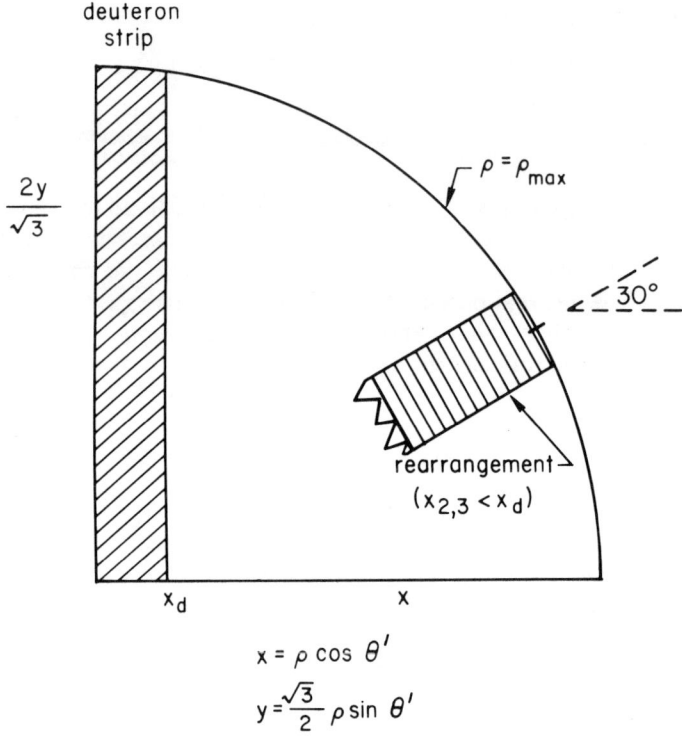

FIGURE 5. The $x-y$ plane for three-nucleon scattering, depicting elastic scattering (the deuteron strip), rearrangement scattering ($\theta' \sim 30°$), and breakup ($90° \geqslant \theta' \geqslant 0°$). Boundary conditions must be specified along $x = 0$, $y = 0$, and $\rho = \rho_{max}$.

ence we also define hyperspherical (polar) coordinates: $(x, y) \to (\rho, \theta')$,

$$\rho^2 = x^2 + \bar{y}^2 \qquad (2.5.5a)$$

$$\tan \theta' = \bar{y}/x \qquad (2.5.5b)$$

where $\bar{y} = \lambda y$, and the $\lambda = 2/\sqrt{3}$ factor arises from $\mu_y/\mu_x = 4/3$. Note the prime on θ' which distinguishes it from θ (the geometric angle between **x** and **y**). Rearrangement scattering corresponds to $\theta' \cong \pi/6$, while the breakup regime asymptotically exists along the entire arc $\rho = \rho_{max}$. Specifying boundary conditions along every portion of that arc, as well as along $x = 0$ and $y = 0$, is necessary in order to specify the solution of our partial-differential equation (the Schrödinger equation) uniquely.

Many years ago Foldy and Tobocman[38] showed that if that equation is converted to an integral equation (the Lippmann–Schwinger equation), whose form guarantees outgoing waves in all channels, the latter equation has no unique solution because the amount of *initial* plane wave in the rearrangement channels is uncontrolled. In order to specify the latter two conditions, two additional equations are required. This can be accomplished with the triad equations,[39] which are not commonly used for practical calculations. The alternative of choice has been the Faddeev equations,[13] which we now discuss in the form introduced by Noyes.[14] Faddeev's work is most commonly presented in momentum space.

5.3. Faddeev–Noyes Equations

The key to understanding the rearrangement problem is the presence of the three two-body potentials in Eq. (2.5.3). One of them creates the bound state for the elastic scattering reaction, while the other two are responsible for asymptotic rearrangement channels. If we could somehow eliminate $V(12)$ and $V(13)$ from the problem (or hide them somewhere), the rearrangement channels would go away and we could resolve our difficulty.

Since rearrangement is an interchange of particles, perhaps particle permutation is the critical mathematical tool we need to employ. In Fig. 4 we picked particles 2 and 3, and then particle 1, to define the two coordinate vectors **x** and **y**. We could have picked any pair to begin with. Any two (nondegenerate) vectors in the plane are linear combinations of **x** and **y** and would also suffice to specify the positions of all particles.

We define three sets of vectors by choosing (i, j, k) to be the cyclic permutations of (1,2,3) [i.e., (1,2,3), (2,3,1), and (3,1,2)]:

$$\mathbf{x}_i = \mathbf{r}_j - \mathbf{r}_k \tag{2.5.6a}$$

and

$$\mathbf{y}_i = \tfrac{1}{2}(\mathbf{r}_j + \mathbf{r}_k) - \mathbf{r}_i \tag{2.5.6b}$$

The pair $(\mathbf{x}_1, \mathbf{y}_1)$ is identical to $(\mathbf{x}, \mathbf{y})$ defined before, and we will use these interchangeably. For later use we also show here relationships among various sets of variables; only two vectors are independent and the others can always be written in terms of those two:

$$(\mathbf{x}_1, \mathbf{y}_1) = (-\tfrac{1}{2}\mathbf{x}_2 - \mathbf{y}_2, \tfrac{3}{4}\mathbf{x}_2 - \tfrac{1}{2}\mathbf{y}_2) = (-\tfrac{1}{2}\mathbf{x}_3 + \mathbf{y}_3, -\tfrac{3}{4}\mathbf{x}_3 - \tfrac{1}{2}\mathbf{y}_3) \tag{2.5.6c}$$

$$(\mathbf{x}_2, \mathbf{y}_2) = (-\tfrac{1}{2}\mathbf{x}_1 + \mathbf{y}_1, -\tfrac{3}{4}\mathbf{x}_1 - \tfrac{1}{2}\mathbf{y}_1) \tag{2.5.6d}$$

$$(\mathbf{x}_3, \mathbf{y}_3) = (-\tfrac{1}{2}\mathbf{x}_1 - \mathbf{y}_1, \tfrac{3}{4}\mathbf{x}_1 - \tfrac{1}{2}\mathbf{y}_1) \tag{2.5.6e}$$

Just as small x_1 and large y_1 specify the usual asymptotic elastic channel, small x_2 (or x_3) and large y_2 (or y_3) specify the asymptotic rearrangement channels. This motivates an ansatz for $\Psi(\mathbf{x},\mathbf{y})$:

$$\Psi(\mathbf{x},\mathbf{y}) = \psi(\mathbf{x}_1,\mathbf{y}_1) + \psi(\mathbf{x}_2,\mathbf{y}_2) + \psi(\mathbf{x}_3,\mathbf{y}_3) \equiv \psi_1 + \psi_2 + \psi_3 \quad (2.5.7)$$

where ψ is the *same* function (for identical particles), but with *different* arguments. Each of these terms appears to refer to different asymptotic channels. We then decompose the Schrödinger equation into three separate equations

$$[E - T - V(23)]\psi_1 = V(23)(\psi_2 + \psi_3) \quad (2.5.8a)$$

$$[E - T - V(13)]\psi_2 = V(13)(\psi_1 + \psi_3) \quad (2.5.8b)$$

$$[E - T - V(12)]\psi_3 = V(12)(\psi_1 + \psi_2) \quad (2.5.8c)$$

which *sum* to the original Schrödinger equation Eq. (2.5.3). These are the differential form of the Faddeev equations.[13,14] At first glance it appears that we have worked in devious ways to replace one equation which is difficult to solve by three equations which are just as intractable. In fact, we have achieved a major simplification at almost no cost. For the case of identical nucleons, once ψ_1 is obtained, ψ_2 and ψ_3 can be obtained from ψ_1 by a simple coordinate permutation. Thus, each of the three equations is a simple coordinate permutation of the others; we need only solve one of them, usually taken to be Eq. (2.5.8a). Moreover, that equation contains $V(23)$, which can only bind particles 2 and 3 asymptotically. Clearly, we have resolved the rearrangement problem. The problem channels are indeed missing in ψ_1, but they are present in Ψ, contained "for free" in ψ_2 and ψ_3.

Several features are more evident if we change our notation slightly to emphasize the symmetry in Eqs. (2.5.8). We adopt an alternative respresentation to specify the two-body potentials: $V_i \equiv V(jk)$. At first glance this appears artificial, since $V_1 = V(23)$ emphasizes the noninteracting particle. Nevertheless, this notation is convenient because the potentials are symmetric under particle interchange and interchanging 2 and 3 leaves 1 unchanged. The three Faddeev equations can now be conveniently written in the generic and elegant form:

$$(E - T - V_i)\psi_i = V_i(\psi_j + \psi_k) \quad (2.5.9)$$

It is the manifest symmetry of V_i under interchange of j and k that leads to this compact equation. Later we will see that if we break that symmetry artificially, a less compact form results.

One immediate benefit of the decomposition we have made is that the Pauli principle is trivial to implement. If we interchange particles 2 and 3, we find after inspecting Eqs. (2.5.6): $(\mathbf{x}_1,\mathbf{y}_1) \to (-\mathbf{x}_1,\mathbf{y}_1)$, $(\mathbf{x}_2,\mathbf{y}_2) \to (-\mathbf{x}_3,\mathbf{y}_3)$, and $(x_3, y_3) \to (-x_2, y_2)$, which leads to $\psi(x, y) \to \psi(-x, y)$, $\psi(x_2, y_2) \to \psi(-x_3, y_3)$, and $\psi(x_3, y_3) \to \psi(-x_2, y_2)$. Thus we have under the preceding transformation:

$$(E - T - V_1)\psi(-x, y) = V_1[\psi(-x_2, y_2) + \psi(-x_3, y_3)] \qquad (2.5.10)$$

which is the original equation with a change of sign in the first argument. Enforcing Bose or Fermi statistics in the variable x of this equation requires $\psi(-x, y) = \pm \psi(x, y)$. Forming the Schrödinger wave function Ψ, we find that the interchange produces $\Psi \to \pm \Psi$. The same result holds for all other permutations as well, and illustrates how easily the appropriate statistics can be implemented in the Faddeev equations.

The Faddeev wave functions (often called amplitudes) ψ_i are unphysical, unlike the Schrödinger wave function Ψ. Theorems proven for Ψ do not necessarily hold for the ψ_i. The situation is analogous to electromagnetic problems, where we wish to solve for the electromagnetic fields, which are physical quantities; this is usually achieved by solving first for the (unphysical) electromagnetic potentials, whose boundary conditions are often easier to implement. From a numerical perspective, Ψ has a more complex structure than ψ and is more difficult to calculate directly.

5.4. Three-Nucleon Forces

Short-range three-body forces are very easy to incorporate into this scheme because they cannot alter the asymptotic configurations. Indeed, it is so easy that there is no unique prescription for including these forces in the Faddeev equations. We now exploit the alternative notation that we introduced for the two-body forces, and extend it to three-body forces, which depend on the coordinates of all three of the particles. The total three-body force W can be written as $W = W_1 + W_2 + W_3$, where $W_i = W(jk; i)$, just as we wrote $V = V_1 + V_2 + V_3$ in the case of the two-body force. This notation emphasizes that particles j and k interact with particle i as an intermediary and, indeed, this is the natural form for most three-nucleon forces (recall that the earth was the intermediary between the moon and the satellite in the classical example we discussed in Section 2). It also requires that $W(jk; i)$ be symmetric in the coordinates of j and k.

We can incorporate W into the first Faddeev equation in a variety of ways.[40] For identical particles, we have

$$(E - T - V_1 - W_1)\psi_1 = (V_1 + W_1)(\psi_2 + \psi_3) \qquad (2.5.11a)$$

$$(E - T - V_1 - W)\psi_1 = V_1(\psi_2 + \psi_3) \qquad (2.5.11b)$$

$$(E - T - V_1 - \tfrac{1}{3}W)\psi_1 = (V_1 + \tfrac{1}{3}W)(\psi_2 + \psi_3) \qquad (2.5.11c)$$

all of which sum with their associated permuted equations to the Schrödinger equation:

$$(E - T - V - W)\Psi = 0 \qquad (2.5.11d)$$

The different forms are therefore identical if treated equivalently, although the function ψ (but not Ψ) will be different in each case. It is only because the potential W has a short range in *every* coordinate that our solution to the arrangement difficulty is not upset by the forms in Eqs. (2.5.11b) and (2.5.11c).

5.5 Coulomb-Modified Faddeev Equations

A similar situation exists when a repulsive Coulomb potential is added to the short-range nuclear force. There is no unique way to accommodate the Coulomb potential, but there is a best way.[16,41] For the moment we ignore any distinctions among the particles, and treat them all the same, leading to three two-body Coulomb interactions. With real nucleons this is trivially accomplished by using isospin operators[4] in V_C to distinguish neutrons from protons. The Coulomb potential between particles 2 and 3 is denoted $V_C(x)$. Clearly, the easiest way to incorporate V_C is just to add it to the short-range interaction $V_1(x) \equiv V(23)$, which we henceforth write as $V(x)$ in the first of the Faddeev–Noyes equations:

$$[E - T - V(x) - V_C(x)]\psi_1 = [V(x) + V_C(x)](\psi_2 + \psi_3) \qquad (2.5.12)$$

Unfortunately, this is the worst possible choice. There is no rearrangement problem, but for the large y and small x corresponding to an outgoing p–d system there is no obvious Coulomb interaction in the y-coordinate, which would lead to outgoing Coulomb waves in ψ_1. Even worse, there is now a long-range interaction in the x-coordinate which couples to the rearrangement channels, and this has no obvious physical interpretation. In order to make the imposition of boundary conditions transparent and tractable, we need a Coulomb potential in the variable y on the left-hand side of Eq. (2.5.12), which will then produce Coulomb waves between the outgoing clusters.

We remarked earlier (and gave examples) of how a short-range three-body force could be added to the Faddeev equations in a nonunique way. Indeed, it is possible to add a fictitious force to the Faddeev equations which vanishes when the various permutations are added together to form the Schrödinger equation. Although this formal introduction of a "distortion" potential has no physical effect, it is a powerful formal device.[41] We introduce a three-body-like distortion potential $X(jk; i) \equiv X_i$, which is required to be symmetric in the indices j and k. More general forms are possible which do not have this restriction and are useful in discussing the Sasakawa–Sawada form[42] of the Coulomb-modified Faddeev equations. Such modifications break the symmetry between ψ_2 and ψ_3 which we see below in Eq. (2.5.13). The modified Faddeev equation,

$$(E - T - V_1 - X_2 - X_3)\psi_1 = (V_1 - X_1)(\psi_2 + \psi_3) \qquad (2.5.13)$$

has the property that adding it to its permuted terms reproduces the Schrödinger equation, which is independent of the X_i. Choosing $X_i = 0$ gives the usual Faddeev equation. Choosing $X_i = V_i$ converts Eq. (2.5.13) back to the Schrödinger equation, which is a poor idea in view of the rearrangement problem. However, if we write $V_1 = V(x) + V_C(x)$, choosing $X_1 = V_C(x)$ completely removes the Coulomb interaction from the right-hand side of Eq. (2.5.13) [leaving $V(x)$], while putting the full Coulomb potential on the left-hand side. This form is often called the Faddeev–Noble equation,[43] and has been exploited successfully for many years by Merkuriev[16,44] and his collaborators:

$$[E - T - V(x) - V_C(x_1, x_2, x_3)]\psi_1 = V(x)(\psi_2 + \psi_3) \qquad (2.5.14)$$

where $V_C(x_1, x_2, x_3)$ is the Coulomb potential between *all* pairs of particles, and has the requisite form for large y. We note that this trick is effective only because the potentials which bind together the asymptotic clusters (i.e., the deuteron) do not contain the Coulomb interaction. It is very easy to simultaneously include both three-nucleon and Coulomb forces.[45]

It has been shown in a variety of numerical calculations[41] that of the several ways that the Coulomb interaction has been introduced into the configuration-space Faddeev equations, the Faddeev–Noble form is optimal because it minimizes the long-range coupling to the permuted terms (i.e., the rearrangement channels) that can cause significant numerical difficulties.

6. BOUNDARY CONDITIONS

In order to obtain unique solutions to the Faddeev (or Schrödinger) equation, we need to specify physically motivated boundary conditions over the entire boundary of the space in which we work. Figure 5 illustrates the x-y plane. Our boundary conditions are:

1. Finiteness.
2. An initial plane wave plus outgoing scattered waves for scattering processes.

The easiest way to implement the finiteness condition along the $x = 0$ and $y = 0$ boundaries is precisely the same as in the two-body case; we use reduced wave functions:

$$\psi_1 \equiv \psi(\mathbf{x}, \mathbf{y}) = \frac{\phi(\mathbf{x}, \mathbf{y})}{xy} \qquad (2.6.1)$$

where $\phi(\mathbf{x}, \mathbf{y}) \equiv \phi(x, y, \mu)$ is required to vanish along $x = 0$ and $y = 0$ in order for ψ to be finite. The reduced wave function satisfies the equation

$$\left[E + \frac{\hbar^2}{M} \left(\frac{\partial^2}{\partial x^2} + \frac{\partial^2}{\partial \bar{y}^2} - \frac{\hat{L}_x^2}{x^2} - \frac{\hat{L}_y^2}{\bar{y}^2} \right) - V(x) - V_C \right] \phi(x, y, \mu)$$

$$= V(x) \left[\phi(x_2, y_2, \mu_2) \frac{xy}{x_2 y_2} + \phi(x_3, y_3, \mu_3) \frac{xy}{x_3 y_3} \right] \qquad (2.6.2)$$

where the quantities $\hat{L}_x^2$ and $\hat{L}_y^2$ are the usual orbital angular momentum operators contained in $\mathbf{V}_x^2$ and $\mathbf{V}_y^2$, respectively, and $\mu_i = \cos \theta_i = \hat{\mathbf{x}}_i \cdot \hat{\mathbf{y}}_i$.

The large-ρ behavior is determined by the particular process.[46] For bound states (with $E = -\hbar^2 \kappa^2 / M$) we require that ψ be vanishingly small ($\sim e^{-\kappa \rho}$), which completes the specification of the boundary conditions for that problem.

6.1. Elastic Scattering

Elastic scattering consists of six different cases:

1. Zero-energy scattering, with $E = E_d$.
2. Scattering below breakup threshold, with $E = E_d + (\hbar^2 k^2 / 2\mu_y) < 0$.
3. Scattering above breakup threshold, with either $E = E_d + (\hbar^2 k^2 / 2\mu_y)$ or $E = \hbar^2 K^2 / M (> 0)$.

The deuteron energy is $E_d = -\hbar^2 \kappa_d^2 / M$. In addition, for each of these categories we have (a) non-Coulomb and (b) Coulomb cases. Nevertheless, all of them require that asymptotically (for large y) a component of the wave function must have the form

$$\phi_{\mathrm{el}}(\mathbf{x}, \mathbf{y}) \sim \phi_d(\mathbf{x}) \chi(\mathbf{y}) \qquad (2.6.3)$$

where $\chi(\mathbf{y})$ is the (relative) N-d wave function, and $\phi_d(\mathbf{x}) = x \psi_d(\mathbf{x})$ is the reduced deuteron wave function which satisfies the equation

$$[E_d - T_x - V(x)] \frac{\phi_d(\mathbf{x})}{x} = 0 \qquad (2.6.4)$$

In addition to the elastic channel, the inelastic or breakup channel (with three free nucleons) contributes to the asymptotic wave function. In general, we have

$$\phi_{asy} \sim \phi_{el}(\mathbf{x}, \mathbf{y}) + \phi_{br}(\mathbf{x}, \mathbf{y}) \tag{2.6.5}$$

Below breakup threshold ($E < 0$), ϕ_{br} decreases exponentially for large values of the hyperradius ρ. We will see below that for $E > 0$, ϕ_{br} decreases $\sim y^{-3/2}$ for small x and large y and this rapid decrease of ϕ_{br} compared to ϕ_{el} always allows us to separate them asymptotically. Thus, we can consider these contributions separately.

Equation (2.6.2) can be solved directly for the asymptotic form of the wave function, ϕ_{asy}. Consider ϕ_{el}. The right-hand side of that equation vanishes for $x > R$, the range of the force. The permuted terms $\phi_{el}(\mathbf{x}_2, \mathbf{y}_2)$ and $\phi_{el}(\mathbf{x}_3, \mathbf{y}_3)$ have the limits $\phi_{el}(\pm \mathbf{y}, -\frac{1}{2}\mathbf{y}) \sim \phi_d(\pm \mathbf{y})\chi(-\frac{1}{2}\mathbf{y})$ for fixed x and large y and vanish exponentially in that limit since $\phi_d(x) \sim e^{-\kappa_d x}$. We can therefore neglect the right-hand side of Eq. (2.6.2). In Fig. 5 the equivalent statement is that direct and rearrangement channels can only overlap for small values of ρ. Thus, the resulting asymptotic form for s-waves for the non-Coulomb case must satisfy

$$\left(\frac{d^2}{dy^2} + k^2\right)\chi(y) = 0 \tag{2.6.6}$$

or

$$\chi^c(y) \sim \frac{e^{i\delta_0}}{k} \sin(ky + \delta_0) \tag{2.6.7a}$$

in complete analogy with the two-body problem, except that δ_0 will be complex above breakup threshold, because an extra channel is open. For $l \neq 0$, Eq. (2.3.40) should be used. At zero energy this reduces to

$$\chi^0(y) \sim y - a \tag{2.6.7b}$$

This completes the non-Coulomb part of this problem.

The Coulomb case requires consideration of the asymptotic form of the Coulomb potential in Eq. (2.6.2):

$$V_C(x_1, x_2, x_3) = \frac{e_1 e_2}{x_3} + \frac{e_1 e_3}{x_2} + \frac{e_2 e_3}{x_1} \tag{2.6.8}$$

We are only interested in the p–d case, where a single pair of the charges, e_i has a nonvanishing product. If the nucleons in the deuteron are given the conventional labels 2 and 3 (and, hence, $e_2 e_3 = 0$), we find that for large y and fixed x we have

$$V_C(x_1, x_2, x_3) \to \frac{(e_1 e_3 + e_1 e_2)}{y} + O(1/y^2) \to \frac{e^2}{y} \tag{2.6.9}$$

This potential generates asymptotic solutions $\chi_C(y)$ for cases (1), (2), and (3) (above) which are given by Eqs. (2.3.36) and (2.3.28), respectively, and by Eq. (2.3.32) for the bound-state case. For scattering with $l \neq 0$, we can use Eqs. (2.3.58) and (2.3.50). Note that for all of the scattering solutions, $|\phi_{el}| \sim$ constant for large y.

We can summarize these results by stating that N–d elastic scattering has boundary conditions which are mere extensions of the two-body problem, and they follow immediately from the asymptotic forms of the Faddeev equations.

6.2. Breakup Scattering

It is most convenient to work out the boundary conditions for breakup scattering in the hyperspherical (polar) coordinates given by Eq. (2.5.5). The reason is that it is very difficult to develop asymptotic forms which simultaneously satisfy the $x = 0$ and $y = 0$ boundary conditions. In order to be as specific as possible, we should also formally include the orbital angular momentum terms in the Faddeev equation, in order to demonstrate that they do not influence the boundary conditions. We first note that as y increases, the angle subtended by the deuterion strip in Fig. 5, x_d/y, becomes vanishingly small, as does R/y, the angle subtended by the range of the strong force. Consequently, we can drop $V(x)$ from Eq. (2.6.2). Although our derivation will not be formally correct for $\theta' = \pi/2$, that result will be justified and extended below. We use $E = \hbar^2 K^2/M$ and $(x_i, \bar{y}_i) \equiv \rho(\cos\theta'_i, \sin\theta'_i)$, with $\theta'_1 \equiv \theta'$ (defined before), to obtain the hyperspherical equation:

$$\left(\frac{\partial^2}{\partial \rho^2} + \frac{1}{\rho} \frac{\partial}{\partial \rho} - \frac{\hat{M}}{\rho^2} + \frac{2K\hat{C}}{\rho} + K^2 \right) \phi_{br}(\rho, \theta', \theta) = 0 \tag{2.6.10}$$

where

$$\hat{M} = \frac{\hat{L}_x^2}{\cos^2\theta'} + \frac{\hat{L}_y^2}{\sin^2\theta'} - \frac{\partial^2}{\partial \theta'^2} \tag{2.6.11}$$

and

$$\hat{C} = -\rho V_C(x_1, x_2, x_3)\frac{M}{2\hbar^2 K} = \frac{-M}{2\hbar K}\left(\frac{e_1 e_2}{\cos\theta'_3} + \frac{e_1 e_3}{\cos\theta'_2} + \frac{e_2 e_3}{\cos\theta'_1}\right) \quad (2.6.12)$$

is the three-body version[16] of the (two-body) Coulomb η-parameter. The additional complexity in Eq. (2.6.12) is that $\hat{C}$ depends on the "angles," θ'_i, which specify the relative velocities of the particles in the CM frame.

We write an outgoing-wave Riccati representation for ϕ_{br} in the form

$$\phi_{br}(\rho, \theta', \theta) = A(\theta', \theta) e^{iK\rho} e^{\int^\rho f(\rho')d\rho'} \quad (2.6.13)$$

where we have suppressed the dependence of the function f on (θ', θ) (which contributes only to higher-order terms in the asymptotic expansion). This leads to

$$f^2 + f' + 2iKf + \frac{iK + f + 2K\hat{C}}{\rho} - \frac{[\hat{M}A]/A}{\rho^2} = 0 \quad (2.6.14)$$

which has an asymptotic solution of the form $f = (-\frac{1}{2} + i\hat{C})/\rho + O(1/\rho^2)$ or

$$\phi_{br}(\rho, \theta', \theta) \sim A(\theta', \theta) \frac{e^{iK\rho}}{\rho^{1/2}} e^{i\hat{C}\ln(2K\rho)}[1 + O(1/\rho)] \quad (2.6.15)$$

where we have arbitrarily added a (constant) phase into the logarithm's argument. This corresponds to the argument leading to Eq. (2.3.28) and has been obtained[15,16,44] by other techniques. Clearly, the square of the breakup amplitude is unaffected by such a phase. We also note that the complicated operator $\hat{M}$ does not contribute to the leading-order term, but rather affects the $O(1/\rho)$ contribution. An expression for the latter term has been worked out in the non-Coulomb case,[47] and is not difficult to obtain using the present techniques. Finally, in order for the reduced wave function to satisfy the boundary conditions at $x = 0$ and $y = 0$, we must have $A(0, \theta) = A(\pi/2, \theta) = 0$.

6.3. Boundary Conditions Redux

We now have all the pieces necessary for a more complete and unified exposition of the asymptotic boundary conditions. We restrict ourselves to the non-Coulomb cases for simplicity, although we will subsequently quote the necessary Coulomb modifications. We wish to solve

$$[E - T_x - T_y - V(x)]\psi(\mathbf{x}, \mathbf{y}) = V(x)[\psi(\mathbf{x}_2, \mathbf{y}_2) + \psi(\mathbf{x}_3, \mathbf{y}_3)] \qquad (2.6.16)$$

for fixed $x < R$ (the range of the nuclear force) and large y, which corresponds to $\theta' \cong \pi/2$ and $\rho \cong \lambda y$, with $\lambda = 2/\sqrt{3}$. From Eqs. (2.5.6) we see that in the large-y limit, we have $(\mathbf{x}_2, \mathbf{y}_2) \to (\mathbf{y}, -\frac{1}{2}\mathbf{y})[\theta_2 = \pi]$ and $(\mathbf{x}_3, \mathbf{y}_3) \to (-\mathbf{y}, -\frac{1}{2}\mathbf{y})[\theta_3 = 0]$. For these arguments we have already developed the required asymptotic form in Eq. (2.6.15), and this immediately leads to

$$[E - T_x - T_y - V(x)]\psi(\mathbf{x}, \mathbf{y}) = V(x)\beta \frac{e^{iK\lambda y}}{y^{5/2}} \qquad (2.6.17)$$

where (to lowest order in y^{-1}) β is a constant determined by the breakup amplitude. The solution of this equation is the sum of the inhomogeneous solution and any homogeneous ones. The elastic-scattered wave that we developed earlier is one homogeneous solution. No homogeneous breakup solution can be written which is an arbitrary product of x- and y-outgoing waves, because such a solution will fail to satisfy the boundary conditions simultaneously along $x = 0$ and $y = 0$. Only the y-dependence exhibited on the right-hand side of Eq. (2.6.17) [and obtained from Eq. (2.6.15)] is acceptable. Therefore, we posit

$$\psi(\mathbf{x}, \mathbf{y}) \sim \beta g(x) \frac{e^{iK\lambda y}}{y^{5/2}} + O(1/y^{7/2}) \qquad (2.6.18)$$

Using $E = \hbar^2 K^2/M$ and $2\mu_y = \lambda^2 M$ and keeping only the leading terms, we arrive at

$$[T_x + V(x)]g(x) = -V(x) \qquad (2.6.19)$$

This remarkable and simple zero-energy equation is quite easy to solve. The homogeneous solution must be proportional to the (regular) scattering-length wave function $\psi^0(x)$, which satisfies the $x = 0$ boundary condition. The arbitrary proportionality constant is denoted by γ. The inhomogeneous solution is obtained by inverting the left-hand side of the equation using a zero-energy Green's function, $G_0 = [T_x + V(x)]^{-1}$:

$$g(x) = -G_0 V \equiv g_{\text{el}}(x) + g_{\text{in}}(x) \qquad (2.6.20)$$

where a spectral decomposition of G_0 allows it to be written in terms of elastic $[\sim \phi_d(x)]$ and inelastic components. Putting it all together, we obtain for small x and large y:

$$\phi_{\text{asy}}(\mathbf{x}, \mathbf{y}) \sim \phi_{\text{el}}(\mathbf{x}, \mathbf{y}) + \frac{e^{iK\lambda y}}{y^{3/2}} x\{\beta[g_{\text{el}}(x) + g_{\text{in}}(x)] + \gamma\psi^0(x)\} \quad (2.6.21)$$

which obviously satisfies the $x = 0$ boundary condition.

The individual terms from left to right are: (1) direct elastic scattering; (2) elastic recombination; (3) inelastic recombination; (4) direct inelastic. The recombination terms arise from breakup, with final-state interactions recombining two nucleons into either a deuteron or a "compact" zero-energy scattering solution. The direct inelastic form also leads to the latter state. Near $x = 0$ the strong nuclear interaction contained in G_0 and ψ^0 can lead to rich structure ("almost elastic" scattering).

If we assume $x > R$, we find $\psi^0(x) \sim 1 + O(1/x)$, and it can be shown that $g(x) \sim O(1/x)$. Consequently, this result immediately matches the result obtained previously with $V(x) \equiv 0$, which satisfies the boundary condition along $y = 0$. The Coulomb case for $x < R$ can be solved after carefully working out the limits of the three-body Coulomb parameter defined in Eq. (2.6.12): $\hat{C} \to (-\alpha Mc\lambda/2\hbar K)$. The obvious result obtains: place a factor of $\exp[i\hat{C}\ln(2K\lambda y)]$ in front of the brace in Eq. (2.6.21) and use Coulomb wave functions for the elastic scattering component. This completes the specification of "Coulomb" boundary conditions. There remains one potentially serious problem with the "correction" terms of order y^{-2} that we ignored in Eq. (2.6.9), and we consider these next.

7. COULOMB POLARIZATION POTENTIAL

7.1. Rayleigh Scattering

The Coulomb polarization potential which has recently been the subject of much controversy, involves the same physical elements as Rayleigh scattering of light. In order to develop insight into this interesting phenomenon, we first briefly review how light scatters from molecules.

One of the great triumphs of nineteenth century physics was the logical formulation of thermodynamics, subject to laws and conservation principles and following from dynamical considerations at the microscopic level. At the same time, classical electron theory was providing a microscopic basis for many optical phenomena. They coincide in providing an explanation for red sunsets and blue skies, which puzzled early scientists.

A perfectly homogeneous (transparent) medium scatters no light. Inhomogeneities (such as density fluctuations) scatter light, and these are determined by the properties of the medium and the microscopic origin of the mechanism for light transmission. At Maxwell's suggestion, Lord Rayleigh investigated[48] whether air molecules could be the origin of the strongly wavelength-dependent scattering of light, which preferentially scatters the blue light from the sun. Rayleigh showed that the attenuation of a beam of light passing through a gas was dependent on the index of refraction, the wavelength of the light ($\sim \lambda^{-4}$) and the density of the gas (which depends on the thermodynamic variables and Avogadro's number N_A). According to Rayleigh, yellow light should have an attenuation length of about 120 km, and he noted that "... Mount Everest appears fairly bright at 100 miles distance as seen from Darjeeling...." Subsequent measurements[49] of starlight attenuation at Mount Wilson verified the λ^{-4} law of Rayleigh scattering and established a value of $N_A = 6.06(4) \times 10^{23}$, which is consistent with measurements made using more accurate techniques.

Classical electrodynamics[34] confirms the microscopic origin of Rayleigh scattering. A long-wavelength electric field applied to a deformable object creates a dipole moment, **p**,

$$\mathbf{p} = \alpha_E \mathbf{E} \tag{2.7.1}$$

proportional to the electric field; the proportionality constant is the electric polarizability. This corresponds to an electromagnetic interaction energy,

$$\Delta H = -\tfrac{1}{2}\alpha_E \mathbf{E}^2 \tag{2.7.2}$$

Because the electric field of a photon is inversely proportional to the wavelength, this interaction leads immediately to a scattering cross section proportional to λ^{-4}.

In quantum theory we can use second-order perturbation theory to obtain Eq. (2.7.2), and this gives[50]

$$\alpha_E = 2\alpha\hbar c \sum_N \frac{|\langle 0|D_z|N\rangle|^2}{E_N - E_0} \tag{2.7.3}$$

where the electromagnetic dipole operator **D** (in units of e) has negative parity and a vanishing ground-state expectation value. Equation (2.7.3) has been evaluated theoretically for the deuteron,[51] and has the value $\alpha_E^d = 0.63 \text{ fm}^3$, with a rather small uncertainty, and this agrees with experimental determinations.[52]

7.2. Polarization Potential

The same physics described in the previous section obtains in nuclear physics. During a collision between charged nuclear fragments, the electric field of each fragment will tend to polarize the charges in the other. This is particularly easy in a nucleus because only the protons are charged, and they separate easily from the neutrons. Indeed, this is the mechanism which produces the giant dipole resonance, a ubiquitous collective excitation in nuclei. Two successive (dipole) stretchings, one which polarizes followed by one which depolarizes, is precisely the physics responsible for Rayleigh scattering of light.

The Hamiltonian in Eq. (2.7.2) describes this phenomenon asymptotically. If we substitute the form of the electric field produced by the (asymptotic) static Coulomb potential of one of the fragments, Ze/r, into that expression, we find

$$V_p(r) = -\frac{Z^2 \alpha \hbar c \alpha_E}{2r^4} \tag{2.7.4}$$

This is a long-range potential, although one which decreases rapidly with increasing r. The size of this interaction relative to the Coulomb potential for p–d scattering ($Z = 1$) is $V_p/V_C = \alpha_E^d/2r^3 = (0.7/r)^3$, using the previously stated value of α_E^d. For $r = 50$ fm (a representative value), this fraction is 2.5×10^{-6}, which is a negligible correction. Note that Eq. (2.7.4) gives only the asymptotic form of V_p. The dipole nature of the potential suggests that it actually vanishes for very small separations of the fragments.

Nevertheless, because $V_p(r)$ is a long-range potential, there are matters of principle which must be confronted. In particular, no matter how small such a force may be, it can cause problems if we integrate the effect of that force over exceptionally large distances. An example of this is the Coulomb force. Even if that force were a million times smaller, the infamous Coulomb divergence would still occur. Nevertheless, in spite of that divergence, standard practice would find this scaled-down Coulomb potential to be negligible in its effects.

Standard practice in configuration space is to integrate the Schrödinger equation inside some boundary r_m, and match these to suitably selected regular and irregular boundary functions, as we have done in this chapter with respect to both two- and three-body problems. If the proper functions are selected, we need only match outside the range of the nuclear forces; any value of r_m must then give the same result. If the long-range force is negligibly weak, matching to ordinary boundary functions (which *do not* include the effect of the long-range force) at modest values of r_m will give the correct answer and exhibit a negligible effect of the long-range force. A problem arises only if we integrate out to enormous distances to compensate for the weak force. This is never done and would be difficult to accomplish in practice.

The weakness of the polarization potential, therefore, means that it can be neglected if standard practice is followed in configuration space, except possibly at zero energy. Equation (2.3.28) exhibits the asymptotic effect of the Coulomb interactions $[\sim \eta \ln(2kr)]$. A negligibly weak V_C would produce a negligible η, except near $k = 0$. In that case we have seen that the form of the solution changes completely: the regular solution is a rising exponential, while the irregular solution is a falling one. This ratio produces a powerful lever arm at large distances, such that even a tiny perturbation could produce an erroneous phase shift. Nevertheless, in the scattering-length regime where the effect is largest, it has been shown[53] numerically that the physical polarization potential will have a negligible effect on the three-nucleon problem, by comparing the results of using boundary functions appropriate to $V_C + V_p$, and those obtained using standard practice (i.e., ignoring V_p and matching at modest distances).

If it is possible to work out the problem "correctly," what is the point of the preceding discussion? Our concern is that the effective polarization Hamiltonian in Eq. (2.7.2) is only the first of an infinite number of increasingly weaker multipole (quadrupole,...) polarization potentials $(\sim r^{-6}...)$ which might have to be taken into account. Because the dominant term is negligible at ordinary values of r_m, we can also ignore the rest. We emphasize that standard practice does not ignore V_p. It is included as a normal part of the Coulomb physics and solution process inside r_m, where it is largest. Outside r_m, where it is negligible, it can be ignored.

In order to illustrate these difficulties more clearly, and the physics associated with them, we next develop the boundary conditions at zero energy for the three-nucleon problem, including the simultaneous effect of the polarization and Coulomb potentials.

7.3. Boundary Conditions for $1/r$- plus $1/r^4$-Potentials

The Coulomb potential used in the Faddeev–Noble equations is rich in the physics which it builds into the asymptotic wave function. For simplicity of presentation we truncate the potential and only allude to what is missing. We first write the complete Coulomb interaction, Eq. (2.6.8), in terms of variables $(\mathbf{x}, \mathbf{y})$:

$$V_C(x_1, x_2, x_3) = \frac{e_2 e_3}{x} + \frac{e_1 e_2}{|\mathbf{y} + \frac{1}{2}\mathbf{x}|} + \frac{e_1 e_3}{|\mathbf{y} - \frac{1}{2}\mathbf{x}|} \qquad (2.7.5)$$

The last two terms can be expanded in a partial-wave series in the angle θ, and we keep only the monopole and dipole terms:

$$V_C = \frac{\alpha \hbar c}{y} + \Delta V_C^s + \beta \frac{\mathbf{x} \cdot \mathbf{y}}{y^3} + \Delta V_C^p + \cdots \qquad (2.7.6a)$$

where

$$\Delta V_C^s = (e_1 e_3 + e_1 e_2)\left(\frac{1}{r_>} - \frac{1}{y}\right) + e_2 e_3 \left(\frac{1}{x} - \frac{1}{y}\right) \qquad (2.7.6b)$$

$$\Delta V_C^p = \beta \mathbf{x} \cdot \mathbf{y} \left(\frac{1}{r_>^3} - \frac{1}{y^3}\right) \qquad (2.7.6c)$$

$\beta = e_1 e_3 - e_1 e_2$, and $r_>$ is the greater of $(y, x/2)$. We have used the fact that $e_1 e_2 + e_1 e_3 + e_2 e_3 = e^2 = \alpha \hbar c$ for the p–d system.

We are only interested in the threshold regime. As argued in Section 6, the rearrangement channels (permuted terms) fall off exponentially in the large-y, finite-x asymptotic region which we investigate. For $y > \frac{1}{2}x$, most of ΔV_C^s and all of ΔV_C^p vanish. Because we have chosen nucleons 2 and 3 to be the bound neutron and proton in the outgoing deuteron, $e_2 e_3$ (and the rest of ΔV_C^s) also vanishes asymptotically.

The region $\frac{1}{2}x > y$ corresponds to the proton projectile inside the deuteron. Because the deuteron has a finite size, its charge distribution modifies the Coulomb potential [cf., Eq. (2.3.20)]. Coulomb scattering by this modified potential is therefore naturally included in the formalism. Friar et al.[4] suggested that this effect (which is quite small) should be subtracted from the directly computed scattering lengths. This, in fact, defines the nuclear scattering with respect to Coulomb scattering by the modified potential, which is not only unconventional but a poor idea. It does not correspond to what is conventionally calculated or measured, which is scattering with respect to a point-Coulomb potential.

We are therefore free to ignore ΔV_C^s, ΔV_C^p,... asymptotically; their entire effect is of short range and is included in the interior solution. We have in the asymptotic region

$$V_C^{\text{asy}} \sim \frac{\alpha \hbar c}{y} + \beta \frac{\mathbf{x} \cdot \mathbf{y}}{y^3} + \cdots \equiv \frac{\alpha \hbar c}{y} + \beta \Delta V \qquad (2.7.7)$$

where the ellipsis indicates long-range potentials of the quadrupole,... type which decrease as y^{-3} and faster.

We are now in a position to develop the behavior of the asymptotic wave function systematically. We wish to include terms up to order β^2, so we write the perturbation expansion

$$\Psi_{\text{asy}}^\beta \sim \Psi_0 + \beta \Psi_1 + \beta^2 \Psi_2 + \cdots \qquad (2.7.8)$$

and

$$(E_d - H_x - H_y - \beta \Delta V) \Psi_{\text{asy}}^\beta = 0 \qquad (2.7.9)$$

where $H_x = T_x + V(x)$ and $H_y = T_y + (\alpha \hbar c/y)$. This leads to the hierarchy of equations:

$$(E_d - H_x - H_y)\Psi_0 = 0 \tag{2.7.10a}$$

$$(E_d - H_x - H_y)\Psi_1 = \Delta V \Psi_0 \tag{2.7.10b}$$

$$(E_d - H_x - H_y)\Psi_2 = \Delta V \Psi_1 \tag{2.7.10c}$$

We immediately see that the s-wave solution is $\Psi_0 = \psi_d(x)F(y)$, where the deuteron wave function is defined in Eq. (2.6.4) and $F(y)$ is either the regular or irregular zero-energy Coulomb solution given by Eqs. (2.3.56). Note that the left-hand side of each equation cannot "vanish" ($E_d < 0$) unless Ψ_i contains a (bound) deuteron.

The next-order solution requires careful consideration. We want only the asymptotic solution in y (not the complete one), and this is easily obtained if we realize that $\Delta V(\mathbf{x}, \mathbf{y})$ falls off for large y as y^{-2}, that it factorizes in $\mathbf{x}$ and $\mathbf{y}$, and that ΔV is itself a solution of Laplace's equation. We try a solution of the form $\Psi_1 \sim f(x)[\Delta V F(y)]$. Operating on the bracketed term with H_y, we find

$$H_y[\Delta V(\mathbf{x}, \mathbf{y})F(y)] = \frac{2\hbar^2}{\mu_y}\left[\frac{F'(y)}{yF(y)}\right]\Delta V F(y) \tag{2.7.11}$$

Moreover, it is easy to show that the two solutions in Eq. (2.3.56) behave as

$$\frac{F'(y)}{F(y)} \sim \pm \sqrt{\frac{\xi}{y}}[1 + O(y^{-1/2})] \tag{2.7.12}$$

and, hence,

$$H_y[\Delta V F(y)] \sim \pm \frac{2\hbar^2 \sqrt{\xi}}{\mu_y y^{3/2}}[\Delta V F(y)][1 + O(y^{-1/2})] \tag{2.7.13}$$

which falls off faster than $[\Delta V F(y)]$ and can be neglected, if desired, when compared to $(E_d - H_x)f\Delta V F$ for large y. We can therefore ignore H_y in Eq. (2.7.10b) and obtain

$$(E_d - H_x)\Psi_1 = \Delta V \Psi_0 + O(y^{-3/2}) \tag{2.7.14}$$

It is clear from the form of this equation that Ψ_1 can be obtained from a two-body Green function's (in the variable x) and, because of the $\hat{\mathbf{x}} \cdot \hat{\mathbf{y}}$ factor in ΔV, it must correspond to the p-wave part of that function. We write

$$\Psi_1 \sim G_d^p \Delta V \psi_d F[1 + O(y^{-3/2}) + O(y^{-1})] \tag{2.7.15}$$

where $(E_d - H_x)G_d^p = 1$, and there is no homogeneous part for p-waves. The $y^{-3/2}$ correction corresponds to Eq. (2.7.13) and can easily be developed, as can the y^{-1} correction which comes from the quadrupole ($l = 2$) part of V_C that we neglected when Eq. (2.7.5) was expanded. This result displays the physics which is significant in the polarization potential: the proton's electric field "stretches" the deuteron preferentially along the $\hat{\mathbf{y}}$ direction ($\Delta V \sim \hat{\mathbf{x}} \cdot \hat{\mathbf{y}}$), which points to the proton projectile. Note that this channel ($l = 1$) has no asymptotically *free* deuteron, and the wave functions fall off much faster in $y (\sim y^{-2})$ compared to Ψ_0.

The second-order solution can be similarly developed. Based on the previous argument, we see that an additional factor of $\Delta V \sim y^{-2}$ will generate a very rapid asymptotic decrease ($\sim y^{-4}$), except for the channel with an outgoing deuteron, where the denominator of the Green's function G_d can become small. Equation (2.7.13) shows that in this case the action of H_y on an inverse power of y increases that power by 3/2. We, therefore, expect the wave function to decrease as $y^{-5/2}$. Using Eq. (2.3.36), we indeed find that

$$\beta^2 \Psi_2 \sim \psi_d(x) \frac{F(y)}{y^{5/2}} \left(\pm \frac{\alpha_E^d \sqrt{\xi}}{10} \right) [1 + O(y^{-1/2})] \tag{2.7.16}$$

This ratio of this to Ψ_0 is

$$\left(\pm \frac{\alpha_E^d \sqrt{\xi}}{10 y^{5/2}} \right) \sim 8 \times 10^{-7}$$

for $y = 50$ fm and $\xi = 0.0463$ fm^{-1}, which is a negligible effect at this distance. However, the ratio of the regular to irregular solutions grows exponentially and will eventually compensate for even this negligible factor. Putting everything together and using $\beta = \alpha \hbar c$, we find to order $\beta^2 y^{-5/2}$ relative to $F(y)$:

$$\Psi_{\text{asy}}^\beta \sim \psi_d(x) F(y) \left(1 \pm \frac{\alpha_E^d \sqrt{\xi}}{10 y^{5/2}} \right) + \frac{\alpha F(y)}{y^2} G_d^p \mathbf{x}' \cdot \hat{\mathbf{y}} \psi_d(x') + \cdots \tag{2.7.17}$$

where we have decomposed V_p in the last term. The upper (lower) sign refers to the regular (irregular) solution. There is an implicit integral over $\mathbf{x}'$ associated with the Green's function G_d^p.

One obvious result from the derivation is that if Eq. (2.7.5) is truncated so that no p-waves are included (i.e., $\beta = 0$), there is no polarization force. This statement depends entirely on how the calculation is organized, but with the Faddeev–Noble method, there is no polarization (of dipole type) if only s-waves are kept, as is commonly done. There is polarization of a different type, but it involves radial excitation of the deuteron (and leads to a relative asymptotic behavior $\sim y^{-4}$).

We summarize this section by noting that the Coulomb polarization force, which has recently been very controversial, is intimately related to Rayleigh scattering of light. The effects of this force can be estimated in several ways, all of them demonstrating that if standard practice in configuration space is followed, it is very tiny and can be ignored. Moreover, leading-order asymptotic boundary conditions have been developed which include this force. Although a fascinating physical phenomenon, polarization in p–d scattering is completely unimportant when treated this way.

8. PRACTICAL THREE-NUCLEON CALCULATIONS

All of the preceding discussion avoided introducing the nucleon's spin and isospin because it was largely unnecessary. In the introduction we argued that the most important component of the nuclear potential was the tensor force, which requires intrinsic nucleon spin for its existence. We are now at the stage where practical calculations can be discussed, and such calculations require the debut of spin and isospin.

8.1. Spin–Isospin–Orbital Coupling Schemes

Each spin-$\frac{1}{2}$, isospin-$\frac{1}{2}$ nucleon has two spin components, while the proton and neutron charge states are the $(+)$ and $(-)$ components of isospin, giving a total of four possible spin–isospin combinations. Typically, these degrees of freedom are treated together, and they are classified according to the representations of $SU(4)$ and the permutation group.

For three nucleons there are $4^3 = 64$ possible states to classify according to total intrinsic spin S and total isospin T. Half of these states correspond to $T = \frac{1}{2}$ and half to $T = \frac{3}{2}$. Of the $T = \frac{1}{2}$ states, which form the basis for the trinucleons, half refer to ^{3}He ($T_z = \frac{1}{2}$) and half to ^{3}H ($T_z = -\frac{1}{2}$). Thus each of the trinucleons contain 16 possible spin–isospin components coupled to various orbital angular momenta. A total angular momentum of $\frac{1}{2}$ and a maximum intrinsic spin of $\frac{3}{2}$ implies that the total orbital angular momentum is restricted to $L = 0, 1, 2$. The large number of spin–isospin components (each with an associated scalar function of the radial variables) greatly increases the computational complexity.

There are a variety of angular momentum coupling schemes which are used.[1] The use of $SU(4)$ spin–isospin states coupled to orbital states is often called a bipolar–harmonic scheme. Closely related to this is the channel-coupling scheme for N–d elastic scattering, where the deuteron spin (1) is coupled to the nucleon spin ($\frac{1}{2}$) to obtain $S = \frac{1}{2}$ or $\frac{3}{2}$, which can subsequently be coupled to the relative orbital angular momentum between these clusters. The total angular momentum of the interacting pair, j, can also be coupled to that of the spectator in the j–J coupling scheme. The intrinsic spins of the pair and the spectator can be coupled together, followed by the orbital waves and, finally, orbital and spin contributions together in the L–S coupling scheme. These alternative procedures are separately useful in varying circumstances and all are used. Standard angular momentum recoupling techniques provide mappings between them.[1]

Standard practice sets up the basis states for practical calculations in $j - J$ coupling, evaluates matrix elements of the N–N potential in L–S coupling, calculates N–d elastic scattering observables in channel coupling, and matrix elements of transition operators using bipolar harmonics. We will present later a schematic description of the j–J coupling basis states, together with examples of the Faddeev equations in such a basis. We will first set up and illustrate the utility of the bipolar harmonic scheme.

8.2. Scales

Under the interchange of particle labels, the spin–isospin wave function of two nucleons can be chosen either symmetric (s) or antisymmetric (a). The Pauli principle is accommodated by restricting the concomitant orbital states. For three particles, however, there are three possible symmetries: s, a, and m (mixed), where the last is more complicated. Complementary orbital states (a, s, and m, respectively) are therefore required.

Basic physics principles can establish their relative importance in the trinucleons. The "sizes" R of the few-nucleon systems range from 1.5–1.75 fm. With use of the uncertainty principle, this corresponds to an "average" momentum $\bar{p} \sim 110$–$130 \text{MeV}/c$. A slightly higher value results if we calculate the kinetic energy per nucleon and equate this to $\bar{p}^2/2M$. A useful paradigm for these estimates is the pion mass $\bar{p} \sim m_\pi c \sim 140 \text{MeV}/c$. We can estimate $(v/c)^2 = (\bar{p}c/mc^2)^2$ to be about 2 percent, which sets the scale for relativistic corrections. We might also expect the angular momentum barrier to dominate when the centrifugal part of the kinetic energy, $\hbar^2 l(l + 1)/2MR^2$, is larger than $\bar{p}^2/2M$, which predicts the dominant waves to be $l < \bar{p}R/\hbar \sim 1$–$2$. It is indeed found that these waves totally dominate the trinucleon structure and binding. Indeed, the bulk of the binding arises from the s-waves and (their) tensor-coupled d-waves.

This is all the information we need in order to understand the sizes of the $SU(4)$ wave function components. Those components with the smoothest spatial (orbital–internal) wave functions will have the least kinetic energy, and therefore will be the largest. This corresponds to a symmetric spatial form, and therefore an antisymmetric spin–isospin wave function. The dominant s-wave component is traditionally denoted the S-state and accounts for roughly 90% of the trinucleon probability. The mixed-symmetry or S'-state has roughly 1–2% probability, while the S''-state (spatially antisymmetric) is minuscule. The strong tensor force introduces D-states (roughly 10%), even though their $(S = \frac{3}{2}, T = \frac{1}{2})$ spin–isospin wave functions are mixed symmetry in nature. The positive-parity P-states are unimportant. Altogether, there are three s-wave components, three d-wave components, and four p-wave components in the traditional wave function decomposition.

Small differences in parts of the nucleon force can play an important role in these components and in physical observables. It is natural to interpret the comparison between ^{3}H and ^{3}He observables (or between n–d and p–d reactions) in terms of the Coulomb interaction in the latter systems, but it should be done with great care, as illustrated by the following example.

The quintessential size of a nucleus is the root-mean-square charge radius, which is defined as the average distance from the nuclear CM to any proton, R_p, and is measured by elastic electron scattering. A *schematic* description of ^{3}He and ^{3}H is provided in Fig. 6, where we have depicted the S-state in (a) as an equilateral configuration. The protons are shaded and the CM is marked by a cross. This is a reasonably accurate first approximation, which fails to take into account that the interaction between two protons or two neutrons is weaker than the interaction between a neutron and a proton (only the latter have a bound state). Therefore, on the average, the protons (neutrons) in ^{3}He (^{3}H) lie further apart than depicted in (a), and this is indicated for ^{3}He in (b) and ^{3}H in (c) as isosceles configurations. Thus, R_p in (b) is greater than R_p in (c), which has practically nothing to do with the Coulomb interaction in (b); ^{3}He therefore has a greater charge radius than ^{3}H. Conversely, ^{3}H has a greater neutron radius than ^{3}He, and this difference can be detected using other probes.[54] The difference between the isosceles and equilateral configurations schematically represents the effect of the S'-state, which enters this process as an amplitude interfering with the S-state. This is a 10% effect, which we will examine later.

8.3. Three-Nucleon Channels and Partial Waves

We showed earlier that the Schrödinger (or Faddeev) equation for three nucleons contains 16 spin–isospin–orbital components which depend on three independent spatial variables. This represents a formidable computa-

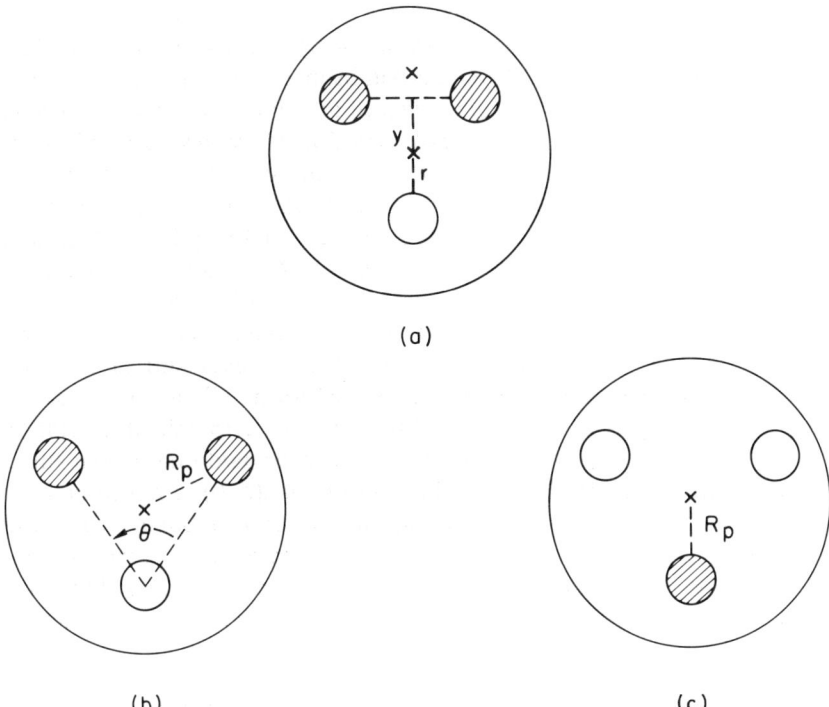

FIGURE 6. Schematic depiction of S- and S'-states. The case of equal n–n (or p–p) and n–p forces is depicted as an equilateral configuration of nucleons. Weaker forces between "like" nucleons produce an isosceles configuration for ^{3}He (b) or ^{3}H (c), with the angle $\theta > 60°$.

tional task. One way to reduce the labor is to make use of the dominance of the angular momentum barrier and decompose the wave function into partial waves in the angle θ between $\hat{x}$ and $\hat{y}$. This has been done in all complete calculations to data. Coupled together with spin and isospin wave functions, each of these three-nucleon partial waves is called a channel, which needs to be distinguished from the same word used to describe different final states.

This scheme allows us to rewrite 16 functions of three continuous variables (x, y, μ) in terms of a modest number of functions of two variables (x, y). Moreover, some realistic nuclear forces, such as the new Nijmegen model (Klomp et al.[11]), are defined only in terms of partial waves, which mitigates use of a partial-wave basis. What is a modest number? Complete calculations[17] for the trinucleons $(J = \frac{1}{2})$ typically keep all N–N "pair" partial waves with $j \leqslant 4$, which corresponds to 34 channels. Analogous scattering calculations[21] for $J = \frac{3}{2}$ require 62 channels for each parity, while larger values of J would require as many as 98 channels.

TABLE I

Spin-Orbital–Angular-Momentum "Channel" Couplings with Total Angular Momentum and Parity $J^\pi = \frac{1}{2}^+$ and $\frac{3}{2}^+$

Channel	Pair	Spectator	J^π
1	$^1S_0$	$s_{1/2}$	$\frac{1}{2}^+$
2	$^3S_1$	$s_{1/2}$	$\frac{1}{2}^+, \frac{3}{2}^+$
3	$^3D_1$	$s_{1/2}$	$\frac{1}{2}^+, \frac{3}{2}^+$
4	$^3S_1$	$d_{3/2}$	$\frac{1}{2}^+, \frac{3}{2}^+$
5	$^3D_1$	$d_{3/2}$	$\frac{1}{2}^+, \frac{3}{2}^+$
6	$^1S_0$	$d_{3/2}$	$\frac{3}{2}^+$
7	$^3S_1$	$d_{5/2}$	$\frac{3}{2}^+$
8	$^3D_1$	$d_{5/2}$	$\frac{3}{2}^+$

This scheme is illustrated in Table I for $J = \frac{1}{2}$ and $\frac{3}{2}$, where positive-parity channels corresponding to the dominant $^1S_0$, $^3S_1 - {}^3D_1$ (pair) components of the N–N force have been kept. This "basic" configuration of five channels for $J = \frac{1}{2}$ and seven channels for $J = \frac{3}{2}$ contains the bulk of the physics.

8.4. Partial-Wave Faddeev Equations

As concrete examples of this scheme we develop the form of the $J = \frac{1}{2}$ and $\frac{3}{2}$ Faddeev equations, including a (isospin-violating) Coulomb interaction. For simplicity we restrict both the pair and spectator to s-waves. Consulting Table I, we see that the doublet case has two channels, in principle, for each isospin ($T = \frac{1}{2}$ or $T = \frac{3}{2}$). The $^3S_1$ pair has total isospin zero, so that it cannot couple with the spectator to form $T = \frac{3}{2}$. Thus, there are three channels for this problem, two with $T = \frac{1}{2}$ and the isospin-violating one with $T = \frac{3}{2}$. The quartet case is restricted to channel 2, which has $T = \frac{1}{2}$. By keeping only "pair" states which satisfy the Pauli principle, we guarantee that the complete Schrödinger wave function also satisfies that principle. Note that the lack of an isospin impurity in this approximation does *not* mean that there is no Coulomb interaction in the (quartet) Faddeev–Noble equations. Had we kept only s-waves in the "standard" Faddeev form, Eq. (2.5.12), this would have been true and is another defect of that form.

The (reduced) quartet wave function $\phi_q(x, y, \mu)$, in this approximation has the form

$$\phi_q(x, y, \mu) = \bar{\phi}_q(x, y) \chi_3 \eta_2 \qquad (2.8.1)$$

where the χ_i and η_i are the spin and isospin wave functions, respectively. These wave functions couple the spin of the spectator to that of the pair:

$$\chi_1 = (\tfrac{1}{2} \otimes 1)_{1/2} \qquad \eta_1 = (\tfrac{1}{2} \otimes 1)_{1/2} \qquad (2.8.2a)$$

$$\chi_2 = (\tfrac{1}{2} \otimes 0)_{1/2} \qquad \eta_2 = (\tfrac{1}{2} \otimes 0)_{1/2} \qquad (2.8.2b)$$

$$\chi_3 = (\tfrac{1}{2} \otimes 1)_{3/2} \qquad \eta_3 = (\tfrac{1}{2} \otimes 1)_{3/2} \qquad (2.8.2c)$$

where the angular momentum coupling scheme for the nucleons is $[1 \otimes (23)]_S$ for historical reasons.[41] These spin–isospin wave functions are orthonormal and complete. We have used an overscore on $\bar{\phi}_q$ to indicate that the s-wave approximation is simply the angular average of $\phi_q(x, y, \mu)$:

$$\bar{\phi}_q(x, y) \equiv \frac{1}{2} \int_{-1}^{1} d\mu\, \phi_q(x, y, \mu) \qquad (2.8.3)$$

Equation (2.6.2) can be used for s-waves by ignoring the angular momentum operators and projecting the equation with respect to the angle θ as well as with $\chi_3 \eta_2$. Note that in obtaining ϕ_2 and ϕ_3 from ϕ_1, we must also permute the spin and isospin labels. We obtain

$$\left[E + \frac{\hbar^2}{M}\left(\frac{\partial^2}{\partial x^2} + \frac{\partial^2}{\partial \bar{y}^2}\right) - \bar{V}_C(x, y) \right] \bar{\phi}_q(x, y)$$
$$= V_t(x) \left\{ \bar{\phi}_q(x, y) + \int_{-1}^{1} d\mu \frac{xy}{x_2 y_2}[-\tfrac{1}{2}\bar{\phi}_q(x_2, y_2)] \right\} \qquad (2.8.4)$$

where V_t is the spin-triplet ($^3S_1$) N–N interaction, and the spin–isospin-angle-projected Coulomb potential is

$$\bar{V}_C = \frac{\alpha \hbar c}{2}\overline{\left(\frac{1}{x_2} + \frac{1}{x_3}\right)} = \frac{\alpha \hbar c}{(y, \tfrac{1}{2}x)_>} \to \frac{\alpha \hbar c}{y} \qquad (2.8.5)$$

for $y > \tfrac{1}{2}x$. Clearly, for $\theta' > 30°$ this has the desired asymptotic form. Note that for $\theta' < 30°$, $\tfrac{1}{2}x$ is greater than y and the incident proton is inside the deuteron, as we discussed earlier. We have used the fact that angle averaging for a function of (x_2, y_2) is exactly the same as for a function of (x_3, y_3) [see Eqs. (2.5.6)]. Spin does not affect this equation because permuting χ_3 always gives χ_3. On the other hand, the isospin is affected and the factors of $\tfrac{1}{2}$ in Eq. (2.8.5) and $-\tfrac{1}{2}$ in Eq. (2.8.4) are the result of permuting particle labels in η_2 while forming ϕ_2 and ϕ_3.

The Scrödinger wave function is obtained from the unreduced Faddeev wave function:

$$\bar{\psi}_q(x, y) = \frac{\bar{\phi}_q(x, y)}{xy} \quad (2.8.6a)$$

which leads to

$$\Psi_q(x, y, \mu) = [\bar{\psi}_q(x, y) - \tfrac{1}{2}\bar{\psi}_q(x_2, y_2) - \tfrac{1}{2}\bar{\psi}_q(x_3, y_3)]\chi_3\eta_2$$
$$+ \frac{\sqrt{3}}{2}[\bar{\psi}_q(x_2, y_2) - \bar{\psi}_q(x_3, y_3)]\chi_3\eta_1 \quad (2.8.6b)$$

This form corresponds to mixed symmetry, which typically couples pairs of radial and spin–isospin wave functions to form a totally antisymmetric wave function.

It is important to note that although $\bar{\psi}_q(x, y)$ does not depend on μ, the permuted variables and permuted terms do, and thus the total wave function Ψ_q also does. We have, in effect, calculated a function of three variables by solving an equation with only two variables!

The doublet case is more complicated because it involves three channels. This is illustrative of the more general case when many channels are coupled together. We write

$$\phi_d(x, y, \mu) = \bar{\phi}_t(x, y)\chi_1\eta_2 + \bar{\phi}_s(x, y)\chi_2\eta_1 + \bar{\phi}_C(x, y)\chi_2\eta_3 \quad (2.8.7)$$

where the t, s, and C subscripts refer to spin-triplet ($T = \tfrac{1}{2}$), spin-singlet ($T = \tfrac{1}{2}$), and spin-singlet ($T = \tfrac{3}{2}$), which is a Coulomb-induced isospin impurity. Again we project Eq. (2.6.2) and suppress the spin-isospin algebra resulting from permutation. We find

$$\left[E + \frac{\hbar^2}{M}\left(\frac{\partial^2}{\partial x^2} + \frac{\partial^2}{\partial \bar{y}^2}\right) - \bar{V}_C^s(x, y)\right]\bar{\phi}_s(x, y) - \Delta\bar{V}_C(x, y)\bar{\phi}_C(x, y)$$
$$= V_s(x)\left\{\bar{\phi}_s(x, y) + \int_{-1}^{1} d\mu \left(\frac{xy}{x_2 y_2}\right)[-\tfrac{3}{4}\bar{\phi}_t(x_2, y_2) + \tfrac{1}{4}\bar{\phi}_s(x_2, y_2)]\right\}$$
(2.8.8a)

$$\left[E + \frac{\hbar^2}{M}\left(\frac{\partial^2}{\partial x^2} + \frac{\partial^2}{\partial \bar{y}^2}\right) - \bar{V}_C^t(x, y)\right]\bar{\phi}_t(x, y)$$
$$= V_t(x)\left\{\bar{\phi}_t(x, y) + \int_{-1}^{1} d\mu \left(\frac{xy}{x_2 y_2}\right)[-\tfrac{3}{4}\bar{\phi}_s(x_2, y_2) + \tfrac{1}{4}\bar{\phi}_t(x_2, y_2)]\right\}$$
(2.8.8b)

$$\left[E + \frac{\hbar^2}{M}\left(\frac{\partial^2}{\partial x^2} + \frac{\partial^2}{\partial \bar{y}^2}\right) - \bar{V}_C^C(x, y)\right] \bar{\phi}_C(x, y) - \Delta \bar{V}_C(x, y) \bar{\phi}_s(x, y)$$

$$= V_s(x) \left\{\bar{\phi}_C(x, y) + \int_{-1}^{1} d\mu \left(\frac{xy}{x_2 y_2}\right)[-\tfrac{1}{2}\bar{\phi}_C(x_2, y_2)]\right\} \quad (2.8.8c)$$

where V_s is the spin-singlet ($^1S_0$) N–N interaction. In addition, the four projected Coulomb potentials are:

$$\bar{V}_C^t = \frac{\alpha \hbar c}{r_>}$$

$$\bar{V}_C^s = \tfrac{1}{3}\alpha \hbar c \left(\frac{1}{r_>} + \frac{2}{x}\right)$$

$$\bar{V}_C^C = \tfrac{1}{3}\alpha \hbar c \left(\frac{2}{r_>} + \frac{1}{x}\right)$$

and

$$\Delta \bar{V}_C = -\frac{\sqrt{2}\alpha \hbar c (1/x - 1/r_>)}{3}$$

where $r_> = (y, \tfrac{1}{2}x)_>$. Note the direct coupling between the (singlet) $T = \tfrac{1}{2}$ and $T = \tfrac{3}{2}$ channels. The coupling between $\bar{\phi}_s$ and $\bar{\phi}_t$ on the right-hand side of these equations is to be expected and, in cases with many partial waves and channels, a completely different notation has to be adopted in order to handle multiple channels. We have tailored our notation in this example with an eye on the underlying physics.

The Schrödinger wave function formed from Eq. (2.8.7) has three components: an antisymmetric spin–isospin term (S-state), and mixed symmetry $T = \tfrac{1}{2}$ (S'-state) and $T = \tfrac{3}{2}$ spin–isospin components. This is left as an exercise for the reader.

8.5. Wave Function Normalization

Bound-state wave functions are normalized so that they have unit norm: $\langle \Psi | \Psi \rangle = 1$. The normalization condition for N–d scattering solutions must conform to the assumed form of the plane-wave normalization. There are many choices for the latter, but all of them are predicated in some way on guaranteeing unit flux for the plane waves. The choice does not affect the direct calculation of phase shifts in any way, but for bound-to-continuum reactions or for making Kohn variational estimates of phase shifts, it is essential to do this consistently. In our case with three identical nucleons there are *three* incident plane waves (one each in ψ_1, ψ_2, and ψ_3) which gives three times the flux for any one of them. Thus, we should divide each Schrödinger wave function by $\sqrt{3}$,

$$\Psi = \frac{\psi_1 + \psi_2 + \psi_3}{\sqrt{3}} \qquad (2.8.9)$$

which again has unit flux. Another equivalent argument is that wave functions must be complete with respect to the phase space. With a conventional two-body phase space, $[d^3p_1 d^3p_2/(2\pi)^6]$, the wave functions in Eq. (2.8.9) are complete; there are three equal contributions contained in $\Psi\Psi^\dagger$, which cancel the factors of $1/\sqrt{3}$ to give the usual completeness δ-function.

Finally, a related problem is the appropriate volume element to use in calculating transition matrix elements or making Kohn variational estimates. An added benefit of Jacobi coordinates is that they preserve volume elements without extraneous factors:

$$d^3r_1 d^3r_2 d^3r_3 = d^3R\, d^3x\, d^3y \qquad (2.8.10)$$

The CM coordinate **R** does not couple to the internal three-nucleon structure, which depends only on **x** and **y**.

8.6. Numerical Modeling

It is far beyond the scope of this chapter to treat the really practical aspects: numerical methods and modeling. Most groups who currently solve the Faddeev equations in configuration space use the spline technique to model the wave function. This technique, which approximates segments of the wave function by polynomials, has proven to be highly flexible and adaptable. The basic ideas and examples for three-nucleon problems are thoroughly presented in Payne.[55]

Finally, we note that most of these groups also convert from (x, y) to (ρ, θ') coordinates before solving the equations. This has the practical advantage that the matrices which result from discretizing the equations are partially banded (i.e., they have many elements which are zero). Again, we refer to Payne.[55] Because there are often many channels involving several variables, the matrices are large and the computational problems are challenging.

9. THE ³HE BOUND STATE

The most visible manifestation of the Coulomb force in the three-nucleon system is the mass difference of ³He and ³H. The trinucleons, ³He and ³H, form an isospin doublet (as do the proton and neutron) and can be viewed as isospin rotations (by π) of each other. Differences between these isospin components, therefore, reflect a very special and restrictive type of isospin-violating interaction[56]: charge-symmetry breaking. The charge-symmetry rotation basically turns protons and neutrons into each other. Because ³He has one p–p pair and two n–p pairs, while ³H has one n–n pair and two n–p pairs, the difference between ³He and ³H is essentially a reflection of the difference between the p–p and n–n interactions. Clearly, the Coulomb interaction in the former is the dominant physics.

The ³He–³H mass difference is the best studied and best understood of the charge-symmetry violations that occur in nuclei. Recent progress in both theoretical and experimental arenas has clarified the roles of what are believed to be the primary mechanisms and, unless something changes to the contrary, the status of our understanding is now satisfactory for the first time. Because differences between the trinucleons immediately imply differences between n–d and p–d scattering (whose states also correspond to an isospin doublet), we will now discuss the status of the trinucleons.

The nuclide ³H is 18.6 keV heavier[57] than ³He and is a β-emitter. When the masses of the constituent neutrons and protons are subtracted, the situation is reversed and ³He has 763.6 keV less binding energy than ³H, denoted ΔE. The primary mechanisms which account for this difference, together with theoretical estimates of their contributions and subjective estimates of their uncertainties[56] are:

1. The Coulomb interaction between the protons in ³He, which accounts for 648(4) keV.
2. Differences in the neutron and proton masses in the kinetic energy, which contributes 14(4) keV.
3. Interactions between the elementary magnets that are part of the neutron and proton, (primarily) magnetic interactions between the moving protons, and other smaller electromagnetic effects,[58] which generate 26(4) keV.
4. Short-range (nonelectromagnetic) differences between the n–n and p–p forces, which account for 67(24) keV.

Adding the various pieces together we arrive at 755(25) keV, which agrees well with the experimental value. We will discuss the Coulomb and short-range mechanisms in more detail.

The initial calculations of the Coulomb effect in ^{3}He had several defects which were gradually eliminated as sophistication increased. Early calculations used the "standard" form of the Coulomb-modified Faddeev equations given by Eq. (2.5.12), which lacks the long-range ($1/y$) Coulomb distortion; instead, it has a long-range coupling to the permuted terms. A consequence is that this form produces a Coulomb effect which is severely affected by truncating the number of three-body channels (partial waves). In other words, the Coulomb partial-waves series converges very slowly if we use Eq. (2.5.12). At the five-channel approximation, which gives most of the binding energy of ^{3}H, a direct comparison of ^{3}He and ^{3}H eigenvalues shows that only 85% of the Coulomb energy is present. The latter fraction can be determined by also performing first-order and second-order perturbation theory calculations[18] with V_C. Defining $\Delta E_C = \Delta E_C^{(1)} + \Delta E_C^{(2)} + \cdots$, we find that $\Delta E_C^{(1)} = 652(4)$ keV and $\Delta E_C^{(2)} = -4(1)$ keV. The partial-wave convergence problem can be solved by using the Sasakawa–Sawada form of the Faddeev equations[58] or, even better, the Faddeev–Noble equations.[18]

The second problem which arises is that most models of the nuclear force underbind the triton, as we discussed before. Typically, N–N forces have 0.6–1.0 MeV too little binding. Underbinding guarantees that the system will have too large a size R and this lowers the Coulomb energy: $V_C \sim 1/R$. Adding a three-nucleon force increases the binding, but the precise amount is determined by features of that force which are not yet subject to experimental testing. What can be done is a scaling plot. Each model calculation of the triton (for example) generates a binding energy (the negative of the energy eigenvalue), E_B, and concomitant wave functions which can be used to calculate $\Delta E_C^{(1)}$. If $\Delta E_C^{(1)}$ is plotted model by model versus the triton binding energy, as in Fig. 7, we find that a smooth curve results, which can be extrapolated to the physical triton binding energy. We emphasize that each symbol on the figure is the result of a theoretical calculation. The curve is a simple fit. The large number of points correspond to various nuclear force models and channel truncations, with solid symbols representing complete or "exact" calculations. Points with $E_B > 7.7$ MeV contain a three-nucleon force. The increase in the Coulomb energy as the binding increases is apparent. There is no significant variation between the eigenvalue differences and the perturbation theory estimates (given above) for the complete calculations if the Faddeev–Noble equation is used.

In the previous section we introduced an isospin impurity ($T = \frac{3}{2}$) into the Faddeev equations. This component, which has a tiny probability ($\sim 10^{-5}$) and typically affects the Coulomb energy by less than 1 keV, can be ignored in almost every circumstance.

The final ingredient which we discuss is the contribution of the short-range charge-symmetry breaking. For many years the experimental evidence indicated that the n–n force was weaker than the Coulomb-

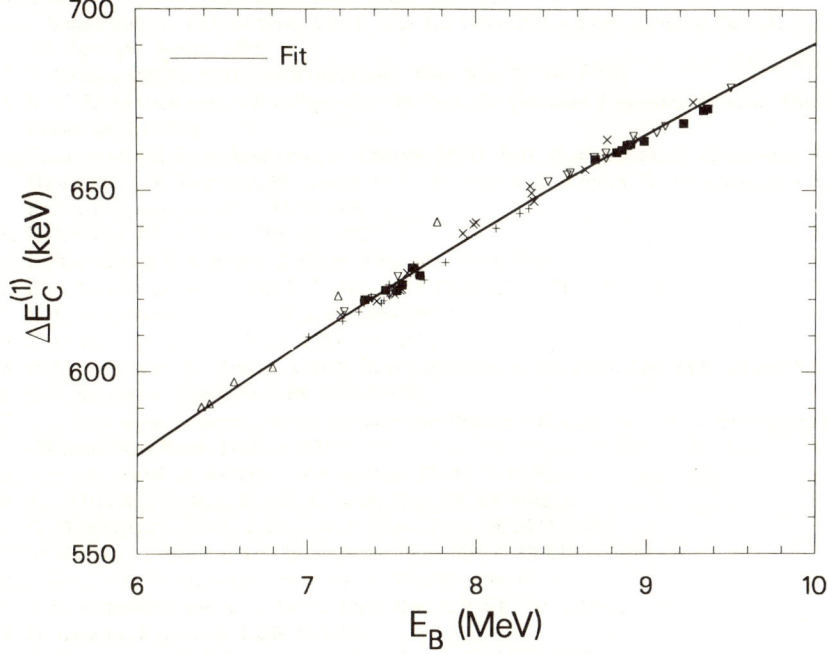

FIGURE 7. Scaling plot of the ^{3}He Coulomb energy in first-order perturbation theory versus triton binding energy. Each point is a separate theoretical calculation. Complete calculations have shaded symbols.

subtracted p–p force. This goes in the wrong direction to explain the ^{3}He–^{3}H binding energy difference.

A recent experiment[59] which directly probed the n–n force at low energies found a scattering length, $a_{nn} = -18.8(3)$ fm, when corrected for small electromagnetic effects. This can be contrasted with the Coulomb-corrected p–p scattering length, $a_{pp} = -17.3(4)$ fm, and compared to the spin-singlet n–p result, $a_{np} = -23.75(1)$ fm. Several calculations[58] have shown that a difference $\Delta a = a_{nn} - a_{pp}$ leads to a change in ΔE which is given empirically by

$$\frac{\Delta E}{\Delta a} \sim -45(5) \text{ keV fm}^{-1} \qquad (2.9.1)$$

The observed difference, $\Delta a = -1.5(5)$ fm, leads to $\Delta E \sim 67(24)$ keV. It is clear that the resolution of the charge-symmetry-breaking problem in the trinucleons depends sensitively on $|a_{nn}| > |a_{pp}|$, and any change in the experimental situation could seriously affect this.

The difference between the spin-singlet a_{np} and $\frac{1}{2}(a_{nn} + a_{pp})$ exemplifies another type of isospin violation: charge dependence. This cannot affect the ^{3}He–^{3}H binding energy difference, but it does play a small role in the absolute binding energy of the triton (or ^{3}He). We use the fact that the trinucleons are dominated by the $^1S_0$ and $^3S_1 - {}^3D_1$ partial waves of the N–N force and that these waves are similar on the average. Relative s-waves for two neutrons or two protons ($T = 1$) means that the total spin must be $S = 0$. An n–p pair will statistically have a 1/4 probability of having $S = 0$, and 3/4 probability of having $S = 1$. Assuming equal probability for each interacting pair gives three interacting pairs in the triton broken down according to $V = \frac{3}{2}V(S = 0) + \frac{3}{2}V_{np}(S = 1)$, with

$$V(S = 0) = \tfrac{2}{3}V_{nn} + \tfrac{1}{3}V_{np}(S = 0) \qquad (2.9.2)$$

Most (but not all) nuclear forces are constructed by fitting the $T = 1$ forces to either p–p data or n–p data, but not both. The aforementioned difference between a_{np} and a_{nn} is usually not taken into account. Studies[30] have indicated that each 1/3 in the "2/3–1/3" rule in Eq. (2.9.2) controls approximately 100 keV in binding. That is, if the weaker (n–n or p–p) force is also used for the n–p pairs, the binding is reduced by 100 keV, while if the stronger n–p force is used for the n–n pair, the binding is increased by 200 keV. These errors are significant on the scale of the current discrepancy. Finally, the deuteron has a single ($S = 1$) n–p pair and our previous argument gave $\frac{3}{2}(S = 1)$ n–p "pairs" for the triton. If the D-state probabilities P_D of the deuteron and triton scale correspondingly we would expect

$$P_D(t) = \tfrac{3}{2}P_D(d) \qquad (2.9.3)$$

and, indeed, this prediction works very well (within a few percent).

Our last example completes the discussion begun in Section 8. Figure 8 displays scaling plots of the rms charge radii of ^{3}He and ^{3}H, similar to the previously displayed plot of ΔE_C, where the appropriate binding energy of each model has been used. The most-bound models contain a three-nucleon force. The finite size of the individual nucleons has been removed, and the size represents only the spread of the wave functions. The agreement between theory and experiment is good.[18] Because the mean-square-radius operator weights the tail of the charge distribution, we might guess that the asymptotic form of the wave function is all that matters. Assuming that this form is given by a properly normalized exponential ($\sim e^{-\kappa\rho}$) multiplied by any power of ρ leads to $\langle r^2 \rangle^{1/2} \sim 1/\kappa \sim E_B^{-1/2}$. The average of ^{3}He and ^{3}H does indeed conform to this dependence. The difference between ^{3}He and ^{3}H largely comes from the mixed-symmetry S'-state, as argued earlier. Direct Coulomb effects are twofold: (a) a change in the binding energy, and

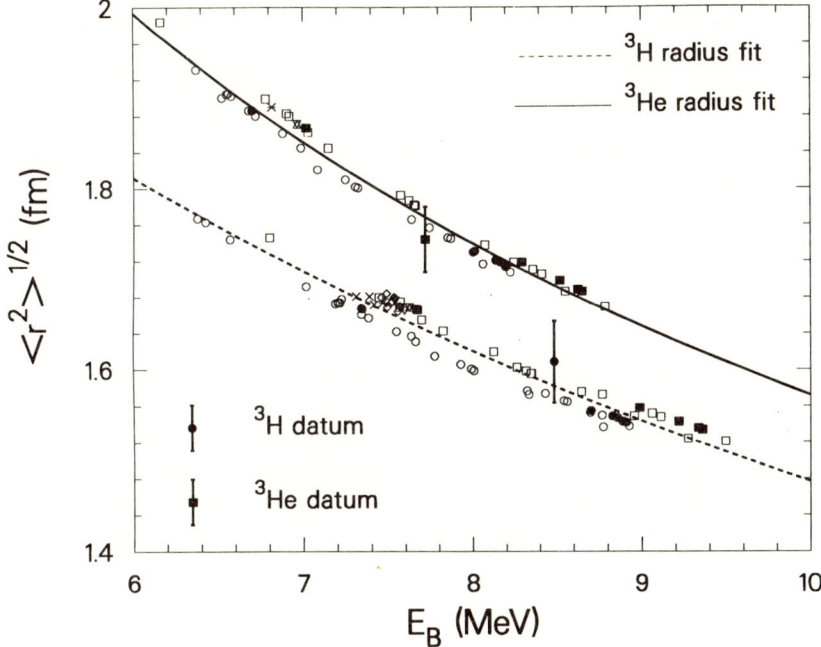

FIGURE 8. Scaling plots of trinucleon rms charge radii (for point nucleons) *vs.* appropriate trinucleon binding energy.

(b) a change in the shape of the tail of the wave function [Eq. (2.3.32)]. These are effects of opposite sign, and the binding effect is twice as large as that due to the change of the shape of the asymptotic wave function. We also see from the figure that the effect of reduced binding is approximately half that of the overall differences. Although Coulomb effects in ^{3}He are not negligible, they do not have a dominating influence.

10. NUCLEON-DEUTERON SCATTERING

10.1. Threshold Scattering: Nucleon-Deuteron Scattering Lengths

After the ground states, the easiest calculations to perform are the scattering calculations below breakup threshold. The only open channels are those which contain a deuteron, and the closed channels exponentially decrease in the asymptotic region like bound-state wave functions. The boundary conditions are therefore two-body-like. Practical difficulties include the slow spatial decrease of the permuted terms (rearrangement channels), which fall off like deuteron wave functions. Whereas $\rho_{max} = 20$ fm is reasonable for a bound-state problem, $\rho_{max} = 50$–70 fm is needed for accurate solutions in the scattering region. Another difficulty pertains to zero energy. At threshold the regular solutions become much larger asymptotically than the irregular ones. This is particularly severe in the p–d case, where the ratio of the two grows exponentially. As discussed in Section 7, either we must not integrate the differential equation out to very large distances ($\gtrsim 1000$ fm) with Coulomb boundary conditions, or we must incorporate a modification of those boundary conditions which includes the effect of Coulomb polarization forces. We chose to do the former.

Our results for the doublet s-wave scattering lengths, a_2, are summarized in Fig. 9. This scaling plot (called a Phillips plot) is the original one of its type. Both n–d and p–d calculations are shown together with the experimental data. The n–d datum lies on the Phillips line and is obscured. The n–d (p–d) cases are plotted versus the ^{3}H (^{3}He) binding energy. The substantial difference reflects the large Coulomb effect. Note that the effect of the Coulomb energy is not simply a reduction in the binding energy of ^{3}He with respect to ^{3}H. This would increase the p–d scattering length with

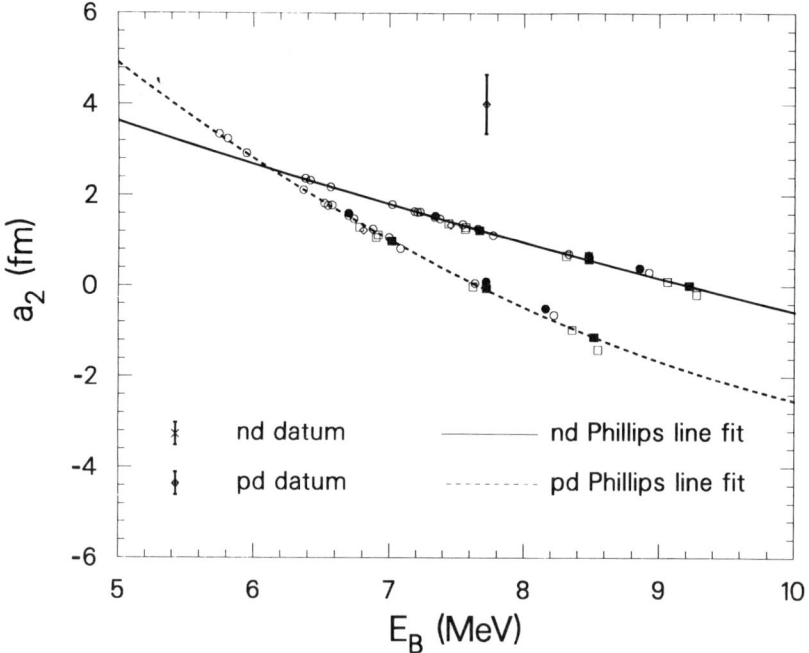

FIGURE 9. Scaling (Phillips) plots of the doublet N–d scattering lengths *vs.* appropriate trinucleon binding energy.

respect to the n–d case. The two curves have different slopes, which reflects the presence in the former of a new scale parameter, $\xi = (4\alpha Mc/3\hbar) = 1/22$ fm. It can be shown using first-order perturbation theory[41] that

$$\frac{1}{a_2^{\text{pd}}} \simeq \frac{\lambda}{a_2^{\text{nd}}} + \delta \qquad (2.10.1)$$

where a_2^{pd} is the Coulomb-modified version of a_2^{nd}, and the (Coulomb) parameters defined by Friar et al.[41] have the properties $\lambda < 1$ and $\delta < 0$. Thus, for large values of a_2^{nd}, δ tends to dominate, while for small values of a_2^{nd}, λ is the important parameter and makes the p–d slope greater than the n–d slope. For quartet scattering, this formula is accurate at the 90% level. The validity of Eq. (2.10.1) was verified numerically for (artificially) small values of ξ, because the Coulomb effect is not linear for physical values of that parameter.

The somewhat obscured n–d data point lies on the Phillips curve, and is a major success for three-nucleon scattering theory, subject to the necessity of extrapolation to the physical binding energy. The p–d data point, however, lies well above the curve and is a major discrepancy. We will discuss this problem, and its likely resolution, in the next section.

The quartet scattering-length results are depicted in Fig. 10, which has a very compressed scale. This scattering problem is not especially sensitive to details of the nuclear force, because its "natural" configuration is an overall s-wave with $S = \frac{3}{2}$, which is Pauli-forbidden if the two protons are in a relative s-wave. This keeps the nucleons apart and reduces sensitivity to the nuclear force. There is essentially no sensitivity to three-nucleon forces. The problem we saw earlier with the doublet scattering length persists here; the experimental values of $a_4^{\text{nd}} = 6.35(2)$ fm and $a_4^{\text{pd}} \sim 11.1(3)$ fm coincide with the calculations for n–d and fail badly for p–d (the experimental point consequently does not fit on the graph).

The calculations reported above were originally controversial, but have thus far stood the test of time. They have been verified using Kohn variational estimates. It has been shown that the elimination of long-range Coulomb coupling to the rearrangement channels is very important for stability and convergence and that the Faddeev–Noble equations are optimal in this respect. It has also been verified that the $T = \frac{3}{2}$ iso-spin-impurity channel has a negligible influence. Finally, we note again that the formalism automatically takes into account the effect of the proton's Coulomb interaction inside the *finite* deuteron, and this is both natural and logical to include as part of the scattering length. We do not recommend that this effect be subtracted from the calculated a_2, as was recommended by Friar et al.[41]

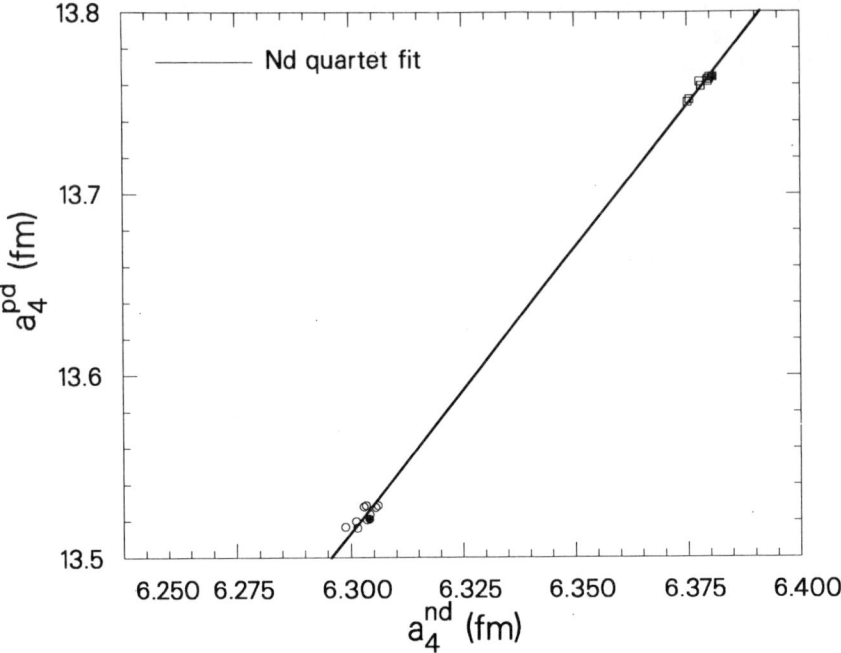

FIGURE 10. Plot of quartet scattering lengths (p–d vs. n–d) for different potential models, indicated by circles and squares.

10.2. Proton-Deuteron Scattering below Breakup Threshold

Although rather extensive momentum-space calculations exist[60] which use a limited number of channels and a separable-potential basis, there are currently no complete Coulomb calculations below breakup threshold (except at zero energy). The existing calculations were performed using a screened Coulomb potential and allowing the screening radius to become very large. A discussion of this technique is far outside the purview of this chapter, but is discussed in Chapter 1 in this volume. The reader is referred to that work and to the original papers.

Nevertheless, s-wave calculations in this energy regime are both interesting and useful because they provide us with some understanding of the discrepancy between the p–d calculations and the experimental data. They also provide us with a physically interesting example of an effective-range function which is not smooth.

The experimental p–d scattering lengths were obtained by measurements[61] at fairly high proton energies [300 keV (CM), or greater], while thermal neutrons were used for the n–d measurements. Thus, the n–d measurements essentially correspond to the calculations, which were performed at zero energy. The proton experiments, on the other hand, require an energy extrapolation of the s-wave effective-range functions for both doublet and quartet spin configurations. Unless very precise data are available over a wide range of energies, the extrapolation[61] requires the *assumption* that the effective range function is smooth (essentially linear) with respect to the energy.

With this in mind s-wave calculations were performed[62] using a simple nuclear potential for both n–d and p–d scattering for the full range of energies below breakup threshold. Phase shifts were extracted and effective range functions were calculated. The quartet cases were smooth and extrapolated easily to the scattering lengths. The doublet cases, however, show a very strong curvature for low energies, which is stronger for the p–d case. If we ignore all Coulomb complications, both doublet cases can be fitted by an effective-range function of the form:

$$K_0(E) \cong \frac{-1/a + \frac{1}{2}r_0 k^2}{1 + k^2/k_0^2} \qquad (2.10.2)$$

which has a pole in the unphysical region ($k^2 < 0$). Note that if the denominator is expanded, it simply adds a term to the effective range, which will be huge if k_0 is small.

The doublet p–d effective-range function from these calculations is shown in Fig. 11. The very strong curvature is due to the pole, which is closer to the physical region for p–d scattering than for n–d scattering. These calculations, which reproduce the measured phase shifts for $E_{CM} > 600$ keV, extrapolate to the zero-energy calculations displayed in Fig. 9, and not to the experimental datum. The reason for this is also clear. There is very little perceptible curvature in the effective-range function for energies greater than 400 keV and this is well below the lowest-energy experimental point. The only possible experimental resolution of this difficulty would be to perform measurements at energies of 300 keV or lower, which would be very difficult.

Thus, we are left with an unsatisfactory experimental situation: the n–d results agree very well with the calculations, but the p–d results do not. Only very difficult experiments can resolve the latter problem.

10.3. Scattering above Breakup Threshold

There are complete calculations above threshold for n–d scattering, but none for p–d scattering. Only a few calculations of the latter exist in any form for local potentials.[16]

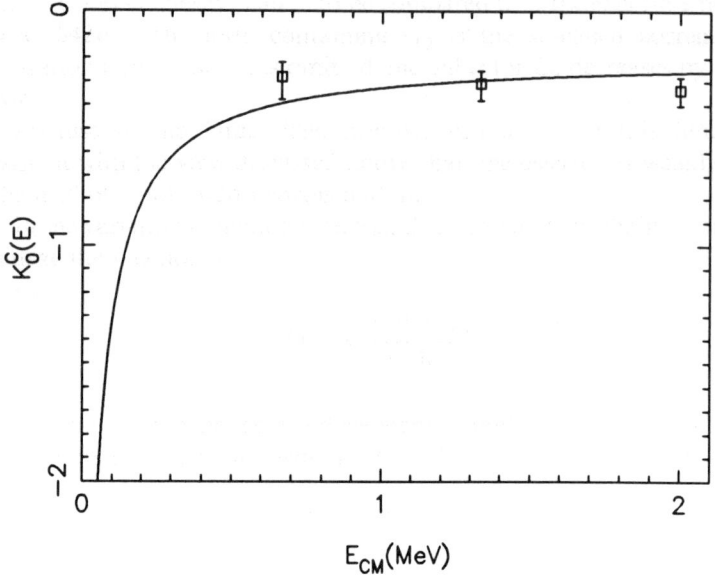

FIGURE 11. Doublet s-wave p–d effective-range function plotted $vs.$ p–d CM energy.

The impressive complete calculations for n–d scattering by the Bochum group (Glöckle et al.[22]) have established that almost all of the experimental data can be explained by accurate calculations with reliable two-nucleon forces. A few examples of discrepancies exist, but at this time it is not clear whether the problems are experimental or theoretical. One area of disagreement which must be resolved arises from comparison of n–d calculations with p–d data. Neutron beams are less common than proton beams and, of course, no neutron targets exist. Much of the accurate scattering data which exists[57] is p–d data.

Although we might suspect that Coulomb effects will become small as energy increases (the Coulomb parameter η becomes smaller), a quantitative evaluation of these effects is an essential calculation which remains to be performed. The boundary conditions developed in Section 6 provide a relatively straightforward way to proceed in configuration space. This calculation is one of the priorities for the field and has the highest priority of any p–d calculation.

Finally, we display in Figs. 12a and 12b, respectively, components of the Faddeev and Schrödinger wave functions for (forward) 14.1 MeV n–d scattering. The coordinate x extends to the right and y to the left. The compact wave running out the y-axis corresponds to elastic scattering, while the faint ripple extending over the entire plot is the effect of breakup. The rearrangement channel ($\theta' = 30°$) is evident only in the Schrödinger wave

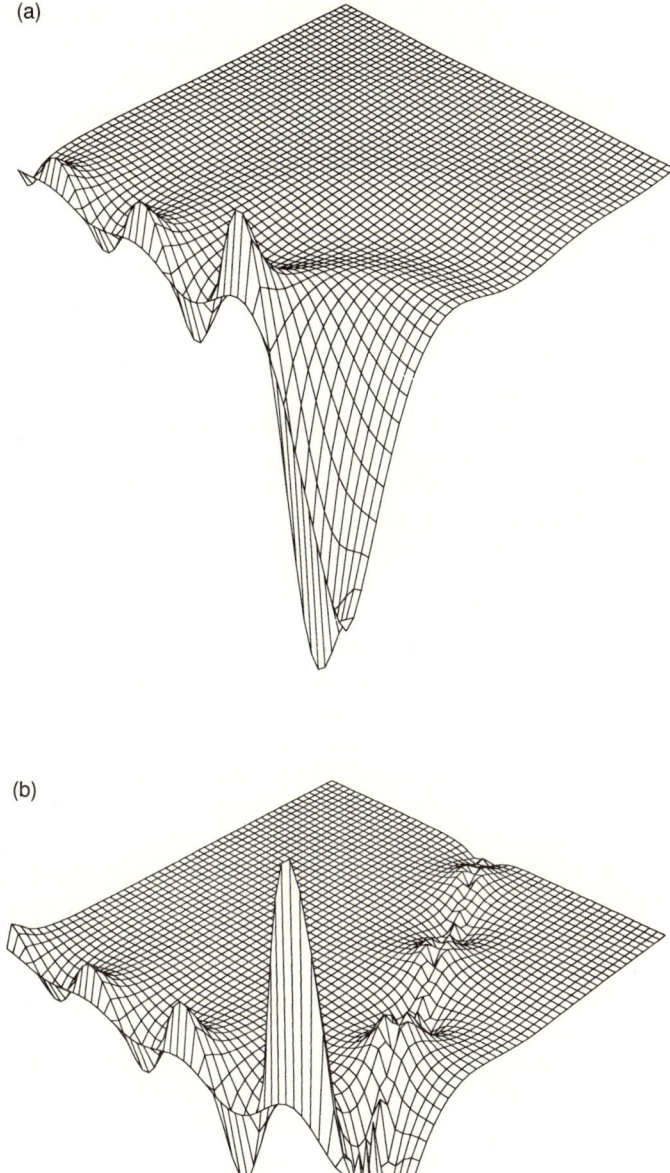

FIGURE 12. Faddeev (a) and Schrödinger (b) wave function components for 14.1 MeV n–d scattering. Only the latter has the prominent rearrangement-scattering ridge, while both display elastic- and breakup-scattering features.

function, as we discussed earlier. The beauty of the configuration space method lies in the direct and close relationship between the wave function and our intuition.

11. PROTON–DEUTERON FUSION REACTIONS

An external interaction (e.g., electromagnetic) is required to break ^{3}He up (into the p–d system, for example) or to coalesce the p–d system into ^{3}He. The traditional way to accomplish this is either by ^{3}He photodisintegration[7] or by p–d radiative capture. The former is very difficult at low energies. The radiative capture reaction has been investigated at relatively low energies in the laboratory in order to extract[8] the (s-wave) astrophysical S-factor. Unfortunately, the Coulomb barrier (epitomized by the Fermi factor C_0^2) greatly suppresses these reactions at low energies and, consequently, the measured S-factor has a large uncertainty. A different and better way to measure this reaction is needed.

11.1. Muon-Catalyzed Fusion

Muon-catalyzed fusion[63] was first observed in bubble-chamber measurements. Negative muons impinging on a liquid hydrogen target (which contained a small admixture of deuterium) were captured by the hydrogen and subsequently transferred to the deuterons, because the atomic reduced-mass effect makes the energy levels of the latter lower than the former. This compact atom eventually collides with ordinary hydrogen and forms a molecule (μ–p–d). The negative charge of the μ^- provides the attraction needed to bind with the proton and deuteron, in spite of the strong Coulomb repulsion between the latter. In addition, the muon's negative charge greatly decreases the Coulomb barrier, and this makes fusion reactions possible within the time it takes for the muon to β-decay.

The first observation of fusion detected internal conversion of the muon. As the proton and deuteron fuse, their joint (transition) charge exerts a Coulomb force on the muon and ejects it from the molecule ($\mu^- + p + d \rightarrow\ ^3\text{He} + \mu^- + 5.5\,\text{MeV}$). This transition is predominantly monopole (i.e., the $E0$ multipole), is extremely clean theoretically, but is not the dominant reaction. The p–d system is typical of few-body electromagnetic reactions at very low energies: it is dominated by magnetic dipole ($M1$) transitions (p + d $\rightarrow\ ^3$He + γ). These processes are interesting because they contain a very substantial contribution from internal nuclear currents,[64] which accompany the exchange of mesons that binds all nuclei together. Unfortunately, these processes are also much more difficult to calculate reliably, and the magnetic dipole fusion is a relatively poor reaction with which to test our knowledge of few-nucleon dynamics. On the other hand, the internal-conversion reaction is nearly ideal.

11.2. Muon Internal Conversion

The scales of the muonic molecule and the range of the nuclear force differ by two orders of magnitude. The molecule is several hundred fm in extent. In order to have a nonvanishing nuclear matrix element, only that part of the initial configuration where the p and d are within a few fm is important, because they must fuse into the compact ^{3}He. This is a tiny part of the molecular wave function and for all practical purposes is a single point in that wave function, exclusive of the muon's coordinate. Because the muon's mass is large, it is ejected from the molecule with a large momentum $|\mathbf{q}|$, which (by the uncertainty principle) only samples very small values of the muon's separation from the fusing ^{3}He. This further means that the probability of coalescence factors into a molecular part $|\psi_{\text{mol}}(0, 0)|^2$ (where the μ, p, and d sit at the same point) and the nuclear matrix elements $|\langle ^3\text{He}|\widehat{E0}|\text{p-d}\rangle|^2$. The energies of p-d motion in the molecule are hundreds of eV, which is effectively zero on the nuclear scale. Thus, the nuclear calculation requires wave functions for the zero-energy p-d system and for the ^{3}He ground state, as well as the $\widehat{E0}$ operator (which is basically the mean-square-radius operator) plus small retarded terms indicated by the ellipsis:

$$\widehat{E0} = -\mathbf{q}^2 \frac{r^2}{6} + \cdots \qquad (2.11.1)$$

This process (to a very good approximation) measures the root-mean-square radius of the transition region for the p + d → ^{3}He fusion.

Complete calculations of this reaction were performed recently.[23] The results for the $E0$ matrix element are shown in Fig. 13. The calculations were performed with many different nuclear models, which are displayed in a scaling plot together with a simple fit. The dashed curve includes the small retardation corrections indicated by the ellipsis in Eq. (2.11.1) and is the preferred result. The extrapolated $E0$ matrix element is in good agreement with the experimental point recently obtained[65] by reanalyzing old bubble-chamber data. The binding-energy dependence ($\sim E_B^{-2}$) in this plot is the strongest seen thus far. The ^{3}He energy calculated using the five-channel approximation, which was the standard ten years ago, is approximately 6.4–6.8 MeV, depending on the nuclear force. This gives a matrix element of 1100–1300 fm$^{7/2}$, instead of the 888(10) fm$^{7/2}$ obtained by extrapolation. Without the extrapolation, calculations would give a very erroneous view of our understanding of this process.

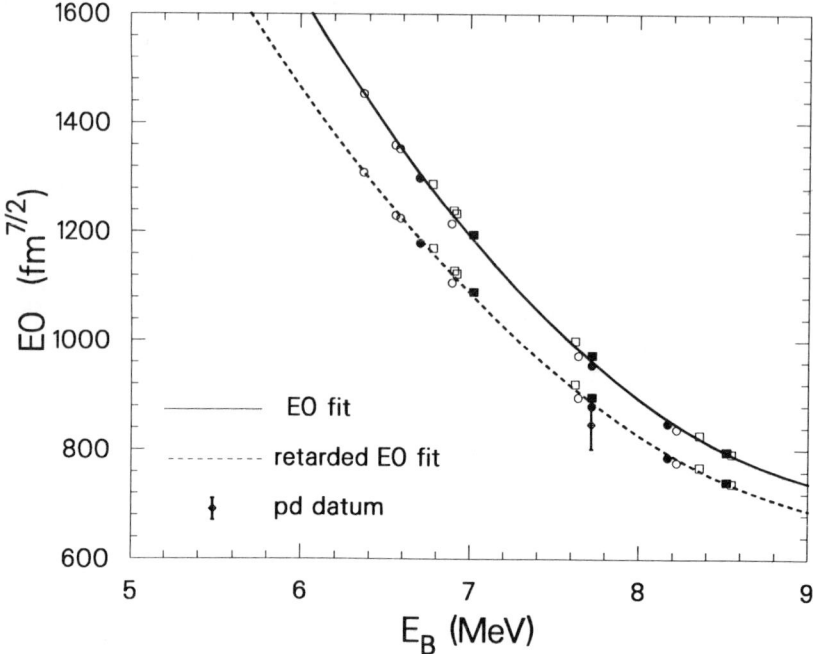

FIGURE 13. Scaling plot of monopole matrix elements for zero-energy p + d → ^{3}He vs. ^{3}He binding energy. This matrix element plays a role in μ-catalyzed p–d fusion.

11.3. Wolfenstein–Gershtein Effect

The previous three decades have produced a great improvement[66] in calculations of the atomic and molecular formation rates which are important in muon-catalyzed fusion. This physics is rich and the details are interesting and physically important. One of the early calculations (by Gershtein[67]) noted that the thermal energy in the hydrogen bubble chambers, kT, was less than the hyperfine splitting in the ground state of the μ–d atom which forms. Collisions quickly thermalize this atom, which becomes trapped in the lowest hyperfine level. Thus, even though the initial spin populations of the μ and d are statistical (random), the freezing of the μ and d spins into a pure (hyperfine) spin state produces a nonstatistical distribution of deuteron spins. The degree of polarization can be changed by changing the deuterium concentration in the bubble chamber. Consequently, the deuteron in the μ–p–d molecule becomes a "polarized target," which has become known as the Wolfenstein–Gershtein (W–G) effect.[67]

This allows the transitions between the various nuclear spins to be resolved. At the very low energies in the molecule, the p–d system is primarily in an s-wave. Consequently, the total nuclear spin J is $\frac{1}{2}$ or $\frac{3}{2}$, which makes a transition to a final spin $\frac{1}{2}$. The allowed multipoles are $E0(\frac{1}{2})$, $M1(\frac{1}{2})$, $M1(\frac{3}{2})$, $E2(\frac{3}{2})$. Disentangling the $M1(J)$ multipoles becomes possible in muonic molecules because of the W–G effect.

Another early calculation estimated that quartet capture should be much smaller than doublet capture because of the Pauli principle. This argument was very influential, and most subsequent analyses of experimental data ignored possible contributions from quartet capture. Early calculations of the W–G effect were in good agreement with the experimental results. Subsequent (better) calculations were in disagreement, however, and a number of exotic atomic processes were examined in attempts to resolve the discrepancy.[66]

The final resolution of this anomaly[23] was provided by the first complete calculations of $p + d \to {}^3\text{He} + \gamma$. The doublet and quartet $M1$ capture matrix elements are comparable and agree with a recent experiment.[9] The argument that the Pauli principle and s-waves between the p and d restrict the size of the quartet spin state is erroneous. The neutron and proton in the bound deuteron are rapidly moving, and s-waves between the p and d do *not* mean s-waves between the two protons. It is favorable for the two protons to be in a ${}^3P$-alignment, with the remaining neutron also in a p-wave, and with the overall system coupled to $L = 0$. This is part of the physics which is implicit in the mixed-symmetry S'-state we discussed in Section 8 and is one reason we worked out the complete wave function for the quartet case.

Finally, although the $M1$ matrix elements are sensitive to the meson-exchange currents, reasonable assumptions concerning the latter produce calculations in reasonable agreement with experiment. If these currents are *fixed* by n–d capture experiments, on the other hand, the agreement between theory and experiment for the p–d matrix elements is excellent.[23] Indeed, these charged-particle reaction rates illustrate how well complete calculations of few-nucleon processes can agree with experimental data.

12. SUMMARY AND PROGNOSIS

In this chapter we have presented a detailed development in configuration space of few-nucleon scattering and reactions with a special emphasis on the p–d system and the inherent complexities of Coulomb physics. We indicated the importance of complete or "exact" calculations, which now allow us to make direct tests of our understanding of the underlying nuclear dynamics. Details of the two-nucleon force can be investigated, as well as the effect of three-nucleon forces. At the present time almost all measured observables agree with these calculations if they are extrapolated (à la the scaling plots) to the proper binding energy. Our understanding of the missing binding energy, however, is deficient, and only circumstantial evidence allows the discrepancy to be attributed to three-nucleon forces. All of the complete calculations mentioned in this chapter included such forces as a natural way to obtain extra binding.

Much remains to be accomplished, however. A number of important calculations need to be completed or performed before we can make a firm assessment of the status of the field. Six such calculations which would greatly advance the field are:

1. Triton and ^{3}He calculations with improved N–N forces which provide a high-quality fit to the N–N data.[29]
2. Improved three-nucleon forces which incorporate important symmetry constraints and provide an adequate phenomenological description of the underlying pion–nucleon physics.
3. Semirelativistic calculations of ^{3}He and ^{3}H which include the proper physics through order $(v/c)^2$.
4. Fully relativistic calculations of ^{3}He and ^{3}H.
5. Complete scattering calculations above breakup threshold which include three-nucleon forces.
6. Complete scattering calculations above breakup threshold which include the Coulomb force.

Various combinations of these ingredients could also be put together. This set of limited goals would, however, teach us much about our field and how well we can describe nature.

Our success over the previous decade has been impressive, and for the first time we have been able to make absolute predictions based on our calculations, without resorting to *ex post facto* "fudge factors." The future holds the promise for more success and looks bright for the field.

REFERENCES

1. J. L. FRIAR, B. F. GIBSON, G. L. PAYNE, AND C. R. CHEN, *Phys. Rev. C* **34**, 1463 (1986).
2. J. L. FRIAR, B. F. GIBSON, G. L. PAYNE, E. L. TOMUSIAK, AND M. KIMURA, *Phys. Rev. C* **37**, 2852 (1988).
3. J. L. FRIAR, B. F. GIBSON, D. R. LEHMAN, AND G. L. PAYNE, *Phys. Rev. C* **37**, 2859 (1988).
4. C. R. CHEN, G. L. PAYNE, J. L. FRIAR, AND B. F. GIBSON, *Phys. Rev. C* **44**, 50 (1991).
5. B. BUDNICK, J. CHEN, AND H. LIN, *Phys. Rev. Lett.* **67**, 2630 (1991).
6. G. A. RETZLAFF *et al.*, *Phys. Rev. C* **49**, 1263 (1994).
7. P. BELLI *et al.*, *Nuovo Cim.* **103A**, 721 (1990).
8. G. M. GRIFFITHS, M. LAL, AND C. D. SCHARFE, *Can. J. Phys.* **41**, 724 (1963).
9. C. PETITJEAN *et al.*, *Muon Catal. Fusion* **5**, 199 (1991).
10. W. J. CUMMINGS *et al.*, *Phys. Rev. Lett.* **68**, 293 (1992).

11. R. A. M. KLOMP, V. G. J. STOCKS, AND J. J. DE SWART, *Phys. Rev. C* **44**, 1258 (1991).
12. H. VAN HAERINGEN, *Charged-Particle Interactions* (Coulomb Press Leyden, Leiden, 1985).
13. L. D. FADDEEV, *Zh. Eksp. Teor. Fiz.* **39**, 1459 (1960) [*Sov. Phys.-JETP* **12**, 1041 (1961)].
14. H. P. NOYES, in *Three-Body Problem in Nuclear and Particle Physics* (J. S. C. McKee and P. M. Rolph, eds.) (North-Holland, Amsterdam, 1970), p. 2.
15. L. ROSENBERG, *Phys. Rev. D* **8**, 1833 (1973).
16. YU. A. KUPERIN, S. P. MERKURIEV, AND A. A. KVITSINSKY, *Yad. Fiz.* **37**, 1440 (1983) [*Sov. J. Nucl. Phys.* **37**, 857 (1983)].
17. C. R. CHEN, G. L. PAYNE, J. L. FRIAR, AND B. F. GIBSON, *Phys. Rev. C* **31**, 2266 (1985).
18. J. L. FRIAR, B. F. GIBSON, AND G. L. PAYNE, *Phys. Rev. C* **35**, 1502 (1987).
19. J. CARLSON, *Phys. Rev. C* **36**, 2026 (1987).
20. J. CARLSON and R. SCHIAVILLA, *Few-Body Systems Suppl.* **7**, 349 (1994).
21. J. L. FRIAR, B. F. GIBSON, G. L. PAYNE, AND C. R. CHEN, *Phys. Lett.* **247B**, 197 (1990).
22. W. GLÖCKLE, H. WITALA, AND TH. CORNELIUS, *Nucl. Phys.* **A508**, 115c (1990).
23. J. L. FRIAR, B. F. GIBSON, H. C. JEAN, AND G. L. PAYNE, *Phys. Rev. Lett.* **66**, 1827 (1991).
24. J. L. FRIAR, B. F. GIBSON, G. L. PAYNE, A. BERNSTEIN, AND T. E. CHUPP, *Phys. Rev. C* **42**, 2310 (1990).
25. D. K. CAMPBELL, in *Nuclear Physics with Heavy Ions and Mesons* (R. Balian, M. Rho, and G. Ripka, eds.) (North-Holland, Amsterdam, 1978), p. 551.
26. R. B. WIRINGA, *Phys. Rev. C* **43**, 1585 (1991).
27. S. A. COON AND J. L. FRIAR, *Phys. Rev. C* **34**, 1060 (1986).
28. J. L. FRIAR, B. F. GIBSON, AND G. L. PAYNE, *Phys. Rev. C* **30**, 1084 (1984).
29. J. L. FRIAR, G. L. PAYNE, V. G. J. STOKS, and J. J. DE SWART, *Phys. Lett.* **B311**, 4 (1993).
30. J. L. FRIAR, B. F. GIBSON, AND G. L. PAYNE, *Phys. Rev. C* **36**, 1140 (1987).
31. G. RUPP AND J. A. TJON, *Phys. Rev. C* **37**, 1729 (1988); *Phys. Rev. C* **45**, 2133 (1992).
32. L. I. SCHIFF, *Quantum Mechanics*, 3rd Ed. (McGraw-Hill, New York, 1968).
33. A. GERSTEN, *Nucl. Phys.* **B103**, 465 (1976).
34. J. D. JACKSON, *Classical Electrodynamics* (Wiley, New York, 1962).
35. E. L. INCE, *Ordinary Differential Equations* (Dover, New York, 1956), p. 23.
36. M. ABRAMOWITZ AND I. A. STEGUN, *Handbook of Mathematical Functions* (Dover, New York, 1965).
37. W. KOHN, *Phys. Rev.* **74**, 1763 (1948).
38. L. L. FOLDY AND W. TOBOCMAN, *Phys. Rev.* **105**, 1099 (1957).
39. W. GLÖCKLE, *Nucl. Phys.* **A141**, 620 (1970).
40. C. R. CHEN, G. L. PAYNE, J. L. FRIAR, AND B. F. GIBSON, *Phys. Rev. Lett.* **55**, 374 (1985).
41. J. L. FRIAR, B. F. GIBSON, AND G. L. PAYNE, *Phys. Rev. C* **28**, 983 (1983).
42. T. SASAKAWA AND T. SAWADA, *Phys. Rev. C* **20**, 1954 (1979).
43. J. V. NOBLE, *Phys. Rev.* **161**, 945 (1967).
44. S. P. MERKURIEV, *Yad. Fiz.* **24**, 289 (1976) [*Sov. J. Nucl. Phys.* **24**, 150 (1976)].
45. C. R. CHEN, G. L. PAYNE, J. L. FRIAR, AND B. F. GIBSON, *Phys. Rev. C* **33**, 401 (1986).
46. S. P. MERKURIEV, C. GIGNOUX, AND A. LAVERNE, *Ann. Phys. (N.Y.)* **99**, 30 (1976).
47. G. L. PAYNE AND W. GLÖCKLE, *Phys. Rev. C* **45**, 974 (1992).
48. LORD RAYLEIGH, *Phil. Mag.* **47**, 375 (1899).
49. F. E. FOWLE, *Astrophys. J.* **40**, 435 (1914).
50. J. L. FRIAR, in *Electron and Pion Interactions with Nuclei at Intermediate Energies* (W. Bertozzi, S. Costa, and S. Schaerf, eds.) (Harwood, New York, 1980), p. 143.
51. J. L. FRIAR AND S. FALLIEROS, *Phys. Rev. C* **29**, 232 (1984).
52. J. L. FRIAR, S. FALLIEROS, E. L. TOMUSIAK, D. SKOPIK, AND E. G. FULLER, *Phys. Rev. C* **27**, 1364 (1983).
53. GY. BENCZE, C. CHANDLER, J. L. FRIAR, A. G. GIBSON, AND G. L. PAYNE, *Phys. Rev. C* **35**, 1188 (1987).

54. W. R. GIBBS AND B. F. GIBSON, *Phys. Rev. C* **43**, 1012 (1991).
55. G. L. PAYNE, *Lecture Notes in Physics* **273**, 64 (1987).
56. J. L. FRIAR, B. F. GIBSON, AND G. L. PAYNE, *Phys. Rev. C* **42**, 1211 (1990).
57. D. R. TILLEY, H. R. WELLER, AND H. H. HASSAN, *Nucl. Phys.* **A474**, 1 (1987).
58. Y. WU, S. ISHIKAWA, AND T. SASAKAWA, *Phys. Rev. Lett.* **64**, 1875 (1990); (E) **66**, 242 (1991).
59. O. SCHORI *et al.*, *Phys. Rev. C* **35**, 2252 (1987).
60. G. H. BERTHOLD, A. STADLER, AND H. ZANKEL, *Phys. Rev. C* **41**, 1365 (1990).
61. E. HUTTEL, W. ARNOLD, H. BAUMGART, H. BERG, AND G. CLAUSNITZER, *Nucl. Phys.* **A406**, 443 (1983).
62. C. R. CHEN, G. L. PAYNE, J. L. FRIAR, AND B. F. GIBSON, *Phys. Rev. C* **39**, 1261 (1989).
63. L. I. PONOMAREV AND G. FIORENTINI, *Muon Catal. Fusion* **1**, 3 (1987).
64. J. L. FRIAR, in *New Vistas in Electro-Nuclear Physics* (E. L. Tomusiak, H. S. Caplan, and E. T. Dressler, eds.) (Plenum, New York, 1986), p. 213.
65. L. N. BOGDANOVA AND V. E. MARKUSHIN, *Muon Catal. Fusion* **5**, 189 (1991).
66. W. H. BREUNLICH, P. KAMMELL, J. S. COHEN, AND M. LEON, *Annu. Rev. Nucl. Part. Sci.* **39**, 311 (1989).
67. S. S. GERSHTEIN, *Zh. Eksp. Teor. Fiz.* **40**, 698 (1961) [*Sov. Phys.- JETP* **13**, 488 (1961)].

CHAPTER 3

TIME-DEPENDENT SCATTERING IN COULOMBIC FEW-BODY SYSTEMS AND THE STRONG OPERATOR APPROXIMATION METHOD

HELMUT KRÖGER

1. INTRODUCTION

The subject of this chapter is nonrelativistic quantum mechanical scattering in Coulombic few-body systems, using the time-dependent formulation. Our ultimate goal is to find efficient and reliable algorithms for numerical computation of scattering observables, such as phase shifts, cross sections, etc. Historically, the notion of a scattering operator was introduced by Heisenberg [1] and Møller.[2] The so-called Møller wave operator maps asymptotic states onto scattering states. At the time when the Møller wave operator was suggested, there was no mathematical proof of its existence. The first proof was given by Cook,[3] formulated in time-dependent language, for a two-body system interacting via a square integrable potential.

Since then scattering theory and numerical computations based upon it have been developed mostly in the time-independent stationary formulation. The connection between the time-dependent and time-independent formulations was established by Gell-Mann and Goldberger.[4] In particular, few-body scattering theory with Coulombic long-range forces has been developed in the time-independent framework (for a review see Alt and Sandhas[5]).

HELMUT KRÖGER • Département de Physique, Université Laval, Québec, Québec G1K 7P4, Canada.
Coulomb Interactions in Nuclear and Atomic Few-Body Collisions, edited by Frank S. Levin and David A. Micha. Plenum Press, New York, 1996.

On the other hand, the development and application of time-dependent methods has grown with the development of supercomputers. Time-dependent methods have been applied in such diverse areas of physics as: (a) atomic/molecular physics, (b) few-nucleon systems, (c) heavy-ion systems, (d) many-body systems and field theories, and also (e) quantum systems with classical chaotic behavior. The application of time-dependent wave packet methods in atomic/molecular and nuclear physics was the topic of an international conference in 1985.[6] An excellent review on time-dependent methods in chemical physics, in particular for reactive scattering, was given in 1988 by Mohan and Sathyamurty.[7] A more recent overview on the use of time-dependent methods in quantum dynamics is to be found in Kulander.[8] A survey of numerical methods which are widely used to compute the time evolution in atomic/molecular systems is presented in Leforestier,[9] where stability, efficiency, and scaling of errors are discussed by comparing the methods applied to a benchmark problem. A general review on time-dependent methods in scattering theory, including the discussion of relativistic models, is given in Kröger.[10]

In the area of atomic/molecular systems much work has been done using time-dependent methods. The reaction probabilities for an atom–molecule exchange reaction have been computed,[11] as has elastic scattering in the hydrogen atom–molecule system.[12] Collision-induced dissociation processes have been computed in the time-dependent framework.[13,14] Also the probability density function and absorption spectra have been determined for collinear reactions.[15] Using the semiclassical wave packet method, short-time dynamical processes have been computed, e.g., photodissociation,[16] photoabsorption and emission,[17] Raman scattering,[18] dynamical tunneling,[19] and atom diffraction by surfaces.[20] Many applications in this area have been discussed by Mohan and Sathyamurthy.[7]

In nuclear few-body systems neutron–neutron (n–n) and proton–proton (p–p) scattering (including the Coulomb potential) have been computed with realistic nucleon–nucleon potentials.[21,22] In the three-nucleon system, the neutron–deuteron (n–d) breakup reaction has been computed and a comparison with standard Faddeev results has been established.[23] Also the p–d breakup has been computed, including the Coulomb potential.[24,25]

Thus it is natural to ask: What are the reasons in favor of the time-dependent formulation of the scattering problem? Why, in particular, should this be useful when treating long-range Coulomb forces? Let us give some reasons:

1. It is "illustrative." That is, in a numerical simulation with a computer, one can "see" the evolution of the wave function and the formation of scattering fragments.

2. Time-dependent methods are successful in applications to atomic/molecular physics. There are many examples. A very successful example is found in collision-induced dissociation processes.[13,14] The atom–diatom

exchange reaction in three dimensions has been computed with time-dependent methods, e.g., for H + H$_2$.$^{(26-28)}$

3. With the advent of present-day supercomputers, i.e., vector machines with multiprocessor architecture, it turns out that the direct solution of the time-dependent Schrödinger equation *may* be more efficient than the solution of the stationary coupled integral equations (Faddeev-type equations). The reason is as follows: Consider an interaction described by local potentials. This corresponds to a Hamiltonian represented by a sparse matrix. The numerical solution of the time-dependent Schrödinger equation means iteration of a sparse matrix, for which very efficient algorithms exist, adapted to be run on a supercomputer. On the other hand, in the stationary method, one has to compute Green's functions (or equivalent objects such as T-matrices). However, the Green's function, being the resolvent of a sparse matrix is *not* sparse. Hence iteration is less efficient. However, as numerical solutions of Faddeev equations show, a small number of iterations (of the order of 10) as summed up by the Padé method is usually sufficient for very good convergence behavior.$^{(29)}$ This issue of comparing efficiency can be settled only by accurate numerical calculations.

4. Monte Carlo methods, and in particular the Monte Carlo Green's function method$^{(30-32)}$ have turned out to be very useful in determining ground state properties of nuclei, e.g. ^{4}He, and atoms, e.g., Li, Be, or N. Inspired by its success, one would like to apply it to the scattering problem as well. However, very little is known about how to use Monte Carlo methods in quantum mechanical scattering. A proposal was made by Kröger *et al.*$^{(33)}$ using the time-dependent formulation, which works for the two-body system. This is a first hint, indicating the usefulness of Monte Carlo for the numerical simulation of scattering reactions.

5. In stationary state few-body ($N \geq 3$) scattering theory, the computation of transition amplitudes proceeds via the Green's functions. The computation of the Green's function in a naive way, e.g., via the N-body Lippmann–Schwinger equation, runs into trouble, insofar as the solution is not unique.$^{(34)}$ This can be cured by writing Faddeev-type integral equations.$^{(35)}$ Some of these are nested equations, which in order to compute the N-body Green's function require the knowledge of the $(N-1)$-body Green's function, $(N-2)$-body Green's function, and so on. This yields a rather complicated structure, in particular for the singularities on the real axis. All those singularities have to be disentangled in a numerical calculation, which is feasible but requires some effort. In the time-dependent framework, the uniqueness problem does not occur. The N-body propagator can be computed, using the appropriately chosen boundary conditions, without knowing the $(N-1)$-body propagator, $(N-2)$-body propagator, and so on.

6. If we consider the few-body scattering problem in the presence of long-range Coulomb potentials, there is another distinction between the stationary and the time-dependent method: In the stationary framework

Coulomb scattering can be described by treating the Coulomb potential as a limit of a screened, i.e., short-range potential. This then allows application of the standard Faddeev-type machinery (plus some kind of renormalization[5]). However, this implies an additional approximation parameter, which is not necessary in the time-dependent framework.

Let us briefly summarize what this contribution is about. We discuss a particular time-dependent method, which is based on the diagonalization of the Hamiltonian. This method has been shown to work very well for the Schrödinger equation with short-range potentials (e.g., the Paris N–N potential) as well as the Coulomb long-range potential, and also for relativistic field theoretical models (scalar ϕ^4 model, Schwinger model, compact electrodynamics in $2+1$ dimensions[10]). We apply it here to proton–proton scattering, where the two-body case is intended to serve as background for the three-particle one. Finally we apply it to proton–deuteron breakup scattering. Other examples where this method could be applied would be reactions such as $e + H \to e + H$, $n + H \to n + p + e$, $e + H \to p + e + e$.

We start our discussion in Section 2 with a review of the basic definitions in scattering theory with long-range forces. We present some of the fundamental relations of one- or two-body scattering theory with Coulomb interactions expressed in the time-dependent language. This includes the time evolution, asymptotic states, Møller wave operator, Dollard's anomalous term, the scattering wave function, S-matrix, T-matrix, and cross section. In Section 3, we generalize this and consider scattering in two- and three-particle systems, interacting via Coulomb-plus-short-range potentials. We classify possible asymptotic situations. In Section 4, we discuss a systematic approximation scheme in Hilbert space. A perturbed, finite-dimensional Hamiltonian is introduced for the purpose of doing numerical computations. In Section 5, we discuss mathematical approximation theory, and in particular introduce the concept of strong resolvent convergence. We present rigorous results, proving convergence (in the sense of strong resolvent convergence) of the perturbed wave operator and S-matrix toward the exact solution. In Section 6, results of numerical computations are presented, namely for elastic proton–proton scattering, as well as proton–deuteron breakup scattering. A summary is given in Section 7.

2. TIME-DEPENDENT SCATTERING THEORY: ONE-BODY SCATTERING

2.1. Short-Range Potential

In this section, we give the basic quantities, such as propagator, wave operator, S-matrix, etc., in time-dependent language. The mathematics of scattering theory, from the physicist's point of view, is excellently described in the books by Amrein et al.,[36] and Reed and Simon.[37] The theory of linear operators in Hilbert spaces and its relation to scattering theory is given in Weidmann's book.[38] Mathematical scattering theory, with emphasis on the two-Hilbert space formulation and spectral theory is the subject of the book by Baumgärtel and Wollenberg.[39] A collection of formulas on scattering with Coulomb forces can be found in van Haeringen's book.[40]

Let us first review the formalism of scattering theory with a short-range potential, which is simpler than the theory with the long-range Coulomb potential. Let us consider a particle moving in the force field given by a short-range potential V. Let H^0 denote the free (noninteracting) Hamiltonian and let $H = H^0 + V$ denote the full Hamiltonian operator. According to the Schrödinger equation,

$$-\frac{\hbar}{i}\frac{\partial}{\partial t}\psi = H\psi \qquad (3.2.1)$$

H is the generator for infinitesimal displacements in time. We define

$$U(t) = \exp(iHt) \qquad (3.2.2)$$

The forward evolution in time or propagator is given by $U(-t)$ (we put $\hbar = 1$ in the following). $U^0(t)$ will denote the corresponding operator for the Hamiltonian H^0.

A scattering process is characterized by the Hamiltonian H of the system, plus the choice of (admissible) initial and final boundary conditions. As an initial boundary condition, one takes a state given by a plane wave which describes a particle moving with a wave vector $\mathbf{k}$, or more realistically a packet of plane waves. The state is completely characterized by giving all the quantum numbers, such as spin, etc. Such a state is called an initial asymptotic state, denoted by $|\phi_{as}^{in}\rangle$. A similar state (with, in general, a different wave number and other quantum numbers) describes the final asymptotic state, $|\phi_{as}^{fi}\rangle$. The asymptotic states can be chosen to be eigenstates of the free Hamiltonian H^0 (by choosing the plane waves), whose eigenvalue is the energy E.

The physical scattering state and the asymptotic state are not identical. It is a priori not obvious that an arbitrary Hilbert state can be approached asymptotically by the physical system. Asymptotic states have to be admissible as boundary conditions. Actually there are prominent examples in nature where this is not the case, i.e., for which there are no scattering solutions where particles can move *asymptotically freely*: (a) confinement of quarks; (b) particles described by a harmonic oscillator interaction; (c) particles interacting via long-range forces, in particular the Coulomb force. Mathematically, the existence of appropriate asymptotic states is expressed by the existence of the Møller wave operator [Eq. (3.2.6) below].

In a scatterring experiment, one prepares the physical system at a time $t = t_{\text{in}}$ (large negative time) to be in the initial asymptotic state. Similarly, one measures the system at the time $t = t_{\text{fi}}$ (large positive time) in the final asymptotic state. The system evolves dynamically, interacting via the potential V. One can ask: What is the probability amplitude for the transition of the physical system from the initial to the final state? The answer to this question involves the definition of the scattering or S-matrix.[41] The physical system is described by a wave function. Scattering states $\psi^{\pm}$ are defined as states, the time-evolution of which asymptotically approaches the (free) time-evolution of the asymptotic states:

$$\|\exp(-iHt)|\psi^{+}\rangle - \exp(-iH^{0}t)|\psi^{\text{in}}_{\text{as}}\rangle\| \xrightarrow[t\to -\infty]{} 0$$

$$\|\exp(-iHt)|\psi^{-}\rangle - \exp(-iH^{0}t)|\psi^{\text{fi}}_{\text{as}}\rangle\| \xrightarrow[t\to +\infty]{} 0 \qquad (3.2.3)$$

where $\|\cdot\|$ denotes the norm of a Hilbert state. Thus, one has essentially two kinds of scattering states, corresponding to the two kinds of asymptotic boundary conditions (initial and final). Note that the assignment of signs is the standard notation adopted from the ε-limit in time-independent scattering theory.[41] Equation (3.2.3) can be rewritten in the form

$$|\psi^{+}\rangle = \lim_{t\to -\infty} U(t)U^{0}(-t)|\psi^{\text{in}}_{\text{as}}\rangle$$

$$|\psi^{-}\rangle = \lim_{t\to +\infty} U(t)U^{0}(-t)|\psi^{\text{fi}}_{\text{as}}\rangle \qquad (3.2.4)$$

One can interpret $|\psi^{+}\rangle$ as the Hilbert state obtained by the following steps: Start from the Hilbert state $|\psi^{\text{in}}_{\text{as}}\rangle$ at $t = 0$, propagate it backward in time until $t = -\infty$ under the free propagator, then switch to the full propagator and propagate it forward in time until $t = 0$. There is a similar interpretation for $|\psi^{-}\rangle$. In particular, for an asymptotic state given by a plane wave $|\mathbf{k}\rangle$, one has

$$|\mathbf{k}^{(\pm)}\rangle \equiv |\psi^{(\pm)}_{\mathbf{k}}\rangle = \Omega^{(\pm)}|\mathbf{k}\rangle \qquad (3.2.5)$$

This leads to the following definition of the Møller wave operators

$$\Omega^{\pm} = \operatorname*{s-lim}_{t\to \mp\infty} U(t)U^{0}(-t) \qquad (3.2.6)$$

The notation s-lim stands for strong operator limit and means that the sequence of operators applied to an arbitrary Hilbert state gives a sequence of Hilbert states which has a limit in Hilbert space, i.e., converges in the norm to some Hilbert state. The s-limit of the sequence of operators is then defined as the mapping of the given Hilbert state to the limit of the sequence of Hilbert states. The existence of the wave operators has been established for the most common nucleon–nucleon short-range potentials. It holds, e.g., for the square-well potential, the Yukawa potential, and a separable potential with square integrable form factors.[36]

An interesting property of the wave operators is expressed by the so-called invariance principle, which states (see Weidmann[42]) that there is a whole class of wave operators whose structure is identical to the wave operator given by Eq. (3.2.6). Let $\vartheta(x)$ denote a twice differentiable, real function, with $\vartheta'(x) > 0$. Let $\Omega^{\pm}(H, H^0)$ denote the wave operator given by the r.h.s. of Eq. (3.2.6), corresponding to the full Hamiltonian H and the asymptotic Hamiltonian H^0. Then the invariance principle takes the form

$$\Omega^{\pm}(H, H^0) = \Omega^{\pm}[\vartheta(H), \vartheta(H^0)] \quad (3.2.7)$$

In the numerical simulations described below, based on the wave operator given by Eq. (3.2.6), one has to compute the nonlinear function $\exp(ix)$, $x = H$. Thus, dealing with another nonlinear function, $\exp[i\vartheta(x)]$ with, e.g., $\vartheta(x) = \arctan(x)$, would not complicate matters further. However it is conceivable that a particular choice of ϑ could improve the convergence properties. To the author's knowledge, no such analysis has been done yet.

Another very important property of the wave operators is the intertwining relation. This turns out to be a very useful tool in numerical simulations in time-dependent scattering theory. It reads

$$H\Omega^{\pm} = \Omega^{\pm} H^0 \quad (3.2.8)$$

According to the definition of the scattering matrix, the probability amplitude can be written as a matrix element of an operator,

$$\langle \psi^- | \psi^+ \rangle = \langle \psi_{as}^{fi} | S | \psi_{as}^{in} \rangle \quad (3.2.9)$$

where

$$S = (\Omega^-)^\dagger \Omega^+ \quad (3.2.10)$$

is called the S-matrix operator. The intertwining relation yields

$$[S, H^0] = 0 \quad (3.2.11)$$

which means that energy is conserved in a scattering reaction.

From the S-matrix, one obtains phase shifts and cross sections, which are measured by experiment, as follows:

$$\langle \mathbf{p}'|S|\mathbf{p}\rangle = \delta(\mathbf{p}' - \mathbf{p}) - 2\pi i \delta(E^{fi} - E^{in})\langle \mathbf{p}'|T|\mathbf{p}\rangle \qquad (3.2.12)$$

where $\langle \mathbf{p}'|T|\mathbf{p}\rangle$ defines the on-shell ($E = p'^2/2m = p^2/2m$) matrix element of the transition matrix or T-matrix. The differential cross section[43] is given by

$$\frac{d\sigma}{d\Omega}(\mathbf{p} \to \mathbf{p}') = |f(\mathbf{p} \to \mathbf{p}')|^2 \qquad (3.2.13)$$

where

$$f(\mathbf{p} \to \mathbf{p}') = -(2\pi)^2 m \langle \mathbf{p}'|T|\mathbf{p}\rangle \qquad (3.2.14)$$

is called the scattering amplitude. Phase shifts are usually introduced by making a partial wave decomposition of the cross section. A partial wave basis $|plm\rangle$ is obtained from the plane wave basis $|\mathbf{p}\rangle$ via

$$|\mathbf{p}\rangle = \sum_{l=0}^{\infty} \sum_{m=-l}^{+l} |plm\rangle Y_{lm}^{*}(\hat{p}) \qquad (3.2.15)$$

The phase shift δ_l is then related to matrix elements of the S-matrix via

$$\langle p', l', m'|S|p, l, m\rangle = \exp[2i\delta_l(p)]\langle p', l', m'|p, l, m\rangle \qquad (3.2.16)$$

In order to establish a connection between the time-dependent and time-independent formulation, one can use the Abelian ε-limit in place of the time limit in Eq. (3.2.4) (see Gell-Mann and Goldberger[4]):

$$\lim_{t \to -\infty} U(t)[U^0(t)]^{-1}|\phi\rangle = \lim_{\varepsilon \to +0} \int_{-\infty}^{0} dt\varepsilon \exp(\varepsilon t) U(t)[U^0(t)]^{-1}|\phi\rangle \qquad (3.2.17)$$

We denote the resolvent of H (propagator in the complex energy plane) by

$$G(z) = (z - H)^{-1} \qquad (3.2.18)$$

and correspondingly the resolvent of H^0 by $G^0(z)$. Putting $|\phi\rangle = |\mathbf{q}\rangle$ (eigenstate of H^0 with energy $E = q^2/2m$), Eq. (3.2.17) leads to

$$\Omega^+|\mathbf{q}\rangle = \lim_{\varepsilon \to +0} G(E + i\varepsilon)[G^0(E + i\varepsilon)]^{-1}|\mathbf{q}\rangle \qquad (3.2.19)$$

This establishes the correspondence between the two formulations for the wave operator Ω^+.

2.2. Coulomb Potential

2.2.1. Coulomb Scattering States. We now discuss the modifications which will occur when we replace the short-range potential by the long-range Coulomb potential, or more generally by a Coulomb-like potential, i.e., a Coulomb plus a short-range potential. The Coulomb potential interacting between two particles of charge e_1 and e_2 is given by

$$V^C(r) = e_1 e_2 / r \tag{3.2.20}$$

and in momentum space by

$$\langle \mathbf{k}' | V^C | \mathbf{k} \rangle = \frac{e_1 e_2}{2\pi^2} \frac{1}{|\mathbf{k}' - \mathbf{k}|^2} \tag{3.2.21}$$

Let us consider the scattering of one charged particle of mass m from the other charge fixed at the origin. In coordinate space $H^0 = -\Delta/2m$. Again, we define the full Hamiltonian $H = H^0 + V^C$. Stationary scattering wave functions for the pure Coulomb potential are known analytically. They were first determined a long time ago by Gordon.[44] They read (following Dollard;[45,46] for more details see Joachain[41] and Mott and Massey[47])

$$\psi_{\mathbf{k}}^{C(+)}(\mathbf{x}) = \exp[i\mathbf{k} \cdot \mathbf{x}] \Gamma(1 + i\gamma) {}_1F_1[-i\gamma, 1, (kx - \mathbf{k} \cdot \mathbf{x})]$$

$$\psi_{\mathbf{k}}^{C(-)}(\mathbf{x}) = [\psi_{-\mathbf{k}}^{C(+)}(\mathbf{x})]^* \tag{3.2.22}$$

where $\gamma = me_1 e_2/k$, $\Gamma(z)$ is the gamma function, ${}_1F_1(a, b, z)$ is Kummer's hypergeometric function, and the superscripts $(+)$ and $(-)$ refer, respectively, to outgoing and incoming waves.

These functions are eigenfunctions of the Hamiltonian H,

$$H \psi_{\mathbf{k}}^{C(\pm)} = E \psi_{\mathbf{k}}^{C(\pm)}, \qquad E = k^2/2m \tag{3.2.23}$$

However, they do *not* behave asymptotically like incoming/outgoing plane waves (characterized by a wave number $\mathbf{k}$). For example, the asymptotic behavior of $\psi_{\mathbf{k}}^{C(+)}$ for $|kx - \mathbf{k} \cdot \mathbf{x}| \gg 1$ is given by

$$\psi_{\mathbf{k}}^{C(+)}(\mathbf{x}) \to \exp\{i[\mathbf{k} \cdot \mathbf{x} + \gamma \log(kx - \mathbf{k} \cdot \mathbf{x})]\} \left[1 + \frac{\gamma^2}{i(kx - \mathbf{k} \cdot \mathbf{x})}\right]$$

$$+ f^C(\theta) \frac{\exp\{i[kx - \gamma \log(2kx)]\}}{x} \left[1 + \frac{(1 + i\gamma)^2}{i(kx - \mathbf{k} \cdot \mathbf{x})}\right] \tag{3.2.24}$$

where

$$f^C(\theta) = -\frac{me_1e_2}{2k^2 \sin^2(\theta/2)} \exp\{i\{2\sigma_0 - 2\gamma \log[\sin(\theta/2)]\}\}$$

$$\exp(2i\sigma_0) = \frac{\Gamma(1+i\gamma)}{\Gamma(1-i\gamma)} \qquad (3.2.25)$$

Note that θ is the angle between $\mathbf{x}$ and $\mathbf{k}$, $f^C(\theta)$ is called the Coulomb scattering amplitude, and σ_0 is the s-wave Coulomb phase shift. In contrast to the asymptotic behavior of a scattering state of a short-range potential, given for $kx \gg 1$ by

$$\psi^+(\mathbf{x}) \rightsquigarrow \exp(i\mathbf{k}\cdot\mathbf{x}) + f(\theta,\phi)\frac{\exp(ikx)}{x} \qquad (3.2.26)$$

in Eq. (3.2.24) there occurs a logarithmic "distortion" of the plane wave, as well as of the spherical wave.

2.2.2. Dollard Wave Operator. If one would be naive and try to construct scattering states from the Møller wave operator, given by Eq. (3.2.6), one would fail. As Dollard[45] showed, the Møller wave operator, defined as the strong limit by Eq. (3.2.6) for a short-range potential, does not exist for the Coulomb potential. Moreover,

$$\underset{t\to\mp\infty}{\text{w-lim}}\, U(t)U^0(-t) = 0 \qquad (3.2.27)$$

Here w-lim stands for weak limit. It means that taking matrix elements of the operator between arbitrary Hilbert states and then taking t to infinity gives a vanishing result. This can be seen as follows: The standard proof of existence of Møller wave operators for short-range potentials proceeds via the Cook–Hack method,[36] in which one writes

$$\exp(iHt')\exp(-iH^0t')|g\rangle|_{t_0}^t = \int_{t_0}^t dt' \exp(iHt')V\exp(-iH^0t')|g\rangle \qquad (3.2.28)$$

and then estimates if the limit of the integral

$$\lim_{t\to\infty}\int_{t_0}^t dt'\,\|V\exp[-iH^0t']|g\rangle\| \qquad (3.2.29)$$

exists. Now let us replace V by the Coulomb potential V^C. According to Dollard,[45] the following heuristic argument applies, which gives a semiclassical idea of the asymptotic behavior: We start from the solution of the free Schrödinger equation, which we multiply by the Coulomb potential $1/x$. For large x and time t, i.e., asymptotically, this corresponds to putting $\mathbf{k} = m\mathbf{x}/t$ and multiplying in momentum space by $m/k|t|$. In coordinate-space k is an eigenvalue of the operator $\sqrt{-\Delta}$ (Δ being the Laplace operator); thus it corresponds to a factor $m/|t|\sqrt{-\Delta}$. From that it is plausible that $\exp(iHt)\exp(-iH^0t)$ behaves like a phase factor built from $m/\sqrt{-\Delta}$. Some analysis shows the following asymptotic behavior

$$\exp(iHt)\exp(-iH^0t) \underset{t \to +\infty}{\sim} = \exp\left(\frac{ime_1e_2 \log|t|}{\sqrt{-\Delta}}\right) \quad (3.2.30)$$

which tends weakly to zero. But this asymptotic behavior suggests that the limit

$$\lim_{t \to +\infty} \exp(iHt)\exp\left(-iH^0t - \frac{ime_1e_2 \log|t|}{\sqrt{-\Delta}}\right) \quad (3.2.31)$$

should exist. Dollard[48] has suggested as Coulomb-modified Møller wave operator,

$$\Omega^{C\pm} = \underset{t \to \mp\infty}{\text{s-lim}} \exp(iHt)\exp[-iH^{0C}(t)]$$

$$H^{0C}(t) = H^0 t + A(t)$$

$$A(t) = \text{sign}(t)\frac{me_1e_2}{\sqrt{-\Delta}}\log\left(\frac{-2|t|\Delta}{m}\right) \quad (3.2.32)$$

which differs slightly in its logarithm from the expression given in Eq. (3.2.31), for the reason explained below. The term $A(t)$ is an anomalous one which does not occur for short-range potentials. Expressing it in terms of H^0, one has

$$A(t) = \text{sign}(t)e_1e_2\sqrt{\frac{m}{2H^0}}\log(4|t|H^0) \quad (3.2.33)$$

Dollard[45,48] has proven firstly that $\Omega^{C(\pm)}$ given by their r.h.s. of Eq. (3.2.32) exists, and secondly that it gives the correct physics, i.e., the Coulomb scattering state. We note the following property of the anomalous term, which will play a role when we discuss the asymptotic Coulomb state,

$$\underset{t \to \mp\infty}{\text{w-lim}} \exp[-iA(t)] = 0 \quad (3.2.34)$$

while the corresponding strong limit does not exist. The anomalous term $A(t)$ can be split up into a time-dependent and a time-independent part, viz,

$$A(t) = \text{sign}(t) e_1 e_2 \sqrt{\frac{m}{2H^0}} \log(|t|) + \text{sign}(t) e_1 e_2 \sqrt{\frac{m}{2H^0}} \log(4H^0) \quad (3.2.35)$$

While the first term, according to Eq. (3.2.30), is necessary to establish convergence in time, the second one obviously does not play a role in the convergence proof. Why has the latter been included? It turns out that it is precisely this term which is necessary to guarantee that the relation

$$|\psi_{\mathbf{k}}^{C(\pm)}\rangle = \Omega^{C(\pm)}|\mathbf{k}\rangle \quad (3.2.36)$$

holds in analogy to Eq. (3.2.5) for a short-range interaction. Here the $|\psi_{\mathbf{k}}^{C(\pm)}\rangle$ are the Coulomb scattering states given by Eq. (3.2.22). This relation is to be understood in the sense of distributions.

2.2.3. Asymptotic Coulomb States. Let us summarize our knowledge about Coulomb scattering. The conventional Møller wave operator does not exist. However, there is the Dollard wave operator, modified by adding a time-dependent anomalous term to the asymptotic time evolution. This wave operator exists, defined in Hilbert space. Moreover, the operator maps an improper plane wave state onto the Coulomb scattering state. One might imagine that the Dollard wave operator $\Omega^{D\pm} [\Omega^D \equiv \Omega^C]$ given by Eq. (3.2.32) could be split into two factors,

$$\Omega^{D\pm} = \Omega^{M\pm} \Omega^{A\pm} \quad (3.2.37)$$

one representing the effect of the conventional Møller wave operator and the other one representing the effect of the anomalous term. Then considering the mapping of the plane wave state onto the Coulomb scattering state, given by Eq. (3.2.36), one might further imagine the existence of an intermediate state, $|\phi_{\mathbf{k}}^{\text{as}\pm}\rangle$, as depicted schematically in Fig. 1. The decomposition given by Eq. (3.2.37), and the state $|\phi_{\mathbf{k}}^{\text{as}\pm}\rangle$ exist, but are defined in the sense of generalized distributions and not in Hilbert space. The state $|\phi_{\mathbf{k}}^{\text{as}\pm}\rangle$, called an asymptotic Coulomb state, was introduced by Nutt[49] and elaborated on by van Haeringen.[50] The relevant relations are

$$|\phi_{\mathbf{k}}^{\text{as}\pm}\rangle = \Omega^{A\pm}|\mathbf{k}\rangle \quad (3.2.38)$$

and

$$|\psi_{\mathbf{k}}^{C\pm}\rangle = \Omega^{M\pm}|\phi_{\mathbf{k}}^{\text{as}\pm}\rangle \quad (3.2.39)$$

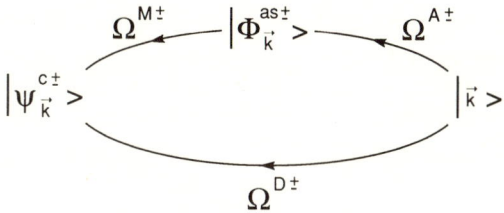

FIGURE 1. Relation between sharp momentum state $|\vec{k}\rangle$, the asymptotic Coulomb state $|\phi_{\vec{k}}^{as}\rangle$, the Coulomb scattering state $|\psi_{\vec{k}}^{C}\rangle$, the mapping of the Dollard wave operator Ω^D, and its (formal) splitting into a Møller wave operator Ω^M and an anomalous wave operator Ω^A.

with proofs being given in the works by Zorbas[51] and van Haeringen.[40,50] Let us now consider how the factorization ansatz, given by Eq. (3.2.37), might apply to the Dollard wave operator written in the form

$$\Omega^{D\pm} = \operatorname*{s-lim}_{t \to \mp\infty} \exp(iHt) \exp(-iH^0 t) \exp[-iA(t)] \qquad (3.2.40)$$

in order to yield Eqs. (3.2.38) and (3.2.39). Naively, one is tempted to decompose $\Omega^{D\pm}$ via the ansatz:

$$\Omega^{M\pm} = \operatorname*{s-lim}_{t \to \mp\infty} \exp(iHt) \exp(-iH^0 t)$$

$$\Omega^{A\pm} = \operatorname*{s-lim}_{t \to \mp\infty} \exp[-iA(t)] \qquad (3.2.41)$$

But we know already that the strong limits on the r.h.s. do not exist. In other words, it is not legitimate to replace the limit of the product [Eq. (3.2.40)] by the product of the limits [Eq. (3.2.41)]. Thus the ansatz given by Eq. (3.2.41) is incorrect. However, there is some truth to the basic idea. It can be made mathematically rigorous by going from the time-dependent formulation to the stationary formulation (replacing the time-limit by an ε-limit) and giving the ε-limit a well-defined meaning in the sense of generalized distributions. Let us define

$$\Omega^D_{\pm\varepsilon} = \pm\varepsilon \int_{\mp\infty}^{0} dt \exp(\pm\varepsilon t) \exp(iHt) \exp(-iH^0 t) \exp[-iA(t)]$$

$$\Omega^M_{\pm\varepsilon} = \pm\varepsilon \int_{\mp\infty}^{0} dt \exp[\pm\varepsilon t] \exp(iHt) \exp(-iH^0 t)$$

$$\Omega^A_{\pm\varepsilon} = \pm\varepsilon \int_{\mp\infty}^{0} dt \exp(\pm\varepsilon t) \exp[-iA(t)] \qquad (3.2.42)$$

That is, we make the transition from the time-dependent to the stationary formulation but do not yet perform the ε-limit.

According to results by Jauch,[52] the existence of the strong time limit implies the existence of the ε-limit and the two are equal. Hence

$$\Omega^{D\pm} = \underset{\varepsilon \to +0}{\text{s-lim}}\, \Omega^{D}_{\pm\varepsilon} \tag{3.2.43}$$

Moreover,

$$\Omega^{D\pm} = \underset{\varepsilon \to +0}{\text{s-lim}}\, \Omega^{M}_{\pm\varepsilon} \Omega^{A}_{\pm\varepsilon} \tag{3.2.44}$$

which is the stationary state analog of Eq. (3.2.40). The operators $\Omega^{M}_{\pm\varepsilon}$ and $\Omega^{A}_{\pm\varepsilon}$ can be computed explicitly. Applied on a plane wave state, $\Omega^{M}_{\pm\varepsilon}$ becomes

$$\Omega^{M}_{\pm\varepsilon} = \pm\varepsilon(E_k \pm i\varepsilon - H)^{-1} \tag{3.2.45}$$

Following van Haeringen[50] and Zorbas,[51] the operator $\Omega^{A}_{\pm\varepsilon}$ [switching to the notation $V^C(r) = -2s/r$, $\hbar = 2m = 1$] is found to be

$$\Omega^{A}_{\pm\varepsilon} = \Gamma(1 \pm is(H^0)^{-1/2})(4H^0/\varepsilon)^{\pm is(H^0)^{-1/2}} \tag{3.2.46}$$

This operator can be represented most easily in momentum space, where it is diagonal. However, because the strong limits on the r.h.s. of Eq. (3.2.41) do not exist, there is no reason that the strong ε-limits of $\Omega^{M}_{\pm\varepsilon}$ and $\Omega^{A}_{\pm\varepsilon}$ individually should exist. Actually they do not. However, if we apply $\Omega^{A}_{\pm\varepsilon}$ on a plane wave state, according to the scheme given in Fig. 1, the ε-limit can be performed in the sense of generalized distributions[40,50,51] and yields the Coulomb asymptotic state. On a suitably defined space of test functions (denoted by $\mathcal{D}_{h_\varepsilon}$ in van Haeringen,[50] which should be consulted for more details), the ε-limit can be performed

$$\lim_{\varepsilon \to +0} \langle h_\varepsilon | \Omega^{A}_{\pm\varepsilon} | \mathbf{k} \rangle = \lim_{\varepsilon \to +0} \langle h_\varepsilon | \phi^{\text{as}\pm}_{\mathbf{k}} \rangle \tag{3.2.47}$$

where h_ε is a test function, implying that

$$|\phi^{\text{as}\pm}_{\mathbf{k}}\rangle = \lim_{\varepsilon \to +0} \Omega^{A}_{\pm\varepsilon} |\mathbf{k}\rangle \tag{3.2.48}$$

The state $|\phi^{\text{as}\pm}_{\mathbf{k}}\rangle$ is called the asymptotic Coulomb state. Explicit expressions can be found in van Haeringen[40,50] e.g.,

$$\langle \mathbf{p} | \phi^{\text{as}+}_{\mathbf{k}} \rangle = \lim_{\varepsilon \to +0} \delta(\mathbf{p} - \mathbf{k}) \left(\frac{p - k - i\varepsilon}{p - k + i\varepsilon} \right)^{-i\gamma} \frac{\exp(\pi\gamma/2)}{\Gamma(1 - i\gamma)} \tag{3.2.49}$$

2.2.4. Connection between the Time-Dependent Formulation and Screening in the Stationary Formulation.

One of the conclusions of the above paragraph is that the Dollard wave operator is the strong time limit of two factors, one being the standard Møller term $\exp(iHt)\exp(-iH^0t)$ and the other being the anomalous term $\exp[-iA(t)]$. The strong time limit of each factor individually does not exist in Hilbert space. This situation is very similar (and actually related) to the situation in the stationary formulation, when Coulomb scattering is treated by screening. Let us briefly outline this similarity and establish the connection (for a review on the screening method see also Alt and Sandhas[5]). Corresponding to the Coulomb potential V^C, let us define a screened potential (Yukawa screening)

$$V_\delta(r) = \exp(-\delta r) V^C(r) \qquad (3.2.50)$$

For $\delta > 0$, V_δ is the short-range Yukawa potential. Hence the apparatus of short-range scattering theory is applicable, yielding a Møller wave operator $\Omega_\delta^\pm$, an S-matrix S_δ, etc. The following questions arise: (a) Does a [strong] limit of $\Omega_\delta^\pm$ and S_δ exist, when $\delta \to 0$? (b) If so, does it give the Coulomb (i.e., Dollard) wave operator and Coulomb S-matrix, respectively? (c) If not, can it be used to obtain information on the latter? The answer to questions (a) and (b) is no! This has been shown explicitly by Dollard.[53] In particular,

$$\underset{\delta \to +0}{\text{w-lim}}\, \Omega_\delta^\pm = 0$$

$$\underset{\delta \to +0}{\text{w-lim}}\, S_\delta = 0 \qquad (3.2.51)$$

and the strong limit does not exist. This is quite analogous to Eq. (3.2.27), stating that the Møller wave operator has the weak limit zero and the strong limit does not exist. In the time-dependent formulation this defect was repaired by taking into account the anomalous term. A similar situation occurs here, which answers questions (c) in an affirmative way: The screened Møller wave operator $\Omega_\delta^\pm$ can be multiplied ("renormalized") by a phase factor $\exp(i\phi_\delta)$ to reconstitute the Dollard wave operator

$$\Omega^{C\pm} = \underset{\delta \to +0}{\text{s-lim}}\, \Omega_\delta^\pm \exp(\pm i\phi_\delta) \qquad (3.2.52)$$

According to Dollard,[53] one has

$$\phi_\delta = \frac{me_1e_2}{\sqrt{-\Delta}} \int_{t_0}^\infty dt\, t^{-1} \exp\left(-\frac{\delta t \sqrt{-\Delta}}{m}\right) + \frac{me_1e_2}{\sqrt{-\Delta}} \log\left(-\frac{2t_0\Delta}{m}\right) \qquad (3.2.53)$$

In momentum space this can be computed to give

$$\phi_\delta(k) = \frac{me_1 e_2}{k}\left[\log\left(\frac{2k}{\delta}\right) - C + O(\delta)\right] \quad (3.2.54)$$

where $C = 0.5772\cdots$ denotes Euler's constant. This renormalization phase factor is in agreement with the result used in stationary scattering theory.[54]

2.2.5. Intertwining Relation, S-Matrix, T-Matrix and Phase Shifts. Considering the structure of the wave operator for the Coulomb potential [Eq. (3.2.32)], one might suppose that the intertwining relation for the Coulomb potential [analogous to Eq. (3.2.8) for the short-range potential] would involve the free Hamiltonian H^0 plus some anomalous term. Actually, as shown in Dollard,[45] this is not the case. The intertwining relation for the Coulomb potential has the same form as the intertwining relation for a short-range potential,

$$H\Omega^{C\pm} = \Omega^{C\pm} H^0 \quad (3.2.55)$$

Also the Coulomb S-matrix is defined in the same way [analogous to Eq. (3.2.10)],

$$S^C = (\Omega^{C-})^\dagger \Omega^{C+} \quad (3.2.56)$$

Thus the S-matrix element corresponding to states which asymptotically behave as "Coulomb distorted plane waves" [Eqs. (3.2.24) and (3.2.25)] is given by

$$\langle \psi_{\mathbf{k}'}^{C(-)} | \psi_{\mathbf{k}}^{C(+)} \rangle = \langle \mathbf{k}' | S^C | \mathbf{k} \rangle \quad (3.2.57)$$

Hence, as with the short-range interaction case [Eq. (3.2.11)], energy is conserved for Coulomb scattering,

$$[S^C, H^0] = 0 \quad (3.2.58)$$

However, there is a difference between short-range and Coulomb interactions when it comes to defining the T-matrix. For short-range potentials, the T-matrix is well defined [Eq. (3.2.12)] by subtracting the momentum conserving δ-function from the S-matrix. However, as Herbst[55] has shown, it is not meaningful for Coulomb scattering to define the T-matrix by Eq. (3.2.12), because such a T-matrix would be even more singular than the momentum conserving δ-function. One can define the Coulomb T-matrix T^C for a complex energy z, off the real axis, i.e.,

Im(z) ≠ 0, in the same way as for a short-range potential via the resolvents $G^0(z)$ and $G^C(z)$:

$$G^C(z) = G^0(z) + G^0(z)T^C(z)G^0(z) \tag{3.2.59}$$

where

$$G^C(z) = (z - H^C)^{-1}$$
$$G^0(z) = (z - H^0)^{-1} \tag{3.2.60}$$

However, then one finds that on-shell ($E = k'^2/2m = k^2/2m$) matrix element

$$\langle \mathbf{k}'|T^C(E + i0)|\mathbf{k}\rangle \tag{3.2.61}$$

is singular and therefore does not exist.

This singularity is related to the fact that the ordinary Møller wave opperator does not exist, but has to be amended by the anomalous term. As we have discussed above, the anomalous term corresponds to mapping the plane wave state $|\mathbf{k}\rangle$ onto the Coulomb asymptotic state $|\phi_\mathbf{k}^{as\pm}\rangle$. Thus it is no surprise that an on-shell matrix element of T^C between Coulomb asymptotic states

$$\langle \phi_{\mathbf{k}'}^{as-}|T^C(E + i0)|\phi_\mathbf{k}^{as+}\rangle \tag{3.2.62}$$

exists.[50,56,57] Putting it differently, one can split off singular factors from $\langle \mathbf{k}'|T^C(z)|\mathbf{k}\rangle$, such that the remaining matrix element has a well-defined on-shell limit [$z = E + i0$, $E = k'^2/2m = k^2/2m$[56,58,59]]:

$$\langle \mathbf{k}'|T^C(z)|\mathbf{k}\rangle = (z - E')^{i\gamma(k')}\langle \mathbf{k}'|T_0^C(z)|\mathbf{k}\rangle(z - E)^{i\gamma(k)} + \langle \mathbf{k}'|T_1^C(z)|\mathbf{k}\rangle \tag{3.2.63}$$

Then both $\langle \mathbf{k}'|T_0^C(z)|\mathbf{k}\rangle$ and $\langle \mathbf{k}'|T_1^C(z)|\mathbf{k}\rangle$ have well-defined on-shell limits, the second one being identically zero. Following Veselova[56] and Bajzer,[59] one finds the following on-shell relation between the S-matrix and the T-matrix:

$$\langle \mathbf{k}'|S^C|\mathbf{k}\rangle = -2\pi i\delta(E' - E)\frac{\exp(-\pi\gamma)}{\Gamma^2(1 - i\gamma)}\left(\frac{2k^2}{m}\right)^{2i\gamma}\langle \mathbf{k}'|T_0^C(E + i0)|\mathbf{k}\rangle \tag{3.2.64}$$

Note that T_0^C has the same dimensions as T^C, which has the same dimensions as the T-matrix for a short-range potential. As shown by Schwinger,[58] the Coulomb Green's function G^C [Eq. (3.2.60)] can be expressed in terms of a one-parameter integral representation, which yields for T_0^C asymptotically the following on-shell relation:

$$\langle \mathbf{k}'|T_0^C(E + i0)|\mathbf{k}\rangle = \langle \mathbf{k}'|V^C|\mathbf{k}\rangle \frac{-2\pi\gamma}{\exp(-2\pi\gamma) - 1} \left(\frac{4k^2}{|\mathbf{k}' - \mathbf{k}|^2}\right)^{i\gamma} [4E]^{-2i\gamma} \quad (3.2.65)$$

This leads to the following analytical expression for the Coulomb S-matrix obtained by Veselova[56] and Bajzer[59]:

$$\langle \mathbf{k}'|S^C|\mathbf{k}\rangle = -2\pi i \delta(E' - E) \frac{e_1 e_2}{2\pi^2} \frac{\Gamma(1 + i\gamma)}{\Gamma(1 - i\gamma)} \frac{(4k^2)^{i\gamma}}{|\mathbf{k}' - \mathbf{k}|^{2 + 2i\gamma}} \quad (3.2.66)$$

As in the case of a short-range potential [see Eq. (3.2.14)], extracting from the r.h.s. of Eq. (3.2.66) the coefficient of $-2\pi i \delta(E' - E)$ and multiplying it by $-(2\pi)^2 m$ gives the scattering amplitude,[43] viz,

$$f^C(\mathbf{k} \to \mathbf{k}') = -\frac{2m e_1 e_2}{|\mathbf{k}' - \mathbf{k}|^2} \exp\left\{2i\left\{\sigma_0 - \gamma \log\left[\sin\left(\frac{\theta}{2}\right)\right]\right\}\right\} \quad (3.2.67)$$

Note that the amplitude $f^C(\mathbf{k}' \leftarrow \mathbf{k})$ is identical to the Coulomb scattering amplitude $f^C(\theta)$, defined in Eqs. (3.2.24) and (3.2.25), if we equate the angle θ between $\mathbf{x}$ and $\mathbf{k}$ with the angle θ between $\mathbf{k}'$ and $\mathbf{k}$. The differential cross section is given by the same relation as in Eq. (3.2.13), which for the Coulomb potential is

$$\frac{d\sigma^C}{d\Omega}(\mathbf{k} \to \mathbf{k}') = |f^C(\mathbf{k} \to \mathbf{k}')|^2 = \left(\frac{2m e_1 e_2}{|\mathbf{k}' - \mathbf{k}|^2}\right)^2 = \frac{m^2 e_1^2 e_2^2}{4k^4 \sin^4(\theta/2)} \quad (3.2.68)$$

the celebrated Rutherford formula (Messiah[60]).

Finally let us discuss the phase shifts. The Coulomb phase shifts are defined by a relation having the same form as for a short-range potential [Eq. (3.2.16)],

$$\langle k'l'm'|S^C|klm\rangle = \exp[2i\delta_l(k)]\langle k'l'm' | klm\rangle \quad (3.2.69)$$

where

$$\delta_l(k) = \arg\{\Gamma[l + 1 + i\gamma(k)]\} \quad (3.2.70)$$

These phase shifts can be computed analytically [see Eq. (3.2.25) for the s-wave].

3. COULOMB-LIKE POTENTIALS: SCATTERING IN CHARGED TWO-BODY AND THREE-BODY SYSTEMS

3.1. Classes of Asymptotic Situations

Among few-body scattering processes with charged particles, one can distinguish the following case: (a) Where a physical scattering process is described by a pure Coulomb potential as, e.g., electron–electron scattering or electron–H^+ scattering. However, in atomic/molecular and nuclear physics one encounters mostly the following types of situations: (b) The interaction is a superposition of the Coulomb potential with a short-range potential. At very large distances only the long-range Coulomb potential survives, since the nucleon–nucleon potential is short range. Example: proton–proton scattering, deuteron–deuteron scattering or ion–ion scattering. (c) In few-body systems with $N \geqslant 3$ the occurrence of positive and negative charges leads to a superposition of attractive and repulsive Coulomb potentials. This results in screening, which is described by an effective short-range potential (note, the electric dipole potential falls off as r^{-2}). One example would be electron–hydrogen scattering.

Until now we have discussed how to describe one-particle scattering in a pure Coulomb potential in the time-dependent formulation. A system to which this would apply approximately is electron–proton scattering [with $m_e/m_p \ll 1$ justifying the static proton approximation]. Now we want to extend the time-dependent formulation to describe one-particle scattering with Coulomb-like potentials, and charged particle scattering in two- and three-particle systems.

3.2. One-Particle Scattering with Coulomb-like Potentials

Let us assume that the total interaction can be written as

$$V = V^C + V^s \qquad (3.3.1)$$

where V^C denotes the Coulomb potential and V^s denotes a short-range potential. As an example, we will discuss in Section 6 numerical results for proton–proton scattering. Let $H = H^0 + V$ denote the full Hamiltonian. As noted above, the asymptotic behavior is dominated by the Coulomb part of the potential, in the same way as for the pure Coulomb potential. Thus one expects the treatment of the boundary conditions in the time-dependent language to be the same as for the pure Coulomb potential. This is actually the case.[45] Hence the formulas for the Dollard wave operator [Eq. (3.2.32], scattering states [Eq. (3.2.36)], S-matrix [Eq. (3.2.56)] are also valid when the pure Coulomb potential V^C is replaced by $V^C + V^s$. Mathematical rigor, of course, imposes some constraints on the short-range potential V^s, but actually not more than those required for the existence of the Møller wave operators for scattering from the short-range potential V^s alone.

An intuitive way to see why the Dollard-type formulation also holds for Coulomb-like potentials is given by the chain rule.[38] The Dollard wave operator is given by

$$\Omega^{C\pm} = \lim_{t \to \mp\infty} \exp[i(H^0 + V^C + V^s)t] \exp[-iH^0 t - iA(t)] \quad (3.3.2)$$

The r.h.s. can be written

$$\lim_{t \to \mp\infty} \exp[i(H^0 + V^C + V^s)t] \exp[-i(H^0 + V^C)t]$$
$$\times \exp[i(H^0 + V^C)t] \exp[-iH^0 t - iA(t)] \quad (3.3.3)$$

The product of terms three and four has a limit, the pure Coulombic Dollard wave operator. The product of terms one and two has a limit, equal to some kind of Møller wave operator. Because of the similarity of the treatment, we subsume in the following the pure Coulomb potential under Coulomb-like potentials.

3.3. Two-Particle Scattering with Coulomb-like Potentials

Let us assume that both of the particles are charged and that the potential acts between them. We proceed in the same way as we would treat two-particle scattering with a short-range interaction, by separating the center-of-mass motion (see, e.g., Glöckle[34]). One introduces Jacobi coordinates for the center-of-mass motion $\mathbf{R}_{cm}$ and the relative motion $\mathbf{r}_{rel}$. One also defines the corresponding momenta $\mathbf{Q}_{cm}$ and $\mathbf{q}_{rel}$. Then the free Hamiltonian can be written

$$H^0 = H^0_{cm} + H^0_{rel} = \frac{Q^2_{cm}}{2M} + \frac{q^2_{rel}}{2\mu_{rel}} \quad (3.3.4)$$

where M is the total mass and μ_{rel} is the reduced mass. This separates the two-particle scattering problem into two independent one-particle scattering problems: (a) the free motion of the center-of-mass (trivial), and (b) the relative motion, which is a one-particle scattering problem as described above. However, there is one modification: the mass is given by the reduced mass of the original two-particle problem. Writing

$$H_{rel} = H^0_{rel} + V \quad (3.3.5)$$

where V is Coulomb-like, the following Dollard wave operator describes the relative motion:

$$\Omega_{\text{rel}}^{\pm C} = \underset{t \to \mp \infty}{\text{s-lim}} \exp(iH_{\text{rel}}t) \exp(-iH_{\text{rel}}^0 t - iA_{\text{rel}}(t))$$

$$A_{\text{rel}}(t) = \text{sign}(t) e_1 e_2 \sqrt{\frac{\mu_{\text{rel}}}{2H_{\text{rel}}^0}} \log(4|t|H_{\text{rel}}^0) \tag{3.3.6}$$

3.4. Three-Particle Scattering with Coulomb-like Potentials

Now we have to be more careful. Let us assume that some of the three particles are charged. We will list different cases of asymptotic situations: (A) There are asymptotic situations where two of the three particles form a composite particle (bound state). Then one can have the following cases: (A1) The composite particle and the third particle are both neutral, e.g., neutron–hydrogen (H is a neutral (p, e^-) bound state]. (A2) Either the composite particle or the third particle is charged, e.g., neutron–deuteron [d is a charged (n, p) bound state], electron–hydrogen. (A3) Both the composite and the third particle are charged, e.g., proton–deuteron. (B) There are asymptotic situations, where *none* of the three particles forms a composite state. Then the following cases can occur: (B1): Only one particle is charged, e.g., p, n, n. (B2) Two particles are charged, e.g., p, p, n. (B3) All three particles are charged, e.g., p, e^-, e^-.

The effect of screening noted above would correspond, e.g., to a proton–hydrogen asymptotic situation. Now we will classify the above list of asymptotic situations into three categories: Class (C1) — None or at most one of the asymptotically occurring particles (constituent or composite) is charged, e.g., Cases A1, A2, B1. Class (C2) — Two of the asymptotically occurring particles are charged, e.g., Cases A3, B2. Class (C3) — Three of the asymptotically occurring particles are charged, e.g., Case B3. For two-particle (elementary particles) scattering reactions, the asymptotic situation falls either into class C1 or C2. The initial asymptotic situation and the final asymptotic situation both belong to the *same* class. But for three-particle scattering the situation is different. The initial and final asymptotic situation can belong to different classes, e.g., atomic breakup reaction

$$n + H \to n + p + e^- \quad (C1 \to C2) \tag{3.3.7}$$

or the nuclear charge transfer reaction (four-body scattering)

$$p + t \to n + {}^3\text{He(nucleus)} \quad (C2 \to C1) \tag{3.3.8}$$

How do we describe scattering? Firstly, we introduce three-body Jacobi coordinates (see Glöckle[34]), which describe the center-of-mass motion, the relative motion between the center-of-mass of a two-body subsystem and the third particle, and the relative motion in the two-body subsystem. Let us denote the corresponding momenta $\mathbf{Q}_{cm}$, $\mathbf{p}_{rel}$, and $\mathbf{q}_{sub}$. This allows us to split the free Hamiltonian

$$H^0 = H^0_{cm} + H^0_{rel} + H^0_{sub} = \frac{Q^2_{cm}}{2M} + \frac{p^2_{rel}}{2\mu_{rel}} + \frac{q^2_{sub}}{2\mu_{sub}} \qquad (3.3.9)$$

Here μ_{rel} is the reduced mass of the subsystem center-of-mass and the mass of the third particle, μ_{sub} is the reduced mass of the two particles of the subsystem. Again the Jacobi coordinates allow us to separate the (trivial) center-of-mass motion. In the following we assume that this has been done, and we treat only the relative and subsystem motion. Thus in the sequel, we use the notation

$$H \equiv H_{rel} + H_{sub} \qquad (3.3.10)$$

Let us assume a potential of the form

$$V = \sum_{ij} V_{ij} + V_{123} \qquad (3.3.11)$$

i.e., a superposition of two-body pair potentials [(12) and (21) being the same pair] plus eventually a three-body potential. The total Hamiltonian is given by

$$H = H^0 + V \qquad (3.3.12)$$

What are now the wave operators corresponding to classes C1, C2, C3? The answer has been given by Dollard.[45] Before discussing it, however, it is useful to introduce channels. If among the constituent particles of the three-body system some are identical, it is useful to firstly treat all particles as distinguishable, introduce channel numbers, and then compute the symmetric or antisymmetric wave functions by a suitable sum over the channels. Example: Let us consider the p, p, n system and suppose we want to describe a p–d scattering state. We assign the numbers p↔1, p↔2, n↔3. Then there are two asymptotic p–d channels: $\alpha = (1, 23)$ and $\beta = (2, 31)$. Corresponding to a p–p–n scattering state, one would associate the channel $\gamma = (1, 2, 3)$.

3.5. Wave Operator for Class C1

With respect to Coulomb effects, this is evidently the simplest category. There is no Coulomb interaction left over at large distances. Thus the wave operator is a Møller-type wave operator,

$$\Omega_\alpha^\pm = \underset{t \to \mp\infty}{\text{s-lim}} \exp(iHt) \exp(-iH_\alpha^{as} t) P_\alpha^{as} \qquad (3.3.13)$$

Here H_α^{as} denotes the asymptotic Hamiltonian corresponding to channel α. Taking as an example the n, n, p system and considering a n–d scattering state, the channel number α runs over $(1, 23)$ and $(2, 31)$, and

$$H_{(1,23)}^{as} = H^0 + V_{23}, \qquad \text{for } \alpha = (1, 23) \qquad (3.3.14)$$

Here V_{23} denotes the potential in subsystem (23), which is a neutron–proton short-range potential. The operator P_α^{as} denotes a projector on the deuteron bound state in the subsystem (ϕ^d being the deuteron wave function),

$$P_{(1,23)}^{as} = \int d^3 p_{rel} |\mathbf{p}_{rel}, \phi^d\rangle_{(1,23)\ (1,23)}\langle \mathbf{p}_{rel}, \phi^d|, \qquad \text{for } \alpha = (1, 23) \qquad (3.3.15)$$

Corresponding to the asymptotic channel $\alpha = (1, 2, 3)$, one has

$$H_{(1,2,3)}^{as} = H^0, \qquad P_{(1,2,3)}^{as} = 1, \qquad \text{for } \alpha = (1, 2, 3) \qquad (3.3.16)$$

Let us pause for a moment to elaborate on a point mentioned in the introduction: Comparing the stationary integral equation approach (Faddeev-type equations) with the time-dependent approach, it turns out that one has to compute more amplitudes in the stationary approach than in the time-dependent approach. Let us consider an example: Suppose we want to compute p–n–n → p–n–n (which is a possible process, although practically hardly measurable by experiment). Then in the stationary integral equation approach, one has to compute all the transition amplitudes $T_{\alpha\beta}$ for $\alpha \in (1, 23)$, $(2, 31)$, $(3, 12)$, $\beta = (1, 23)$, and finally $T_{\gamma\gamma}$ for $\gamma = (1, 2, 3)$. However, in the time-dependent approach one needs to compute only the S-matrix element $S_{\gamma\gamma}$ for $\gamma = (1, 2, 3)$. This difference is even more pronounced, e.g., in a four-body system.

3.6. Wave Operator for Class C2

With respect to Coulomb effects, this case is analogous to two-particle scattering with a Coulomb-like potential. Let us consider as an example the p, p, n system. The p–d scattering state is obtained via the following Dollard-type wave operator:

$$\Omega_\alpha^\pm = \underset{t \to \mp \infty}{\text{s-lim}} \exp(iHt) \exp[-iH_\alpha^{as}t - iA_\alpha(t)]P_\alpha^{as} \qquad (3.3.17)$$

where for $\alpha = (1, 23)$,

$$H_{(1,23)}^{as} = H^0 + V_{23}$$

$$H^0 = H_{\text{rel}(1,23)}^0 + H_{\text{sub}(23)}^0$$

$$A_{(1,23)}(t) = \text{sign}(t) e_p e_d \sqrt{\frac{\mu_{\text{rel}(1,23)}}{2H_{\text{rel}(1,23)}^0}} \log(4|t|H_{\text{rel}(1,23)}^0) \qquad (3.3.18)$$

One should note that the Dollard anomalous term $A_{(1,23)}(t)$ takes care of the Coulomb "distortion," expressed in terms of the kinetic energy of the relative motion between the proton and the center-of-mass of the charged composite particle, i.e., the deuteron. On the other hand, the Coulomb potential V_{12}^C acts between both protons, but the proton inside the deuteron is not located (classically) at the center-of-mass of the deuteron.

The p–p–n scattering state is generated from the Dollard wave operator given by Eq. (3.3.17), but now for $\alpha = (1, 2, 3)$,

$$H_{(1,2,3)}^{as} = H^0$$

$$A_{(1,2,3)}(t) = \text{sign}(t) e_p e_p \sqrt{\frac{\mu_{\text{sub}(12)}}{2H_{\text{sub}(12)}^0}} \log(4|t|H_{\text{sub}(12)}^0) \qquad (3.3.19)$$

3.7. Wave Operator for Class C3

Now we have three charged constituent particles, where asymptotically none of the constituents forms a bound state. As an example, consider the e, e, p system forming the electron–hydrogen breakup (ionization) state. Mathematically, Dollard's theory[48] allows one to construct wave operators by generalizing the case of two charged particles, to treat three or even N charged particles. For the case of the e, e, p system, the Dollard wave operator for the breakup state is given by

$$\Omega_{(1,2,3)}^\pm = \underset{t \to \mp \infty}{\text{s-lim}} \exp(iHt) \exp[-iH^0 t - iA_{(1,2,3)}(t)]$$

$$A_{(1,2,3)}(t) = \text{sign}(t) \sum_{1 \leq i < j \leq 3} e_i e_j \sqrt{\frac{\mu_{\text{sub}(i,j)}}{2H_{\text{sub}(i,j)}^0}} \log(4|t|H_{\text{sub}(i,j)}^0) \qquad (3.3.20)$$

From the point of view of mathematics, the existence of this wave operator can be rigorously established. From the point of view of physics, we know that the above construction gives the correct physical scattering states when applied to the two-body Coulomb problem. However, N-body stationary Coulomb scattering states are not known, and hence it is an open question if for $N = 3$ Dollard's construction will give the analog of Eq. (3.2.36). It is possible that another phase factor would occur. It should be clear that this remark applies only to the case where asymptotically there are three charged particles, i.e., case C3.

4. HILBERT SPACE APPROXIMATION METHOD

4.1. Construction of Finite-Dimensional Operators

In the above sections we have reviewed the mathematical framework of time-dependent scattering theory. In this section we discuss a systematic approximation scheme that converts the mathematical theory into algorithms for the purpose of doing numerical simulations. We present a Hilbert space method, which relies on the property that any state can be expanded in terms of basis states. Hence all the quantities we compute are matrix elements of Hilbert states. Usually the amplitudes for scattering reactions are expressed in terms of matrix elements between sharp momentum states, which are *not* square integrable wave packets, i.e., are not Hilbert states. In our approach, the asymptotic states are represented by wave packets. However, they can be chosen quite narrow in order to approximate the sharp momentum states (in the sense of distributions). Besides, in a scattering experiment, the asymptotic states are wave packets, i.e., the sharp momentum states are an idealization.

In physics one encounters the necessity to compute long-time evolutions, e.g., when studying quantum systems with a classically chaotic behavior or scattering processes in quantum mechanics. A time scale in a scattering process is set by $t_{scatt} = \hbar/E_{scatt}$, where E_{scatt} is the (conserved) energy of a scattering reaction, i.e., the quantum expectation value of the asymptotic Hamiltonian in the initial asymptotic state. A long-time evolution can be computed in a one-step process, i.e., essentially via diagonalization of the Hamiltonian. This has the following advantages: (a) it is simple, (b) symmetries are conserved, like unitarity and time-reversal symmetry, (c) it allows one to treat Coulomb forces in a straightforward way, and (d) it can be considered as a starting point to develop stochastic methods to compute the time-evolution of a many-body Hamiltonian system. However, the numerical effort for complete diagonalization goes like $O(N^3)$, compared to $O(N^2)$ for an iterative scheme or even $O(N \log N)$ using the fast Fourier transform,[61] where N is the dimension of the system.

Thus we seek approximations of states (asymptotic and scattering) and operators (Hamiltonian, time evolution, wave operator, S-matrix, etc.) in the topology of the Hilbert space. Let us assume that the Hamiltonian H is a self-adjoint operator in the Hilbert space. We start by replacing the Hamiltonian:

$$H \to H(N) = P(N) H P(N) \qquad (3.4.1)$$

where $P(N)$ is the orthogonal projector onto the first N states of a complete orthonormal basis in the Hilbert space. Thus $H(N)$ is a finite-dimensional operator. Moreover, $H(N)$ is self-adjoint, because H is self-adjoint. The Hamiltonian $H(N)$ can be diagonalized,

$$H(N)|\psi_\alpha\rangle = E_\alpha |\psi_\alpha\rangle, \qquad \alpha = 1, 2, \ldots, N \qquad (3.4.2)$$

As H is the generator of displacement in time, one can consider $H(N)$ as a perturbed generator. Thus it is natural, corresponding to the (backward time evolution) operator $U(t) = \exp(iHt)$, to introduce a perturbed (backward time-evolution) operator

$$U(N, t) = \exp[iH(N)t] \qquad (3.4.3)$$

Once the perturbed Hamiltonian is diagonalized, it is easy to compute it:

$$U(N, t) = \sum_{\alpha=1}^{N} |\psi_\alpha\rangle \exp(iE_\alpha t)\langle\psi_\alpha| \qquad (3.4.4)$$

This method of computing the exponential of a matrix, if the exponent is self-adjoint, has been recommended by Moler and van Loan in their review article "Nineteen Dubious Ways to Compute the Exponential of a Matrix."[62] Thus, we have accomplished several things at one time: (a) $U(N, t)$ can be computed exactly (up to numerical errors occurring in the diagonalization) for all values of t. It is a one-time-step calculation. Nearly all the numerical effort goes into the diagonalization of $H(N)$ [Eq. (3.4.2)]. This being done once, it takes a negligible effort to compute $U(N, t)$ [Eq. (3.4.4)] for a variety of values of t. (b) Because of the perturbed generator $H(N)$ is self-adjoint, the perturbed operator $U(N, t)$ belongs to a one-parameter unitary group. In other words, the perturbation conserves the dynamical group properties.

The conservation of the symmetry of the dynamical group is a very important property which deserves some further comment. It implies that the S-matrix operator $S(N, t)$ to be defined below is exactly unitary. Let us give two examples, one from classical mechanics and the other one from quantum mechanics, which underscore the importance of conserving the symmetry of the dynamical group. (a) Example from classical mechanics: Suppose one wishes to compute the classical trajectories of charged particles in an accelerator ring. The particles are forced into orbital trajectories by magnetic fields, which determine the Hamiltonian of the problem. In particular one wishes to compute the trajectory of a particle during many orbits, i.e., one wants to know the time-evolution for a long period. This kind of problem occurs in the design of accelerator machines. In classical mechanics the underlying dynamical symmetry group is the symplectic one. For the purpose of computing the classical trajectories of particles in an accelerator, Dragt and his collaborators[63] have formulated classical dynamics in the language of the Lie algebra of the symplectic group. They have managed to conserve the dynamical group under perturbation of the generator, i.e., the introduction of an approximation to the original Hamiltonian (given by the magnets in the accelerator ring). Thus they have obtained an approximated but exactly symplectic Hamilton flow. This approach has worked very well numerically for calculating many orbits and is used in the design of new machines.[64] (b) Example from quantum mechanics: If one wants to solve the time-dependent Schrödinger equation by discretization in space and time [$\psi_j^n \equiv \psi(x = j\varepsilon, t = n\delta)$], and one does it in a naive way, one obtains a one-step forward Euler scheme,

$$\psi_j^{n+1} = (1 - i\delta H)\psi_j^n + O(\delta^2) \tag{3.4.5}$$

which violates time-reversal symmetry and unitarity and is numerically unstable. On the other hand, if one writes the elementary time step as a Cayley transform, which is unitary, one has

$$(1 + i\delta H/2)\psi_j^{n+1} = (1 - i\delta H/2)\psi_j^n \tag{3.4.6}$$

This is the Crank–Nicholson algorithm, which is unitary and stable (Varga[65]).

In scattering reactions we wish to compute the wave operator and the S-matrix. Then the asymptotic propagator is involved. In most cases, the asymptotic Hamiltonian can be diagonalized and hence the asymptotic propagator can be computed exactly without introducing any finite-dimensional projector. However, as numerical calculations have shown[21], the numerical results for the S-matrix are often better if the same kind of approximation is also applied to the asymptotic generator (allowing for some kind of cancellation of errors). Hence we write in analogy to Eqs.

(3.4.1) and (3.4.3),

$$H^0(N) = P(N) H^0 P(N) \qquad (3.4.7)$$

and

$$U^0(N, t) = \exp[iH^0(N)t] \qquad (3.4.8)$$

For the case of short-range potentials, we can then write the perturbed Møller wave operator

$$\Omega(N, \pm t) = U(N, \mp t)U^0(N, \pm t) \qquad (3.4.9)$$

and the perturbed S-matrix

$$S(N, t) = \Omega(N, -t)^\dagger \Omega(N, t) \qquad (3.4.10)$$

For the case of Coulomb-like potentials, one has to define a perturbed anomalous term. In analogy with Eq. (3.4.7), one defines

$$H^{0C}(N, t) - P(N) H^{0C}(t) P(N) \qquad (3.4.11)$$

where $H^{0C}(t)$ is defined by Eq. (3.2.32). Because $H^{0C}(t)$ is a nonlinear function of H^0, and H^0 is diagonalized by momentum eigenstates, it is natural when constructing finite-dimensional perturbations to work with a basis which diagonalizes H^0, i.e., a basis in momentum space (see Sections 5 and 6). Then the perturbed Dollard-type wave operator is given by

$$\Omega^C(N, \pm t) = U(N, \mp t) \exp[iH^{0C}(N, \pm t)] \qquad (3.4.12)$$

Using the eigen representation of the Hamiltonians $H(N)$ and $H^0(N)$, this reads

$$\Omega^C(N, t) = \sum_{\alpha,\beta=1}^{N} |\psi_\alpha\rangle \exp(-iE_\alpha t)\langle\psi_\alpha|\psi_\beta^0\rangle$$

$$\times \exp\left[iE_\beta^0 t + i \operatorname{sign}(t) e_1 e_2 \sqrt{\frac{m}{2E_\beta^0}} \log(4|t|E_\beta^0)\right]\langle\psi_\beta^0| \qquad (3.4.13)$$

Finally, the perturbed S-matrix is given by

$$S^C(N, t) = \Omega^C(N, -t)^\dagger \Omega^C(N, t) \qquad (3.4.14)$$

In order to give a physical meaning to the perturbed wave operators and S-matrix, one has to address the crucial question of convergence. The unperturbed wave operators and S-matrix are defined as a time limit, i.e., the time does not occur any more. Do the corresponding perturbed operators converge as a function of the time parameter t and the dimension N? A little thought shows immediately that the following time-limits do *not* exist:

$$\lim_{t \to \pm \infty} \Omega^C(N, t) \quad \text{for } N \text{ fixed} \quad (3.4.15)$$

It has neither a strong nor a weak limit, but it behaves like a superposition of a finite number of harmonics with logarithmic phase factors. At a first glance this seems to be a drawback. On the other hand, one can show (for a rigorous proof see Section 5) that the following limit exists:

$$\underset{N \to \infty}{\text{s-lim}}\, \Omega^C(N, t) \quad \text{for } t \text{ fixed} \quad (3.4.16)$$

In principle, we want both t and N to go to infinity. The last two equations indicate that it is not possible to obtain convergence of $\Omega^C(N, t)$ *uniformly* in both N and t. However, there is a path in the plane of parameters N and t where both N and t go to infinity and $\Omega^C(N, t)$ converges to the physically correct limit. This path can be thought of as a valley in the error mountains of the (N, t)-plane. The path means that N is a function of t and vice versa. In Section 5, establishing the mathematical proof of the existence of such a path, basically one will find N as a function of t. In a numerical calculation, one searches on t as a function of N. How to do this in practice is our next question.

4.2. Approximation Parameters

What value should one associate with the time parameter t for a given dimension N? The typical behavior of the imaginary part of the S-matrix as a function of t is shown in Fig. 2 for the case of a short-range potential. For $t = 0$, one has $S(N, 0) = 1$, hence $\text{Im}\langle S(N, 0) \rangle = 0$. For small values of t, $\text{Im}\langle S(N, t) \rangle$ increases linearly in t. For intermediate values of t, one observes a region of stability, where $\text{Im}\langle S(N, t) \rangle$ is close to the reference value of the exact solution. For large values of t, one observes an oscillatory, i.e., a singular behavior. One sees clearly that $\lim_{t \to \infty} \text{Im}\langle S(N, t) \rangle$ does not exist. The picture is qualitatively similar for the Coulomb case, except for the (irrelevant) behavior near the origin. It is important to note that the region of stability in Fig. 2 becomes broader and in this region the error

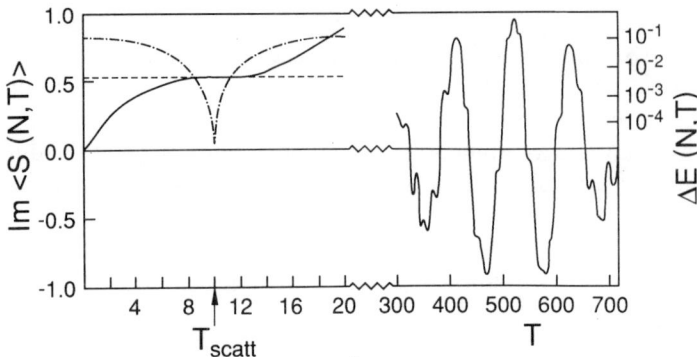

FIGURE 2. Typical behavior of the S-matrix as a function of time T for a short-range potential. Solid line: imaginary part of S-matrix $\text{Im}\langle S(N, T)\rangle$, dashed line: imaginary part of exact S-matrix $\text{Im}\langle S\rangle$, dashed-dotted line: violation of energy conservation $\Delta E(N, T)$.

decreases when N is increased. How do we then choose the time parameter, in the following called T_{scatt}, from the region of stability? The natural definition would be: T_{scatt} is the value t, where the relative error of the S-matrix,

$$\Delta S^C(N, t) = \left| \frac{\langle \phi^{\text{fi}}|S^C(N, t) - S^C_{\text{ref}}|\phi^{\text{in}}\rangle}{\langle \phi^{\text{fi}}|S^C_{\text{ref}}|\phi^{\text{in}}\rangle} \right| \quad (3.4.17)$$

takes a minimum as a function of t. Here, S^C_{ref} is a reference value of the S-matrix against which $S^C(N, t)$ is measured. Note that this value depends on the dimension N. However, when the reference value S^C_{ref} of the S-matrix is unknown, i.e., in most of the cases of interest, then this definition is useless. For those cases we have to have a working definition. One could define T_{scatt} such that the variation of $\text{Re}\langle S^C(N, t)\rangle$ and $\text{Im}\langle S^C(N, t)\rangle$ becomes minimal. A better definition, which we will adopt, is based on conserved observables. For instance, the energy is conserved, i.e., the expectation value of the asymptotic Hamiltonian in the asymptotic state is identical to the expectation value of the full Hamiltonian in the scattering state. However, when going over to the finite-dimensional perturbed Hamiltonian, one can introduce a function corresponding to $\Delta S^C(N, t)$ that measures the error in the violation of energy conservation in the finite-dimensional system:

$$\Delta E(N, t) = \left| \frac{\langle \Omega^C(N, t)\phi^{\text{as}}|H(N)|\Omega^C(N, t)\phi^{\text{as}}\rangle - \langle \phi^{\text{as}}|H^0(N)|\phi^{\text{as}}\rangle}{\langle \phi^{\text{as}}|H^0(N)|\phi^{\text{as}}\rangle} \right| \quad (3.4.18)$$

We define T_{scatt} as the value of t, where $\Delta E(N, t)$ has its first minimum as a function of t (see Fig. 2). Firstly, this definition does not require knowledge of the exact solution. Secondly, Fig. 2 shows that the domains where $\Delta S^C(N, t)$ and $\Delta E(N, t)$ become small coincide to a large extent. One should note that we are only interested in the first minimum for each of those functions. The subsequent minima are caused by the periodicity of the approximate solutions and do not correspond to physics. The property that the error in the S-matrix has a minimum in the same domain as the error of the violation of energy has been investigated and confirmed for a large number of physical models.[10]

We note that T_{scatt}, called the lattice scattering time, depends on the physical model, on the parameters of the finite-dimensional approximation, and on the physical boundary conditions, i.e., on energy, quantum numbers, etc. In scattering calculations, T_{scatt} is generally found to be a long time,

$$\langle E \rangle T_{\text{scatt}}/\hbar \gg 1 \quad (3.4.19)$$

which clearly shows that it is prohibitive to apply, e.g., a Taylor expansion in t in order to compute the propagator $U(t)$. When going toward the continuum limit ($N \to \infty$), then T_{scatt} also goes to infinity.

In stationary scattering theory the wave operator, given by Eq. (3.2.19), is less singular than either of the Green's functions $G(E + i0)$ and $G^0(E + i0)$. Analogously in the time-dependent framework, the wave operator is less singular than either one of the time-evolution operators $U(t)$ or $U^0(t)$. [$A(t)$ is said to be more singular than $B(t)$ when $A(t)$ oscillates more rapidly in time than $B(t)$.] This property of $\Omega(N, t)$ compared to $U(N, t)$ and $U^0(N, t)$ can be nicely demonstrated using the energy eigen representation of the Hamiltonians (let us consider the case of a short-range interaction; for a Coulomb-type interaction the behavior is qualitatively the same). We consider a typical scattering process (low-energy nucleon–nucleon scattering), characterized by $\langle E \rangle = 1.5\,\text{fm}^{-1}$ and $T_{\text{scatt}} = 2.7\,\text{fm}$, such that the action $\langle E \rangle T_{\text{scatt}} = 4.05$ [compare Eq. (3.4.19)]. A typical numerical example is shown in Table I. One observes the following behavior of the eigenvalues

TABLE I
Difference of Eigenvalues [Eq. (3.4.20)], and of Eigenvectors [Eq. (3.4.21)] between the Full and the Free Hamiltonian ($N = 10$).

α	1	2	3	4	5	6	7	8	9	10
E_α	0.084	0.501	1.36	2.64	4.33	6.44	8.96	11.9	15.3	19.1
ΔE_α	0.0419	0.0570	0.0352	0.0219	0.0146	0.0103	0.00761	0.00583	0.00460	0.00372
$\Delta \psi_\alpha$	0.120	0.128	0.649	0.322	0.174	0.103	0.0654	0.0437	0.0302	0.0204

and eigenvectors at the upper end of the spectrum, i.e., $\alpha, \beta \simeq N$,

$$\Delta E_\alpha = E_\alpha - E_\alpha^0 \xrightarrow[\alpha \to N]{} 0 \qquad (3.4.20)$$

$$\Delta \psi_\alpha = \|\psi_\alpha - \psi_\alpha^0\| \xrightarrow[\alpha \to N]{} 0 \qquad (3.4.21)$$

Consequently, one has

$$\langle \psi_\alpha | \psi_\beta^0 \rangle \xrightarrow[\alpha, \beta \to N]{} \delta_{\alpha, \beta}$$

$$\exp(i \Delta E_\alpha T_{\text{scatt}}) \xrightarrow[\alpha \to N]{} 0 \qquad (3.4.22)$$

Let us consider $\Omega(N, T_{\text{scatt}})$,

$$\Omega(N, T_{\text{scatt}}) = \sum_{\alpha, \beta = 1}^{N} |\psi_\alpha\rangle \exp(-iE_\alpha T_{\text{scatt}}) \langle \psi_\alpha | \psi_\beta^0 \rangle \exp(iE_\beta^0 T_{\text{scatt}}) \langle \psi_\beta^0 | \quad (3.4.23)$$

Equations (3.4.20) and (3.4.21) imply that the contributions coming from the part of the spectrum corresponding to the large eigenvalues asymptotically approach the unit operator. Consequently, (a) the time dependence of the wave operator is under control and one is far from a singular, oscillatory behavior; and (b) the S-matrix for large energy values approaches unity and the phase shifts go to zero.

5. STRONG RESOLVENT CONVERGENCE

In this section we present some mathematically rigorous statements on the convergence properties of the perturbed time-evolution operator, wave operators, and S-matrix. We will see that the time-evolution operator, the Dollard wave operators, the scattering states, and the S-matrix in the finite-dimensional approximation as defined in Section 4 converge toward the exact time evolution, wave operators, scattering states, and S-matrix, respectively. The proof is based on the notion of strong resolvent convergence. In the following we outline some ideas of the proof. Let us consider two-body scattering (after splitting off the center-of-mass motion) in a Coulomb-like potential.

Let us start with some definitions. For proper definitions of mathematical terms in this section see Weidmann.[38] Let $\mathcal{H} = \mathcal{L}^2(\mathcal{R}^3)$ denote the Hilbert space of square integrable functions (more precisely the vector space of equivalence classes of Lebesgue measurable complex valued functions defined on $\mathcal{R}^3$ which are absolutely square integrable]. Let $H = H^0 + V^S + V^C$ denote the total Hamilton operator, H^0 the free Hamiltonian, V^S a short-range potential, and V^C the Coulomb potential. For the purposes of our proof we need to work with self-adjoint Hamiltonians. The operator H^0 is defined by

$$(H^0 f)(\mathbf{k}) = \frac{\mathbf{k}^2}{2m} f(\mathbf{k}), \qquad f \in D(H^0) \tag{3.5.1}$$

where the domain of H^0 is

$$D(H^0) = \left[f \in L^2(\mathcal{R}^3) \,\Big|\, \int d^3k |\mathbf{k}^2 f(\mathbf{k})|^2 < \infty \right] \tag{3.5.2}$$

Then $D(H^0)$ is dense in $\mathcal{H}$. H^0 is self-adjoint on $D(H^0)$ as a consequence of the following proposition[36]:

PROPOSITION 1 (AMREIN–JAUCH–SINHA): *Let Δ be a measurable set in $\mathcal{R}^n$ and ψ a real valued measurable function defined on Δ which is finite almost everywhere (with respect to the Lebesgue measure). Define an operator A in $L^2(\Delta)$ by*

$$(Af)(\mathbf{k}) = \psi(\mathbf{k}) f(\mathbf{k}) \quad \text{for } f \in D(A) \tag{3.5.3}$$

where

$$D(A) = [f \in L^2(\Delta) \,|\, \psi(\mathbf{k}) f(\mathbf{k}) \in L^2(\Delta)] \tag{3.5.4}$$

Then A is self-adjoint.

Also, we want H to be self-adjoint. There are quite general theorems to guarantee this, e.g., the theorem by Rellich–Kato.[36]

THEOREM 1 (RELLICH–KATO): *Let A be a self-adjoint operator and B a symmetric operator. If B is A-bounded with A-bound less than 1, then $A + B$ is self-adjoint with $D(A + B) = D(A)$. If B is A-compact, then $A + B$ is self-adjoint and is bounded below when A is so bounded.*

Here A and B are linear operators in $\mathcal{H}$. B is A-bounded means $D(A) \subseteq D(B)$ and that there are $\alpha, \beta \geq 0$ such that for all $f \in D(A)$

$$\|Bf\| \leq \alpha \|f\| + \beta \|Af\| \tag{3.5.5}$$

The smallest β is called the A-bound. B is A-compact means that $D(A) \subseteq D(B)$ and $B(z - A)^{-1}$ is a compact operator for some $z \in \rho(A)$ (resolvent set of A).

Based on this is the following proposition[36]:

PROPOSITION 2 (AMREIN–JAUCH–SINHA): *Let H^0 be the free Hamiltonian defined by Eq. (3.5.1) and $V(\mathbf{x})$ be a real local potential. (a) If $V \in L^2(\mathcal{R}^3)$ then V is H^0-compact and $H^0 + V$ is self-adjoint with $D(H^0 + V) = D(H^0)$. (b) If $V \in L^\infty(\mathcal{R}^3)$ (space of the essentially bounded measurable functions), then V is H^0-bounded with H^0-bound 0 and $H^0 + V$ is self-adjoint. If, moreover, $V(\mathbf{x})$ converges to 0 as $|\mathbf{x}| \to 0$, then V is H^0-compact.*

One should note that H^0-compactness implies H^0-boundedness. As a consequence of this proposition, one has: (a) A short-range Yukawa potential, given in coordinate space by $V^Y(\mathbf{x}) = \lambda \exp[-\mu|\mathbf{x}|]/|\mathbf{x}|$, belongs to $L^2(\mathcal{R}^3)$ and is H^0-compact. It yields a Hamiltonian $H = H^0 + V^Y$, which is self-adjoint, with $D(H) = D(H^0)$. (b) The Coulomb potential V^C is H^0-compact. It yields a Hamiltonian $H = H^0 + V^C$, which is self-adjoint, with $D(H) = D(H^0)$. (c) The Coulomb-like potential $V = V^Y + V^C$ is H^0-compact and yields a self-adjoint Hamiltonian $H = H^0 + V^Y + V^C$, with $D(H) = D(H^0)$. In the remainder of this section we assume that the total Hamiltonian H is self-adjoint on $D(H) = D(H^0)$.

As indicated in the above section, we introduce a finite-dimensional approximation (perturbation) of the Hamiltonian, via orthogonal projection [Eq. (3.4.1)] onto a basis. As an example of such a basis, one could take the Hermite functions, although this choice is not optimal with respect to numerical simulations. A more suitable basis will be discussed below. Thus we construct a finite-dimensional Hamiltonian $H(N)$, which is also self-adjoint. We hope that $H(N)$ will approximate H in some sense. Keeping in mind our goal to approximate the wave operators, we firstly aim to approximate the time-evolution operator. Thus we pose our next question: What kind of approximation of the Hamiltonian H by $H(N)$ will guarantee that

$$\operatorname*{s-lim}_{N \to \infty} \exp[iH(N)t] = \exp(iHt) \quad (3.5.6)$$

holds? We cannot give a necessary condition, but state a sufficient condition, following Weidmann.[38]

PROPOSITION 3 (WEIDMANN): *Let H and $H(N)$ be self-adjoint and let $H(N)$ converge toward H in the sense of strong resolvent convergence. Then one has*

$$\operatorname*{s-lim}_{N \to \infty} F[H(N)] = F(H) \quad (3.5.7)$$

for every continuous bounded function F defined on $\mathscr{R}$, in particular for $F(x) = \exp(ixt)$.

Here $H(N)$ is said to converge toward H in the sense of strong resolvent convergence, if for some complex number z with $\mathrm{Im}(z) \neq 0$ the following limit holds:

$$\operatorname*{s\text{-}lim}_{N \to \infty} [z - H(N)]^{-1} = (z - H)^{-1} \tag{3.5.8}$$

In practice, this condition is not very useful because it is often complicated to compute the resolvent $(z - H)^{-1}$ of an operator H. Thus it would be desirable to have a sufficient condition for strong resolvent convergence, but expressed in terms of H and $H(N)$. Such a condition is formulated in the next proposition.[38]

PROPOSITION 4 (WEIDMANN): *Assume that there is a core $C(H)$ of H. For every $f \in C(H)$ let $f \in D[H(N)]$ and*

$$H(N)f \xrightarrow[N \to \infty]{} Hf \tag{3.5.9}$$

Then $H(N)$ converges toward H in the sense of strong resolvent convergence.

A core $C(H)$ is a subset of the domain $D(H)$, with the property

$$\overline{H|_c} = H \tag{3.5.10}$$

where $\bar{X}$ denotes a closure of an operator X.

We want to point out that all the mathematical subtleties involved are necessary only if H is an unbounded operator, as it is the case for the nonrelativistic description of the nucleon–nucleon system. If H were bounded the matter would be trivial. However, the statement of Proposition 3 means that a bounded function of an unbounded operator can be as nicely approximated as if the operator itself were bounded.

The question arises: For what kind of finite-dimensional approximation will Proposition 4 be applicable? In Kröger[66] the following momentum-space basis functions have been considered:

$$e_I(\mathbf{q}) = h_n(q) Y_{lm}(\hat{q}) \tag{3.5.11}$$

where I denotes the linearized multi-index (n, l, m), h_n are step-functions corresponding to a partition $0 = q_0 < q_1 < \cdots < q_{n_{\max}} = \Lambda$ of the interval $(0, \Lambda)$, and the index of the spherical harmonics runs over $0 \leq l \leq l_{\max}$ and $-l \leq m \leq +l$. Let $P(N)$ denote the orthogonal projector onto the space spanned by those basis functions. Let us suppose the basis is such that $\Lambda \to \infty$, $\Delta q \to 0$, $l_{max} \to \infty$, when $N \to \infty$. Then this basis is complete in $L^2(\mathcal{R}^3)$, i.e.,

$$\underset{N \to \infty}{\text{s-lim}} P(N) = 1 \tag{3.5.12}$$

Then one can show[66] using this basis that for every $f \in D(H^0)$

$$H^0(N)f \xrightarrow[N \to \infty]{} H^0 f \tag{3.5.13}$$

Because H^0 is self-adjoint and $D(H^0)$ is dense in the Hilbert space, the domain $D(H^0)$ is also a core, $C(H^0) = D(H^0)$. Thus the assumptions of Proposition 4 are satisfied, and hence Proposition 3 implies

$$\underset{N \to \infty}{\text{s-lim}} U^0(N, t) = U^0(t), \quad \text{for every } t \in \mathcal{R} \tag{3.5.14}$$

Next we consider (for $t \neq 0$) Dollard's modified asymptotic Hamiltonian $H^{oc}(t)$, given by Eq. (3.2.32). Its domain is

$$D[H^{oc}(t)]$$
$$= \left\{ f \in L^2(\mathcal{R}^3) \Big| \int d^3k \left\| \left[\frac{\mathbf{k}^2}{2m} t + \text{sign}(t) \frac{m e_1 e_2}{|\mathbf{k}|} \log\left(\frac{2\mathbf{k}^2 |t|}{m}\right) \right] f(\mathbf{k}) \right\|^2 < \infty \right\} \tag{3.5.15}$$

As $D(H^0)$, $D[H^{oc}(t)]$ is also dense in $L^2(\mathcal{R}^3)$. Proposition 1 implies that $H^{oc}(t)$ is self-adjoint on $D[H^{oc}(t)]$. In analogy to Eq. (3.5.13), one can show[66] using the basis defined in Eq. (3.5.11) that for every $f \in D[H^{oc}(t)]$,

$$H^{oc}(N, t)f \xrightarrow[N \to \infty]{} H^{oc}(t)f \tag{3.5.16}$$

Again $C[H^{oc}(t)] = D[H^{oc}(t)]$ is a core and Propositions 3 and 4 imply

$$\underset{N \to \infty}{\text{s-lim}} \exp[iH^{oc}(N, t)] = \exp[iH^{oc}(t)] \quad \text{for every } t \in \mathcal{R}, t \neq 0 \tag{3.5.17}$$

Finally, we want to establish that the analog of Eq. (3.5.14) also holds for the time-evolution of the total Hamiltonian. Then we have to make some assumptions about the potential. In Proposition 2 we have made some assumptions in order that $H = H^0 + V$ be self-adjoint. In all cases, V was H^0-bounded. Let us consider here two kinds of assumptions: (a) V is bounded, (b) V is H^0-bounded (for the definition of H^0-boundedness see Theorem 1 above). Boundedness is stronger than H^0-boundedness, i.e., (a) implies (b). Let us define $V(N) = P(N)VP(N)$. In case (a) one can show[66] using the expansion functions defined in Eq. (3.5.11) that

$$\operatorname*{s-lim}_{N\to\infty} V(N) = V \qquad (3.5.18)$$

In case (b) one can show for every $f \in D(H^0)$[66] that

$$V(N)f \xrightarrow[N\to\infty]{} Vf \qquad (3.5.19)$$

By Proposition 2 and following, we have $D(H) = D(H^0)$. Then Eqs. (3.5.13) and (3.5.19) imply for all $f \in D(H)$ that

$$H(N)f \xrightarrow[N\to\infty]{} Hf \qquad (3.5.20)$$

Again, $D(H)$ is a core and Propositions 3 and 4 imply

$$\operatorname*{s-lim}_{N\to\infty} U(N, t) = U(t) \qquad (3.5.21)$$

This establishes strong convergence of the perturbed time-evolution operators generated by the finite-dimensional total Hamiltonian toward the time-evolution operator generated by the original total Hamiltonian.

We can ask now: What does this imply about the convergence of the approximate Dollard wave operators, defined by Eq. (3.4.12)? Let us introduce the notation

$$\Omega^C(\pm t) = \exp(\mp iHt)\exp[iH^{0C}(\mp t)], \quad t \neq 0 \qquad (3.5.22)$$

We have established above the strong convergence of the perturbed time-evolution of the total Hamiltonian [Eq. (3.5.21)], and the strong convergence of the perturbed Dollard modified asymptotic time evolution [Eq. (3.5.17)]. This, together with the property that all time-evolution operators are bounded operators on Hilbert space, implies strong convergence of the product,[38] i,.e.,

$$\operatorname*{s-lim}_{N\to\infty} \Omega^C(N, \pm t) = \Omega^C(\pm t), \quad t \neq 0 \qquad (3.5.23)$$

In order to guarantee the existence of the Dollard wave operator (i.e., the time limit on the r.h.s.), one has to make some assumptions about the potential; e.g., the following condition can be imposed: V^S is a local potential which is square integrable. Then the Dollard wave operator for the Coulomb-like potential $V^S + V^C$ exists[45]:

$$\operatorname*{s-lim}_{t \to \pm\infty} \Omega^C(t) = \Omega^{C\pm} \qquad (3.5.24)$$

Equations (3.5.23) and (3.5.24) imply strong convergence of the perturbed Dollard wave operator in N and t, but not a uniform convergence. This result is formulated in Theorem 2.

Finally we consider the convergence of the S-matrix. Let us introduce the notation

$$S^C(t) = \Omega^C(-t)^\dagger \Omega^C(t) = \exp[iH^{0C}(t)]\exp(-i2Ht)\exp[iH^{0C}(t)] \qquad (3.5.25)$$

Strong convergence of $\Omega^C(t)$ in the time parameter [Eq. (3.5.24)] and the fact that $\Omega^C(t)$ are defined in the whole Hilbert space and obey the bound $\|\Omega^C(t)\| \leq 1$ imply at least weak convergence of its adjoint in the time parameter, and hence,

$$\operatorname*{w-lim}_{t \to \infty} S^C(t) = S^C \qquad (3.5.26)$$

Together with the strong convergence of the perturbed wave operators [Eq. (3.5.23)], this implies weak convergence of the perturbed S-matrix, given by Eq. (3.4.14). The results are summarized in the following theorem:

THEOREM 2: *Let $V = V^S + V^C$, where V^S is a short-range potential, such that $H = H^0 + V$ is self-adjoint with $D(H^0) = D(H)$. Let V^S be bounded or relatively H^0-bounded. V^C is the Coulomb potential. We assume that V^S is such that Dollard's modified wave operators $\Omega^{C(\pm)}$ exist. Let $P(N)$ denote the orthogonal projector corresponding to the set of expansion functions, given by Eq. (3.5.11). Then, for each $\varepsilon > 0$ and each $f, g \in \mathcal{H}$, one can find a $T > 0$ and an integer N_0, such that for every $N > N_0$*

$$\|\Omega^{C(\pm)}f - \Omega^C(N, \pm T)f\| < \varepsilon \qquad (3.5.27)$$

and

$$|\langle f|S^C|g\rangle - \langle f|S^C(N, T)|g\rangle| < \varepsilon \qquad (3.5.28)$$

A generalization of the proof from the two-body to an N-body system has been discussed in Kröger.[66]

6. NUMERICAL RESULTS

6.1. Proton–Proton Scattering

In order to test the method, let us consider the proton–proton system, and in particular compute singlet s-wave scattering. The p–p interaction can be described phenomenologically by a nucleon–nucleon short-range potential plus the Coulomb potential. For the N–N potential we have chosen a so-called realistic interaction, namely the Paris potential, which is derived from a field theoretical model of mesons and excited nucleons. In particular we have used a separable representation of this potential, i.e., the Graz potential (Schweiger[67]) (a potential is called separable if it has the form $V = \Sigma_{i,j=1}^{I} |\chi_i > \lambda_{ij} < \chi_j|$). In the singlet s-wave channel, the Graz potential is given by a rank-two separable potential. Its parameters are given in Schweiger.[67] The usefulness of a separable potential is that it allows one to obtain an analytical solution of the T-matrix and S-matrix. This is also true for a separable-plus-Coulomb potential.[59] The analytical solution for the Graz-plus-Coulomb potential has been used in the following as a reference solution.

Because ours is a Hilbert space method, we build asymptotic states from wave packets. We have chosen the following wave packet (to be normalized to unity),

$$\langle \mathbf{q}|\phi\rangle = \begin{cases} 1 - \cos\left[2\pi \dfrac{q_{up} - q}{q_{up} - q_{low}}\right], & \text{if } q_{low} \leq q \leq q_{up} \\ \text{and } 0 \text{ elsewhere} \end{cases} \quad (3.6.1)$$

This is a bell-shaped wave packet, which is symmetrical with respect to the peak at $q_M = \frac{1}{2}(q_{up} + q_{low})$, having a half-width $q_W = \frac{1}{2}(q_{up} - q_{low})$. In the calculations presented below we have used a wave packet with $q_M = 0.2 \text{ fm}^{-1}$ and $q_W = 0.02 \text{ fm}^{-1}$ ($\hbar = c = 1$). This is denoted as a standard wave packet [SWP]. Its width is narrow enough so that the relative deviation between the S-matrix element $\langle\phi^{SWP}|S|\phi^{SWP}\rangle$ and the corresponding sharp momentum S-matrix element is smaller than 10^{-3}. As basis functions we have worked with step functions in momentum space, as described in Eq. (3.5.11). The step functions are characterized by choosing a cut-off Λ and a particular partition of the interval $(0, \Lambda)$.

The following numerical results are taken from Batinić.[21] For the pure Coulomb potential, the Graz N–N potential and the Graz-plus-Coulomb p–p potential, the relative error of the S-matrix element, denoted by $\Delta S(N, T)$, [Eq. (3.4.17)], is presented in Fig. 3. One observes a wide shallow valley around the minimum for the Graz potential, and a relatively shallow valley around the first minimum for the Coulomb potential as well as for

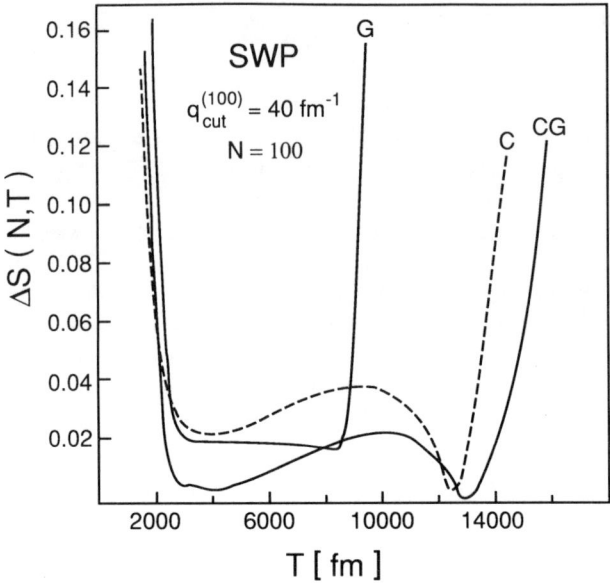

FIGURE 3. Proton–proton scattering. Relative error $\Delta S(N, T)$ of the S-matrix compared to the exact value (computed analytically) for the pure Coulomb potential (C), Graz potential (G), and Graz-plus-Coulomb potential (GC) as a function of the time parameter T. Number of nodal points (dimension of basis) $N = 100$.

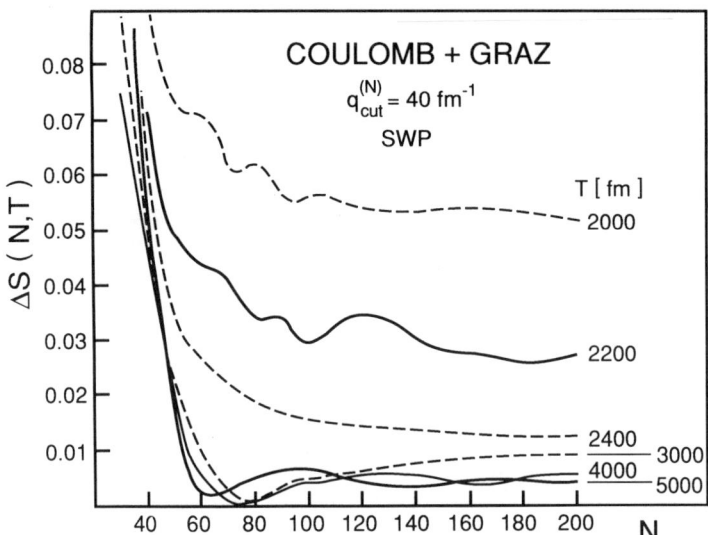

FIGURE 4. Proton–proton scattering. Graz-plus-Coulomb potential. Relative error $\Delta S(N, T)$ as a function of nodal points N and the time T. Convergence for large N and T is clearly seen.

the Graz-plus-Coulomb potential. For the Graz-plus-Coulomb potential, $\Delta S(N, T)$ is displayed in Figs. 4 and 5 as a function of N and T. Figure 5 shows that there is a well-defined first minimum as a function of T and that the valley around the minimum becomes wider as N increases. As discussed above, in cases where the reference S-matrix is not known, we need an auxiliary function to determine the scattering time parameter T_{scatt}. As suggested above, we use the minimum in the violation of energy conservation, $\Delta E(N, T)$ [Eq. (3.4.18)]. Figure 6 shows that the minimum of $\Delta E(N, T)$ and the minimum of $\Delta S(N, T)$ lie close together, thus justifying the definition of T_{scatt}. Comparison of Fig. 6 with Fig. 7 shows that not only is there a region where the minima of $\Delta E(N, T)$ and $\Delta S(N, T)$ nearly coincide, but also that in this region the S-matrix element $\langle S(N, T) \rangle$ is stable as a function of T. Finally, in Fig. 8, the cut-off dependence is studied. It shows for this particular case that $\Lambda = 20\,\text{fm}^{-1}$ is sufficiently large. The above results have been obtained with an equidistant distribution of nodal points in the partition of $(0, \Lambda)$. With this kind of basis it is quite straightforward also to accommodate a nonequidistant distribution. Some possibilities have been explored by Batinić.[21] One finds that in order to find numerically converged results it is necessary to distribute enough nodal points in the domain of the wave packet, i.e., where the wave packet is nonzero.

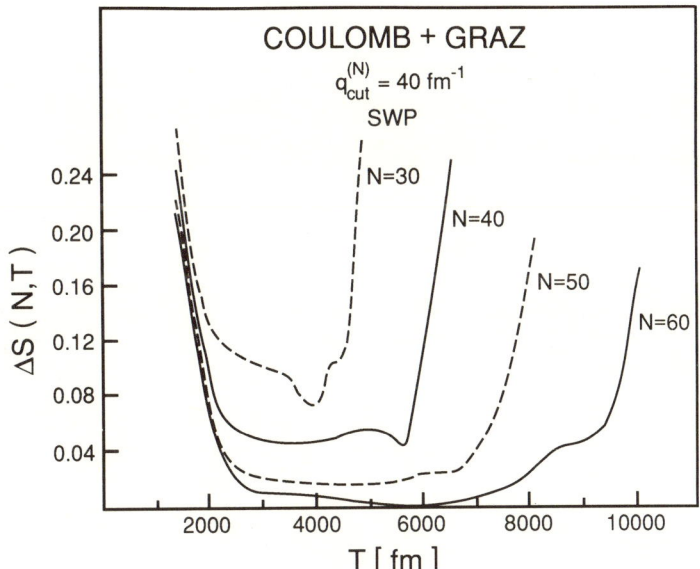

FIGURE 5. Proton–proton scattering. Graz-plus-Coulomb potential. $\Delta S(N, T)$ as a function of time T and nodal points N. By increasing N, the depth and the width of the valley of stability is increased.

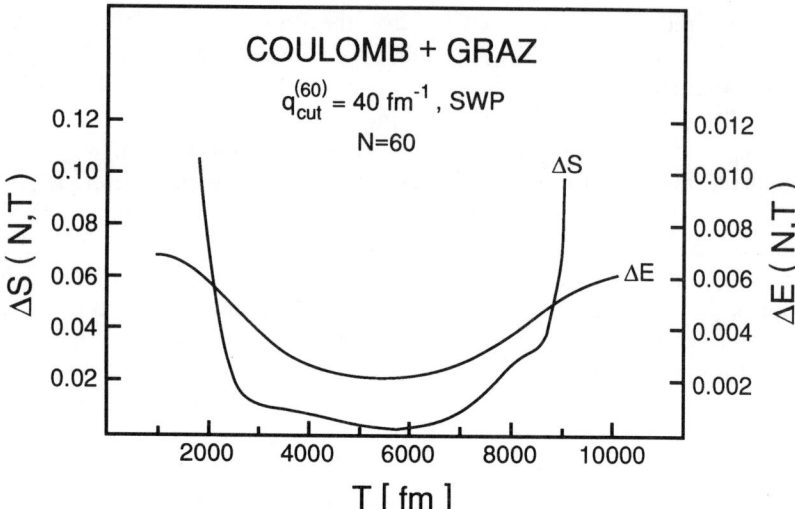

FIGURE 6. Proton–proton scattering. Graz-plus-Coulomb potential. $\Delta S(N, T)$ and $\Delta E(N, T)$ (violation of conservation of energy) as a function of T. Number of nodal points $N = 60$. The position of the minima of $\Delta S(N, T)$ and $\Delta E(N, T)$ nearly coincide.

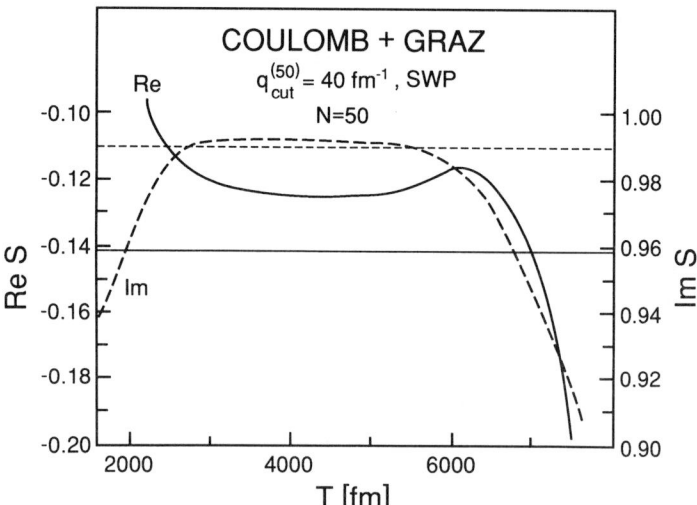

FIGURE 7. Proton–proton scattering. Graz-plus-Coulomb potential. Real part (solid line) and imaginary part (dashed line) of $\langle S(N, T) \rangle$ as a function of T. $N = 50$. The straight lines give the corresponding reference values.

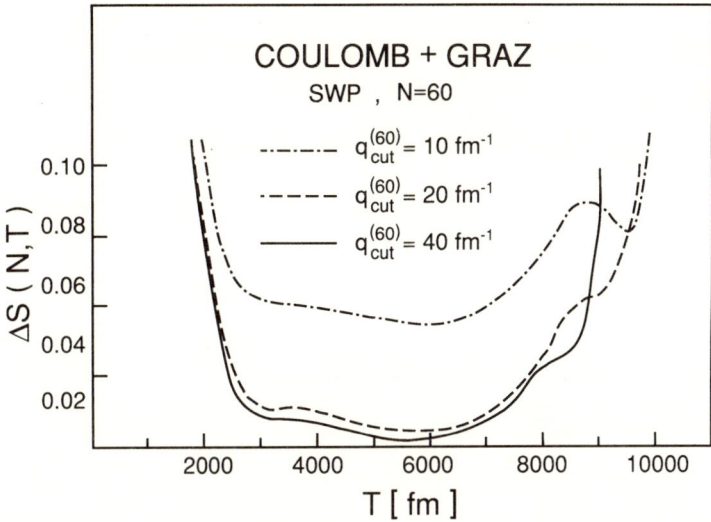

FIGURE 8. Proton–proton scattering. Graz-plus-Coulomb potential. $\Delta S(N, T)$ as a function of the cut-off $\Lambda = q_{cut}$ and time T. $N = 60$.

From the numerical point of view, in the p–p scattering calculation most of the effort is spent in computing $\exp[iH(N)t]$ via diagonalization of $H(N)$. There are other alternatives for computing $\exp[iH(N)t]$ (for a review se Kröger[10]). Let us discuss briefly one particular method, which has been also recommended in Moler and van Loan's review on the computation of exponentials of matrices.[62] As noted above, the problem is to compute the exponential for large values of the time parameter, i.e., $\langle E \rangle T/\hbar \gg 1$. This can be done as follows: One computes $\exp[iH(N)\tau]$ for a small value of τ, using the Taylor expansion or the Padé approximation.[62] Then one generates $\exp[iH(N)T]$ for $T \gg \tau$ by iterative squaring of $\exp[iH(N)\tau]$. This is done by using the scaling property of the exponential function,

$$\exp[iH(N)T] = \left\{ \exp\left[\frac{iH(N)T}{m} \right] \right\}^m \qquad (3.6.2)$$

and choosing $m = 2^{n_{sq}}$, such that $\tau = T/m$ is small. In the case of p–p scattering, a sufficient number of squaring operations n_{sq} was found to be of the order of 10. This method has been tested by Girard et al.[68] for N–N scattering with a short-range Yamaguchi potential and by Batinić[69] for the Graz-plus-Coulomb potential. In Fig. 9 some results from the iteration method are shown for p–p scattering with the Graz-plus-Coulomb potential. Displayed are the relative deviations between the diagonalization

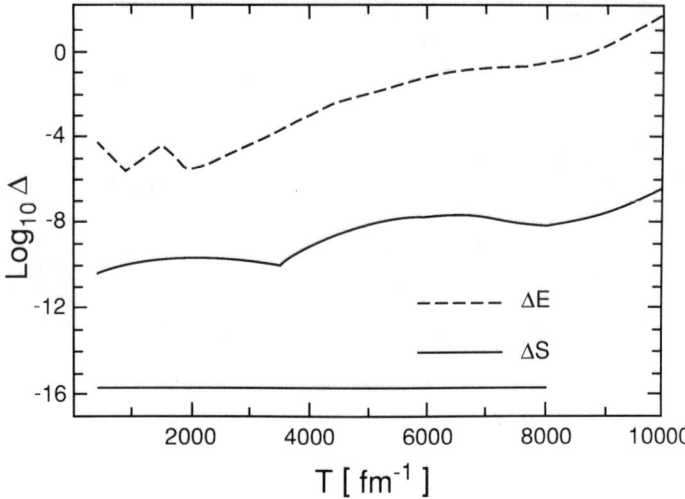

FIGURE 9. Proton–proton scattering. Graz-plus-Coulomb potential. Comparison between the diagonalization method and the iterative squaring method. The results from the diagonalization method are taken as a reference. Relative deviation for the S-matrix $\langle S(N, T)\rangle$ and for the expectation value of the energy $\langle H(N)\rangle$ as a function of T. The results from iterative squaring correspond to 7 terms in the Taylor series and 18 squaring operations.

method and the iteration method in the expectation values of the S-matrix and the energy, respectively. The relative deviation in the S-matrix is found to be of the order of 10^{-10}. Comparing this result with the relative error in the S-matrix between the diagonalization method and the exact value, Figs. 3 to 6, which is of the order of 10^{-2} to 10^{-4}, one concludes that the iteration method is equivalent to the diagonalization method with respect to numerical accuracy.

6.2. Proton-Deuteron Breakup Scattering

When studying the three-nucleon system, one of the most important open questions concerns the role of three-nucleon (NNN) forces.[70] There seems to be a sizable effect for the 3N bound state, i.e., for ^{3}He and ^{3}H.[73,74] For ^{3}H the Argonne V_{14} nucleon–nucleon potential yields 7.68 MeV binding energy, while inclusion of the Tucson 3N potential yields 8.42 MeV, while the 3N potential from the Brazilian group yields 8.44 MeV. On the other hand, various NN potentials, such as Reid soft core, Argonne, Paris, and Nijmegen yield between 7.36 and 7.62 MeV. However, according to the Bonn potential,[71,72] which yields 8.29 MeV, there is little space left for a 3N force. One should be careful, however, to disentangle 3N forces from effects hidden in the NN potential. For example, the Δ excitation of the nucleon in the medium gives rise to a three-body force due to π and ρ exchange with different nucleons. On the other hand, for elastic electron scattering form factors of ^{3}He and ^{3}H no substantial sensitivity to the 3N force has been found. For ^{3}He and ^{3}H charge, form factors and charge radii have been determined based on the new Bonn potential without 3N forces.[75] One finds r_c(He) = 1.89 fm and r_c(^{3}H) = 1.72 fm, in very good agreement with experimental data.

Slaus et al.[76] explain the difference in the scattering length a_{nn} extracted from the reaction ^{2}H(p, n)2n and from ^{2}H(π^-, γ)2n in terms of a 3N force. On the other hand, Faddeev-type calculations of the n–d breakup process at incident neutron energies of 7.5, 14.4, and 18.5 MeV which include a 3N force gives effects of maximal 10% in the final-state region.[77] A maximal effect of 2–3% is estimated from the p–d breakup reaction at E_p = 14.1 MeV in Karus et al.[78] An investigation to estimate the possible effect of a 3N force has been carried out by Witala et al.,[79] who did a Faddeev calculation based on the Paris potential and the new Bonn potential to search for 3N effects in the ^{2}H(p, pp)n reaction at 14.1 MeV but without treating Coulomb effects. They looked into final-state interaction configurations and collinear configurations, and found a significant discrepancy between experiment and calculation for the cross section of collinear configurations which is a possible sign of a 3N force. Also there is a big difference in the final-state region as well as for collinear configurations for the analyzing power.

In order to estimate the effect of a 3N force, it is very important to incorporate rigorously the Coulomb force, in particular for the charged p–d reaction at low energy. Nachabe[80] reported kinematically complete measurements of the d–p breakup reaction close to the breakup threshold, and a very crude bound of 30% for 3N effects has been estimated. However, in the low-energy scattering region Coulomb effects are important.[81] While at higher energies, the Coulomb effects can be taken into account by some corrections to the neutral amplitudes,[82] at very low energies the Coulomb force has to be taken into account exactly.

This is possible in the time-dependent framework. Below we will present results from such a calculation applied to the d–p breakup reaction at E_d = 7.4 MeV. The potential employed is a sum of NN short-range pair potentials plus the Coulomb potential between the protons. The strong NN potential used in the calculation is of separable form, rank two (corresponding to the singlet and triplet channel) with s-wave form factors of Gaussian type. The parameters were chosen to reproduce the effective range parameters: deuteron binding energy, singlet and triplet scattering length, and singlet effective range. The potential and its numerical values are given in Kröger and Slobodrian.[25] We have used the momentum-space step functions as a basis, as we did for p–p scattering. Denoting by $\mathbf{p}_{rel}$ the relative momentum between one nucleon and the center-of-mass of the complementary two-nucleon subsystem, and by q_{sub} the relative momentum between the two nucleons in the subsystem, we have used $N = 10$ step functions to describe the $|p_{rel}|$ dependence and $N = 10$ step functions for the $|q_{sub}|$ dependence. For each of the total angular momenta, the relative angular momentum and the subsystem angular momentum, s-, p-, and d-waves have been taken into account. Also the full spin dependence has been accounted for. Results have been reported in Kröger[24] and Kröger and Slobodrian[25] including only s- and p-waves in Kröger et al.[83] also including d-waves.

We have calculated at $E_d = 7.4$ MeV the breakup cross section for a set of p–p detection angles. However, only the relative shape of the curve has been determined; the absolute normalization has been adjusted to the peak of the experimental data. The results are shown in Figs. 10 and 11. In order

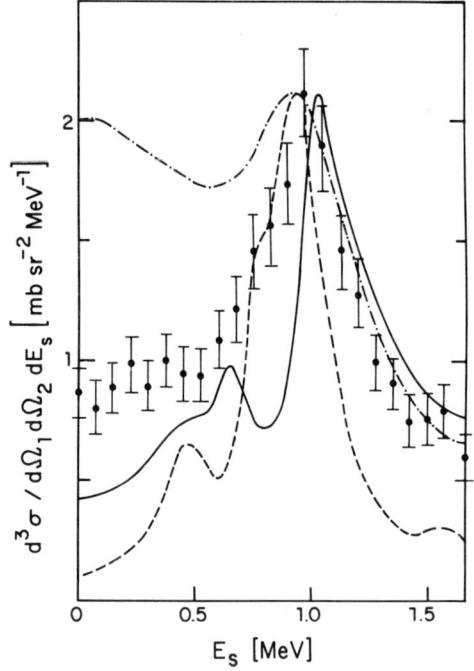

FIGURE 10. Kinematically complete d–p breakup cross section at $E_{lab}^d = 7.4$ MeV. E_s is measured along the kinematical curve of the outgoing protons. $E_s = 0$ corresponds to the point where the protons have the same energy (close to the origin). The p–p detection angles are 13°–13°. The dashed and the solid lines correspond to the time-dependent calculation. The dashed curve corresponds to inclusion of s- and p-waves for total, relative, and subsystem angular momenta. The full line also includes the d-wave. Both curves are normalized to the height of the experimental peak. For comparison, the dashed-dotted line shows a d–n cross section obtained by a Faddeev calculation, also normalized to the peak of the d–p reaction.

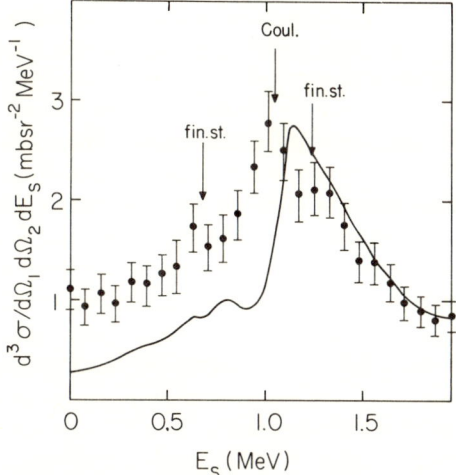

FIGURE 11. Same as Fig. 10, but at 12°–12°.

to demonstrate the effect of the Coulomb force at low energy, we have included the result of a d–n Faddeev calculation.[84] Also in this case, the calculated curve has been adjusted to the height of the experimental d–p peak. The shape of the d–n (Faddeev) curve differs largely from the d–p curve resulting from the time-dependent calculation as well as from the d–p experimental data. In Kröger and Slobodrian[81] we have pointed out that the position of the peak of the cross section as a function of the kinematical curve lies very close to the point where the relative energy of the outgoing protons has a maximum. Thus the general shape of the d–p cross sections having a pronounced peak and a sharp fall-off behavior could be explained in terms of the repulsive Coulomb force in the outgoing channel, which tends to suppress configurations with a small relative momentum between the protons and enhance configurations with a large relative momentum. It should be noted that in the final state the center-of-mass energy of the three nucleons is only 0.241 MeV. On the other hand, the pronounced peak is not found in the neutral d–n case; only a weak maximum is found there, at nearly the same position. It is possible that in the d–p as well as in the d–n cross sections a peak reminiscent of the quasi-free scattering peak is present, although strictly speaking quasi-free scattering can only occur above the threshold $E_d > 4|B_d|$.

These calculations have been limited by the computing facilities available at the time. It would be desirable to redo these computations with a larger number of nodal points and, most importantly, a higher number of partial waves. Also the use of redundant basis functions in different channels, as proposed and applied by Kuruoglu and Levin[23] in a time-dependent n–d calculation might lead to further improvement.

7. SUMMARY

We have reviewed time-dependent scattering theory, in particular in the presence of long-range Coulomb forces. In Section 2 we collected the basic results, based on Dollard's wave operators. We outlined the relation between time-dependent and time-independent stationary scattering theory. In Section 3 we discussed the Dollard wave operators for charged-particle scattering in two- and three-body systems. In Section 4 we discussed how the time-dependent formalism can be approximated and discretized in order to render it amenable to numerical calculations. Some rigorous mathematical results were given in Section 5, proving convergence of the approximated S-matrix toward the exact solution. In Section 6, some examples of numerical calculaions were presented from p–p scattering and p–d breakup scattering. Here p–p scattering serves as a benchmark problem for the time-dependent method. We have studied convergence properties as a function of all the approximation parameters involved. Finally we presented an application to the p–d breakup scattering reaction close to the breakup threshold. Let us stress again that the p–d breakup reaction has been investigated when searching for effects of a 3N force at low energy, i.e., close to the breakup threshold. For such an analysis, a correct treatment of Coulomb effects as presented here is indispensable. Similarly in atomic/molecular reactions, the method presented here should be adequate to treat reactions with two or three charged particles in the incoming/outgoing state, e.g., electron–hydrogen elastic scattering and ionization.

While there are many ways in which time-dependent methods have been applied to scattering problems in atomic/molecular physics, e.g., $H + H_2$ and $F + H_2$[85] (for an overview see Kulander[8]), there are only a few cases where time-dependent methods have been applied to three-nucleon scattering (for review see Kröger[10]). In our view the virtues of the time-dependent methods lie in their conceptual simplicity and a formulation which lends itself to computations using vector processors and multi-parallel-processor architectures.

ACKNOWLEDGMENT. This work has been supported by the NSERC Canada and FCAR Québec.

REFERENCES

1. W. HEISENBERG, *Z. Phys.* **120**, 513 (1943).
2. C. MØLLER, *Danske Vid. Selsk. Mat.-Fys. Medd.* **23**, 1 (1945).
3. J. M. COOK, *J. Math. Phys.* **36**, 82 (1957).
4. M. GELL-MANN AND M. L. GOLDBERGER, *Phys. Rev.* **91**, 398 (1953).

5. E. O. ALT AND W. SANDHAS, Chapter 1, this volume.
6. J. BROEKHOVE, L. LATHOUWERS AND P. VAN LEUVEN (eds.), *Lecture Notes in Physics, Vol. 256* (Springer, Berlin, 1985).
7. V. MOHAN AND N. SATHYAMURTHY, *Comp. Phys. Rep.* **7**, 214 (1988).
8. K. C. KULANDER (ed.), *Time-Dependent Methods for Quantum Dynamics, Comput. Phys. Comm.* **63**, 1 (1991).
9. C. LEFORESTIER, R. H. BISSELING, C. CERJAN, M. D. FEIT, R. FRIESNER, A. GULDBERG, A. HAMMERICH, G. JOLICARD, W. KARRLEIN, H. D. MEYER, N. LIPKIN, O. RONCERO, AND R. KOSLOFF, *J. Comput. Phys.* **94**, 59 (1991).
10. H. KRÖGER, *Phys. Repts.* **210**, 45 (1992).
11. J. MAZUR AND R. J. RUBIN, *J. Chem. Phys.* **31**, 1395 (1959).
12. E. A. MCCULLOUGH JR. AND R. E. WYATT, *J. Chem. Phys.* **51**, 1253 (1969).
13. K. C. KULANDER, *J. Chem. Phys.* **69**, 5064 (1978).
14. K. C. KULANDER, *Nucl. Phys.* **A353**, 341 (1981).
15. P. M. AGRAWAL, V. MOHAN, AND N. SATHYAMURTHY, *Chem. Phys. Lett.* **114**, 343 (1985).
16. E. J. HELLER, *J. Chem. Phys.* **68**, 2066 (1978).
17. E. J. HELLER, in *Potential Energy Surfaces and Dynamics Calculations* (D. G. Truhlar, ed.) (Plenum, New York, 1981), p. 103.
18. S. Y. LEE AND E. J. HELLER, *J. Chem. Phys.* **71**, 4777 (1979).
19. E. J. HELLER AND M. J. DAVIS, *J. Chem. Phys.* **85**, 307 (1981).
20. G. DROLSHAGEN AND E. J. HELLER, *J. Chem. Phys.* **79**, 2072 (1983).
21. M. BATINIĆ, Ž. BAJZER, AND H. KRÖGER, *Phys. Rev. C* **33**, 1187 (1986).
22. J. HOLZ AND W. GLÖCKLE, *Phys. Rev. C* **37**, 1386 (1988).
23. Z. C. KURUOGLU AND F. S. LEVIN, *Phys. Rev. Lett.* **64**, 1701 (1990).
24. H. KRÖGER, *Phys. Lett.* **135B**, 1 (1984).
25. H. KRÖGER AND R. J. SLOBODRIAN, *Phys. Rev. C* **30**, 1390 (1984).
26. D. NEUHAUSER, M. BAER, R. JUDSON, AND D. J. KOURI, *J. Chem. Phys.* **90**, 5882 (1989).
27. D. NEUHAUSER, M. BAER, R. JUDSON, AND D. J. KOURI, *J. Chem. Phys.* **93**, 312, (1990).
28. R. JUDSON, D. J. KOURI, D. NEUHAUSER, AND M. BAER, *Phys. Rev. A* **42**, 351 (1990).
29. H. KRÖGER AND W. SANDHAS, *Phys. Rev. Lett.* **40**, 834 (1978).
30. M. H. KALOS, *Monte Carlo Methods in Quantum Problems*, Paris, France, 1982, *Proc. NATO Adv. Res. Workshop*, NATO ASI Ser. C, Vol. 125 (Reidel, Dordrecht, 1984).
31. K. E. SCHMIDT AND M. H. KALOS, *Applications of the Monte Carlo Method in Statistical Physics* (K. Binder, ed.) (Springer, Berlin, 1984).
32. M. H. KALOS AND P. A. WHITLOCK, *Monte Carlo Methods* (John Wiley, New York, 1986).
33. H. KRÖGER, K. J. M. MORIARTY, AND J. POTVIN, *Phys. Rev. A* **42**, 2661 (1990).
34. W. GLÖCKLE, *The Quantum Mechanical Few-Body Problem* (Springer, Berlin, 1983).
35. L. D. FADDEEV, *Mathematical Aspects of the Three-Body Problem in the Quantum Scattering Theory* (Israel Program for Scientific Translation Jerusalem, 1965).
36. W. O. AMREIN, J. M. JAUCH, AND K. B. SINHA, *Scattering Theory in Quantum Mechanics, Lecture Notes and Supplements in Physics* (Benjamin, Reading, MA., 1977).
37. M. REED AND B. SIMON, *Methods of Modern Mathematical Physics, Vol. III: Scattering Theory* (Academic Press, New York, 1979).
38. J. WEIDMANN, *Linear Operations in Hilbert Spaces* (Springer, New York, 1980).
39. H. BAUMGÄRTEL AND M. WOLLENBERG, *Mathematical Scattering Theory* (Birkhäuser Verlag, Basel, 1983).
40. H. VAN HAERINGEN, *Charged-Particle Interactions* (Coulomb Press, Leiden, 1985).
41. C. J. JOACHAIN, *Quantum Collision Theory* (North Holland, Amsterdam, 1975).
42. J. WEIDMANN, *Linear Operations in Hilbert Spaces* (Springer, New York, 1980), Theorem 11.13, p. 352.
43. J. R. TAYLOR, *Scattering Theory* (John Wiley, New York, 1972).

44. W. Gordon, *Z. Phys.* **48**, 180 (1928).
45. J. D. Dollard, *Rocky Mountain J. Math.* **1**, 5 (1971).
46. J. D. Dollard, *Rocky Mountain J. Math.* **2**, 317 (1972).
47. N. F. Mott and A. S. W. Massey, *The Theory of Atomic Collisions* (Oxford University Press, 1949).
48. J. D. Dollard, *J. Math. Phys.* **5**, 729 (1964).
49. G. L. Nutt, *J. Math. Phys.* **9**, 796 (1968).
50. H. van Haeringen, *J. Math. Phys.* **17**, 995 (1976).
51. J. Zorbas, *Nuov. Cim. Lett.* **10**, 121 (1974).
52. J. M. Jauch, *Helv. Phys. Acta* **31**, 127 (1958).
53. J. D. Dollard, *J. Math. Phys.* **9**, 620 (1968).
54. E. O. Alt, W. Sandhas, and H. Ziegelmann, *Nucl. Phys.* **A445**, 429 (1985).
55. I. W. Herbst, *Comm. Math. Phys.* **35**, 181 (1974).
56. A. M. Veselova, *Theor. Math. Phys.* **13**, 368 (1972).
57. Z. Bajzer, in *Few-Body Nuclear Physics*, (G. Pisent, V. Vanzani, and L. Fonda, eds.) (IAEA, Vienna, 1978), p. 365.
58. J. Schwinger, *J. Math. Phys.* **5**, 1606 (1964).
59. Z. Bajzer, *Z. Phys.* **A278**, 97 (1976).
60. A. Messiah, *Quantum Mechanics, Vol. I* (Elsevier, Amsterdam, 1961).
61. A. A. Samarskii and E. S. Nikolaev, *Numerical Methods for Grid Equations* (Birkhäuser, Basel, 1989).
62. C. Moler and C. Van Loan, *SIAM Rev.* **20**, 801 (1978).
63. A. J. Dragt, in *Lectures on Nonlinear Orbit Dynamics* (Fermilab. 1982), Proceedings of the Conference of High Energy Particle Accelerators, AIP Conf. Proc. No. 87 (R. A. Carrigan and F. R. Huson, eds.) (AIP, New York, 1982), p. 147.
64. E. Forest, Superconducting Super Collider Central Design Group Publication No. SSC-111 (1987).
65. R. S. Varga, *Matrix Iterative Analysis* (Prentice Hall, Englewood Cliffs, N.J. 1980).
66. H. Kröger, *J. Math. Phys.* **25**, 1875 (1984).
67. W. Schweiger, W. Plessas, L. P. Kok, and H. van Haeringen, *Phys. Rev. C* **27**, 515 (1983).
68. R. Girard, H. Kröger, P. Labelle, and Ž. Bajzer, *Phys. Rev. A* **37**, 3195 (1988).
69. M. Batinić, Ž. Bajzer, and H. Kröger, *Phys. Rev. C* **38**, 2955 (1988).
70. E. Hadjimichael, *Nucl. Phys.* **A508**, 161c (1990).
71. R. Machleidt, K. Holinde, and C. Elster, *Phys. Repts.* **149**, 1 (1987).
72. R. Machleidt, in *Advances in Nuclear Physics, Vol. 19.* (J. W. Negele and E. Vogt, eds.) (Plenum, New York, 1989), p. 189.
73. Ch. Hajduk and P. E. Sauer, *IX International Conference on Few-Body Problems,* (M. J. Moravcsik and F. S. Levin, eds.) University of Oregon, Eugene, 1980), Paper II-16.
74. R. B. Wiringa, *Proc. 3rd International Conference on Recent Progress in Many-Body Theory* (H. Kümmel and M. Ristic, eds.) (Springer, Berlin, 1983).
75. K. T. Kim, Y. E. Kim, D. J. Klepacki, R. A. Brandenburg, E. P. Harper, R. Machleidt, *Phys. Rev. C* **38**, 2366 (1988).
76. I. Slaus, Y. Akaishi, and H. Tanaka, *Phys. Rev. Lett.* **48**, 993 (1982).
77. W. Meier and W. Glöckle, *Phys. Lett.* **B138**, 329 (1984).
78. M. Karus, M. Buballa, J. Helten, B. Laumann, R. Melzer, P. Niessen, H. Oswald, G. Rauprich, J. Schulte-Uebbing, and H. Paetz gen. Schiek, *Phys. Rev. C* **31**, 1112 (1985).
79. H. Witala, W. Glöckle, and T. Cornelius, *Phys. Rev. C* **39**, 384 (1989).
80. A. M. Nachabe, R. J. Slobodrian, B. K. Sinha, R. Roy, and H. Kröger, *J. Phys.* (Paris) **47**, 1141 (1986).
81. H. Kröger and R. J. Slobodrian, *Phys. Lett.* **B144**, 19 (1984).

82. V. V. KOMAROV, A. M. POPOVA, AND V. V. ZATEKIN, Proc. X. International Conference on Few-Body Problems in Physics, Karlsruhe (B. Zeidnitz, ed.) (North Holland, Amsterdam, 1984), Vol. 2, p. 387.
83. H. KRÖGER, A. M. NACHABE, AND R. J. SLOBODRIAN, *Phys. Rev. C* **33** 1208 (1986).
84. P. DOLESCHALL, private communication.
85. D. NEUHAUSER, M. BAER, R. S. JUDSON, AND D. J. KOURI, *Comput. Phys. Comm.* **63**, 460 (1991).

CHAPTER 4

H⁻ SPECTROSCOPY

H. C. BRYANT AND M. HALKA

1. INTRODUCTION

Among the simplest three-body two-electron atomic systems, the negative hydrogen ion, H⁻, is the most interesting because the electron correlations are especially strong; the usual approximation that the two electrons move in hydrogen-like orbits with their mutual interaction modeled by a screening constant is grossly inadequate. In fact, this independent particle approximation predicts H⁻ to be unbound—evidence that the correlations between the two electrons as they move about the proton must be carefully considered.[1] These correlations make possible the existence of doubly-excited states, where both electrons have a principal quantum number greater than one.

Since the late 1960s the study of photodetachment of negative ions has identified many unique properties of correlated electron systems. Experimental work has facilitated a steadily increasing understanding of the processes involved. Theory has so far been able to explain or predict the general behavior of cross sections and resonance positions and widths, but mysteries remain, especially when external fields take part.

The correlations between interacting atomic electrons are intricate, making calculation difficult. The dynamics, especially for high excitations, is not completely understood, and controversy still flares concerning the mechanisms of double escape. The general case of the stability of bound states in three-body Coulomb systems with unit charge has been treated most recently by Bishop and Frolov.[2] In the following discussion we review some basic experimental techniques and results, and attempt to impart an elementary understanding of the theory as it stands at this time.

H. C. BRYANT • Department of Physics and Astronomy, University of New Mexico, Albuquerque, NM 87131-1156. M. HALKA • Department of Physics, Portland State University, Portland, Oregon 97207-0751.

Coulomb Interactions in Nuclear and Atomic Few-Body Collisions, edited by Frank S. Levin and David A. Micha. Plenum Press, New York, 1996.

Section 2 describes the basic H^- system. Section 3 illustrates what has been the most useful technique for the study of a wide range of photodetachment energies (principally in the vacuum ultraviolet): the crossing of laser beams with a relativistic H^- beam. In Section 4 we examine the behavior of cross sections near the one-electron detachment thresholds with and without applied fields. The current controversy regarding the presence or absence of observable cross-section modulations in two-electron detachment thresholds is also presented in Section 4 along with experimental measurements. Convenient coordinate systems are introduced in Section 5, which also describes some of the earliest experimental work, including the measurement of the $n = 2$ Feshbach and shape resonance photodetachment cross sections with and without applied static fields, as well as higher-energy doubly-excited states converging from below to the H ($n = 3, 4, 5, 6, 7$, and 8) production thresholds. Some of these have been observed in applied fields, and the results are examined qualitatively as calculations have not been available.

A brief review of experimental and theoretical work on the angular distribution of photoelectrons from H^- is presented in Section 6. Section 7 offers a brief overview of multiphoton experiments and theories as applied to the H^- spectrum.

Some very curious results have been seen when relativistic H^- ions are passed through thin foils. These are presented in Section 8. Section 9 contains the summary and an outlook on future developments, including what we consider to be the most important unsolved questions inherent in the H^- system.

Before going further we wish to caution the reader that several systems of units commonly used in theoretical or experimental atomic physics will be encountered. Our computations will mostly be done in SI units, but some results will also be presented in atomic units. We shall try to be clear about this as we go along.

2. H^- STRUCTURE

2.1. Simple Considerations

To begin, we consider an electric charge brought near a conducting neutral sphere. The electric field from the charge will induce a dipole on the sphere which will then attract the charge. If the hydrogen atom is the sphere and an electron a distance r from the sphere is the charge, the force on the electron is attractive and goes as r^{-5}, a rather short-range influence. The potential goes as r^{-4}. Classically, at least insofar as the atom may be regarded as a conducting sphere, the electron will be attracted and H^-

negative ions can be expected to form. Thus the formation of an induced dipole, essential for binding, illustrates the importance of correlations.

To solve the Coulombic three-body problem, of which H^- is a good example, requires approximate methods, even though the Hamiltonian is known exactly. The simplest quantum mechanical approximation treats the ion as a two-body problem,[3] in which the binding potential of the outer electron due to the inner electron and proton is considered to have a very short range (e.g., a δ-function). For r greater than the range of the force, the Schrödinger equation is ($\hbar = 1$)

$$-\frac{1}{2\mu}\frac{d^2}{dr^2}(r\Psi) = -E_B(r\psi) \qquad (4.2.1)$$

where E_B is the binding energy and μ is the reduced mass. (The deuteron problem is similar to this.) The solution of (4.2.1) is

$$\Psi = \varepsilon^{1/4}\frac{\exp(-r\varepsilon^{1/2})}{4\pi r} \qquad (4.2.2)$$

where $\varepsilon = 2\mu E_B = 1/(2.24\,\text{Å})^2$ for $E_B = 0.7542\,\text{eV}$, the measured binding energy of the H^- ion's outer electron.[6] Note that the characteristic distance of the outside electron of H^-, 2.24 Å, is 4.25 times the Bohr radius, the atomic unit of distance or characteristic dimension of the neutral hydrogen atom. The rms radius of the outer electron is 1.58 Å or three Bohr radii.

Fano and Rau[4] use this function to create a two-electron wave function:

$$\Psi(\mathbf{r}_1,\mathbf{r}_2) = \frac{1}{\sqrt{2}}(1 + P_{12})\frac{C}{r_2}\exp(-k_B r_2)U_0(r_1) \qquad (4.2.3)$$

where $k_B^2 = \varepsilon$, U_0 is the hydrogenic ground state, and P_{12} permutes 1 and 2. This wave function begs the question of binding by setting ε equal to the experimental binding energy.

In order to address the question of binding, one typically uses the variational technique, described below. The early, famous trial wave function,

$$\psi = (e^{(-ar_1-br_2)} + e^{(-br_1-ar_2)})(1 + cr_{12}) \qquad (4.2.4)$$

yields minimum energy when $a = 1.075$, $b = 0.478$, and $c = 0.312$ when r is in atomic units, giving the ionization potential $E_B = 0.0518\,\text{Rydberg} = 0.704\,\text{eV}$.[1] A Rydberg is the binding energy of hydrogen and is one-half an atomic unit of energy. This is to be compared with the experimental value $E_B = 0.7542\,\text{eV}$. The term containing r_{12} is the so-called correlation or polarization term. When it is omitted, the value for E_B decreases by a factor of two.

Because of the large difference between a and b this function is consistent with the view discussed above that one electron is weakly bound in the field of a polarized hydrogen atom.

In the variational method (see, e.g., Bransden and Joachain[7]) one must minimize the functional

$$E(\phi) = \frac{\langle \phi | \mathbf{H} | \phi \rangle}{\langle \phi | \phi \rangle} \qquad (4.2.5)$$

$\mathbf{H}$ is the nonrelativistic, spin-independent Hamiltonian for two electrons, written in its most general form (see Fig. 1),

$$\mathbf{H} = -e^2 \left(\frac{Z}{r_1} + \frac{Z}{r_2} - \frac{1}{r_{12}} \right) - \frac{\hbar^2}{2} \left(\frac{\nabla_1^2}{m} + \frac{\nabla_2^2}{m} + \frac{\nabla_p^2}{M} \right) \qquad (4.2.6)$$

The last term containing the nuclear mass M is the mass polarization term. The charge on the nucleus is taken to be Ze.

The problem is that ϕ is unknown. One can show that the correct value of the ground state E_0 for the Hamiltonian $\mathbf{H}$ is the minimum value of $E(\phi)$, and that ϕ approaches the correct wave function as $E(\phi) \to E_0$. Thus one simply parametrizes ϕ with as many parameters as the computer can handle and minimizes $E(\phi)$. By an equivalent procedure, Pekeris[8] found the electron affinity (EA) of hydrogen to be $0.7542\,\text{eV}$.

H^- photodetachment experiments dating from the early 1970s have attempted to measure the EA.[6] Most of these provided very good accu-

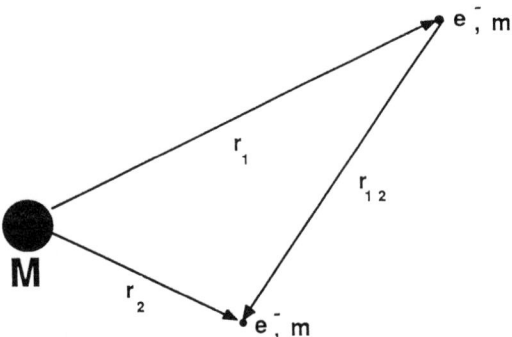

FIGURE 1. Variables used in Eq. (4.2.6) shown schematically. The nucleus M is at the origin. r_1 is the distance to one electron, r_2 is the distance to the other electron, and r_{12} is the distance between the two electrons.

racies with errors of less than 12 meV. Lykke et al.[6] have performed the most sensitive experiment to date, using a tunable F-center laser ($\lambda = 1.45 \to 1.75\,\mu$m) intersecting a 2.5-keV H$^-$ beam. They found an EA of 6082.99 ± 0.15 cm^{-1} [0.754195(19) eV], which is in excellent agreement with Pekeris's value. To our knowledge, this measurement is the only one sensitive to the hyperfine interaction [not included in Eq. (4.2.6)] in the final state.

2.2. Electron Correlations

Along with studies of He doubly-excited states, studies of the H$^-$ ion have been largely responsible for the current understanding of strong electron correlations. While the H$^-$ photodetachment spectrum displays evidence of many doubly-excited states, it is believed that it does not harbor any singly-excited bound states. "Folk wisdom" says that these states — so abundant in He — all disappear into the continuum and lose their couplings as Z goes to 1. In support of this expectation, Tang[9] has calculated the binding energy of the $1s2s\,^3S^e$ state for a two-electron system vs. Z. The results are shown in Table I. Extrapolation to $Z = 1$ gives an energy right at the single-electron continuum. Since this state is the lowest of singly-excited states, we conclude that no other states can be bound. Hill[10] has shown rigorously that H$^-$ has only one bound state.

Because of their near degeneracies, the excited states of neutral hydrogen behave as if they had permanent electric dipole moments even in weak electric fields. Thus, when an electron approaches an excited hydrogen atom, it feels the tug of a *permanent* dipole organized by its own field. The

TABLE I

Energies in Atomic Units for the State $2\,^3S^e(1s2s)$ in the Two-Electron Atom as a Function of the Hypothetical Charge on the Nucleus for Z between 1 and 2

Z	E (a.u.)
2.0	−1.77
1.8	−1.738479
1.6	−1.35198
1.4	−1.016034
1.2	−0.7310938
1.1	−0.6083818
1.0	unbound

*Calculations of Tang.[9] This progression illustrates the disappearance of this lowest excited state in the two-electron system as Z approaches unity.

direction of the tug depends on the alignment of the dipole with the number of possible orientations increasing with principal quantum number.

As an example, the $n = 2$ state, for which the $2s$ and $2p$ states are mixed for very small fields, behaves as if it has a permanent dipole moment of $3ea_0$ where a_0 is the Bohr radius for atomic hydrogen. Since the states are not exactly degenerate, a small field is required to mix them. The $2s_{1/2}$ (and $2p_{1/2}$)–$2p_{3/2}$ splitting (neglecting Lamb shifts) is about $0.36\,\text{cm}^{-1}$ or $45\,\mu\text{eV}$. Mixing of the two states should occur for a field F of the order of

$$F = \frac{45(10^{-6})\,\text{eV}}{3ea_0} = 3 \times 10^5\,[\text{V/m}] \tag{4.2.7}$$

Since the field of an electron goes as $e/(4\pi\varepsilon_0 r^2)$ in SI units, this field is present when the distance from the electron to the neutral atom is

$$r = \left[\frac{9(10^9)1.6(10^{-19})}{3(10^5)}\right]^{1/2}\,\text{m} = 692\,\text{Å} = 1310 a_0 \tag{4.2.8}$$

Therefore, when an electron approaches a hydrogen atom in its $n = 2$ state, it feels a long-range dipole field, a potential that falls as $1/r^2$:

$$V = \frac{3e^2 a_0 \cos\theta}{4\pi\varepsilon_0 r^2} = \frac{3\cos\theta}{r^2} \tag{4.2.9}$$

where θ is the angle of the approaching electron's velocity with the direction of the dipole. At large distances an electron is much more strongly attracted if the atom is excited than if it is not. Note that when angular momentum is present, there is an additional $1/r^2$ angular momentum barrier term in the radial equation.

When one realizes that the binding mechanism must depend on the excited atom presenting a dipole to the incoming electron, it is clear that correlation must be very important in the resonant states, where both electrons are excited. These are discussed in Section 5.4.

3. EXPERIMENTAL TECHNIQUES

The usual procedure for the study of submicroscopic objects is to probe them and observe the outcome. The use of photons as probes in the study of atomic systems has been very productive. The mode we wish to discuss here is the use of a tunable photon beam to produce one or two free electrons. We review the study of ions moving at very high speeds ($v \approx c$) through the laboratory. As we shall see, such a situation provides several experimental advantages that apparently cannot be obtained in any other way with current technology. We review the kinematics of special relativity, show how this can be applied to an atomic system, and discuss the experimental results.

3.1. Relativistic Kinematics

When an atom moving near the speed of light is probed with laser beams, electric and magnetic fields, or thin foils, the description of the interaction can be quite different in the atom's frame from that of the laboratory. The connection between these two descriptions is given by the Lorentz transformation. The assumption of special relativity is entirely adequate[11] for the measurements we describe below.

Let us assume two parallel Cartesian coordinate systems, one fixed in the laboratory and the other fixed on the atomic system, such that the relative velocity v is along the x-axes of both and they have a common origin in space-time. We take these frames to be inertial, in the sense that Newton's first law (the law of inertia) is valid in both. Following the Minkowski convention, for which time and time-like components are treated as imaginary, with $\beta = v/c$ and $\gamma = (1 - \beta^2)^{-1/2}$, the tensor form of the Lorentz transformation may be written

$$\mathbf{L} = \begin{bmatrix} \gamma & 0 & 0 & i\beta\gamma \\ 0 & 1 & 0 & 0 \\ 0 & 0 & 1 & 0 \\ -i\beta\gamma & 0 & 0 & \gamma \end{bmatrix} \quad (4.3.1)$$

The invariance of the square of the length of the space-time position four-vector can be easily demonstrated using (4.3.1). That is, if two events are separated by projections x, y, z, and t in space-time, then

$$s^2 = c^2 t^2 - x^2 - y^2 - z^2 \quad (4.3.2)$$

where s is known as the interval, has the same value in both reference frames. This relationship leads immediately to the phenomenon of time dilation, in which the time intervals of a moving clock appear to an observer in the stationary frame to dilate by a factor of γ and to the Lorentz contraction, in which a rod would appear to contract by a factor of $1/\gamma$ along its line of motion.

The Lorentz transformation enables us to sort out the relationships between kinematical variables compounded from the space-time coordinates. In particular, the energy–momentum four-vectors of a particle as determined in the two frames are related by

$$\begin{bmatrix} p'_x \\ p'_y \\ p'_z \\ iE'/c \end{bmatrix} = \mathbf{L} \begin{bmatrix} p_x \\ p_y \\ p_z \\ iE/c \end{bmatrix} \qquad (4.3.3)$$

In the case where the particle is a photon, the connection is particularly simple because of the Lorentz invariant relation (similar to interval invariance),

$$p^2 - \frac{E^2}{c^2} = p'^2 - \frac{E'^2}{c^2} = 0 \qquad (4.3.4)$$

since the photon mass is zero. This relationship follows simply from the fact that the length of a four-vector squared, as in Eq. (4.3.2), (its dot product with itself) is an invariant under the Lorentz transformation. In the case of the energy–momentum four-vector of a particle, the length is $(mc)^2$, where m is the particle's rest mass.

In order to relate a photon's kinematics as seen in the two frames, we take the nonconstraining definition that $p_z = 0$, and explicitly use $p = E/c$, so that

$$\begin{bmatrix} p' \cos \theta' \\ p' \sin \theta' \\ 0 \\ ip' \end{bmatrix} = \mathbf{L} \begin{bmatrix} p \cos \theta \\ p \sin \theta \\ 0 \\ ip \end{bmatrix} \qquad (4.3.5)$$

where θ and θ' are the angles the photon's momentum makes with the x- and x'-axis in the xy- and $x'y'$-planes, respectively.

From the fourth component of (4.3.5) we have

$$ip' = \gamma(-i\beta p \cos \theta + ip) \qquad (4.3.6)$$

or

$$E' = \gamma(1 - \beta \cos \theta)E \qquad (4.3.7)$$

a relationship known as the "Doppler shift." As we shall discuss later, this shift can be substantial for presently available H$^-$ beams. At LAMPF, the linear accelerator at Los Alamos National Laboratory, with an 800-MeV H$^-$ beam, upward shifts by as much as a factor of 3.4 can be achieved.

The relative motion can also produce an enhancement in the intensity of the laser. The intensity $I(\text{W}/\text{cm}^2)$ is proportional to the magnitude of the Poynting vector $|\mathbf{E} \times \mathbf{H}|$.[12] Transforming $\mathbf{E}$ and $\mathbf{H}$ according to the Lorentz field transformation formulas (see, e.g., Rindler[13]), one obtains

$$\frac{I_{\text{CM}}}{I_{\text{lab}}} = \gamma^2(1 - \beta\cos\theta)^2 \qquad (4.3.8)$$

where I_{CM} is the intensity in the atom's center of mass, and I_{lab} is the measured intensity of the laser beam in the laboratory. Thus for $\beta = 0.842$ and $\theta = \pi$, an order of magnitude (factor of 11.6) is also gained in laser intensity.

3.2. An Experimental Application of Time Dilation

When an atom is traveling with a velocity near the speed of light, time in its frame passes appreciably slower from our point of view than it would if the atom were at rest in the laboratory. Thus atomic lifetimes of short duration become more accessible to measurement. Moreover, a mean lifetime τ in the atom's frame becomes a mean decay distance from its production point $d = \beta\gamma c\tau$ in the laboratory, and the atom would decay with an e-folding distance of d.

Suppose we wanted to measure the width Γ of a spectral line approximately $10\,\mu\text{eV}$ wide ($10\,\mu\text{eV} = 2.42\,\text{GHz} = 0.08\,\text{cm}^{-1}$). By the uncertainty principle

$$\tau \approx \frac{\hbar}{\Gamma} = \frac{6.582(10^{-16})\,\text{eV}\,\text{s}}{10^{-5}\,\text{eV}} \cong 6.6(10^{-11})\,\text{s} \qquad (4.3.9)$$

In the laboratory the mean flight path d of this atom, from its production point to its decay, would therefore be $d = 3.08\,\text{cm}$ when $\beta = 0.842$. This length is much easier to measure by Doppler-shifted laser spectroscopy than the 10-μeV width. Note that a narrower (in energy) state has a longer lifetime and a longer flight path, so that "flight-path spectroscopy" becomes more precise as the line narrows.

3.3. Aberration of Light

By taking the ratio of the first two space-like components in Eq. (4.3.5), we can also derive a relationship between θ and θ':

$$\tan\theta' = \frac{\sin\theta}{\gamma(\beta - \cos\theta)} \qquad (4.3.10)$$

This connection, first observed in a nonrelativistic context by Bradley (see Stewart[14]), is known as the "aberration of light," and must be considered in relativistic experiments, as its effect can be considerable.

3.4. The Electromagnetic Tensor

The electromagnetic field can be compactly expressed as an antisymmetric tensor **F**, whose Minkowski-space form is

$$\mathbf{F} = \begin{bmatrix} 0 & -B_z & B_y & iF_x/c \\ B_z & 0 & -B_x & iF_y/c \\ -B_y & B_x & 0 & iF_z/c \\ -iF_x/c & -iF_y/c & -iF_z/c & 0 \end{bmatrix} \quad (4.3.11)$$

where $\mathbf{F} = (F_x, F_y, F_z)$ is the electric field and $\mathbf{B} = (B_x, B_y, B_z)$ is the magnetic induction. **F** transforms from one frame to another according to the rule

$$F' = \mathbf{L}F\mathbf{L}^{-1} \quad (4.3.12)$$

The transformations of fields from the laboratory to the atom's frame can result in substantial changes which are exceedingly advantageous from an experimental point of view. Consider for example the consequence of the application of a modest magnetic field in the laboratory transverse to a relativistic beam. From (4.3.12) we have

$$F'_z = \gamma(F_z + \beta B_y c) \quad (4.3.13)$$

We apply this to the case of a hydrogen atom with an energy of 800 MeV ($\beta = 0.842$, $\gamma = 1.853$) moving through a magnetic field of 1 T (10^4 gauss) — readily obtainable in the laboratory using an ordinary electromagnet — and $F_z = 0$. Then the atom experiences a field of $F'_z = 4.7$ MV/cm, an extraordinarily high field by most laboratory standards. Thus we have a tool for the study of atoms in very strong electric fields. Note that, of course, there is still a magnetic field in the atom's frame,

$$B'_y = \gamma\left(B_y + \frac{\beta F_z}{c}\right) = 1.853 \, \text{T} \quad (4.3.14)$$

but this is not a large field by atomic standards.

3.5. Application to Photoabsorption: Doppler Tuning

Let us now turn in more detail to the specific situation in which a photon beam (e.g., a laser beam) is directed at a beam of atoms or ions moving at nearly the speed of light in the laboratory. In the H⁻ work at LAMPF, the angle α between the laser beam and the particle beam (for historical reasons) is taken to be zero when the two approach each other head-on, that is, when the gap between the particle and the photon is closing at the maximal rate. Thus $\alpha = -\theta$ and $\alpha' = -\theta'$. The primed system is the center-of-mass (CM) frame of the atomic system and the unprimed system is the laboratory. Thus a photon of energy E_{lab} in the laboratory has an energy E_{CM} in the CM of the atomic system [from Eq. (4.3.7)] of

$$E_{CM} = \gamma E_{lab}(1 + \beta \cos \alpha) \qquad (4.3.15)$$

Figure 2 illustrates how E_{CM} varies with angle α for various pulsed lasers of experimental interest for the relevant case of the beam at LAMPF with $\beta = 0.842$ for the 800 MeV ion beam energy. With these conditions the laser photon energy can be tuned nearly continuously with a range from about 0.2 to 20 eV using a variety of lasers (CO_2, N_2, Nd:YAG, ArF).

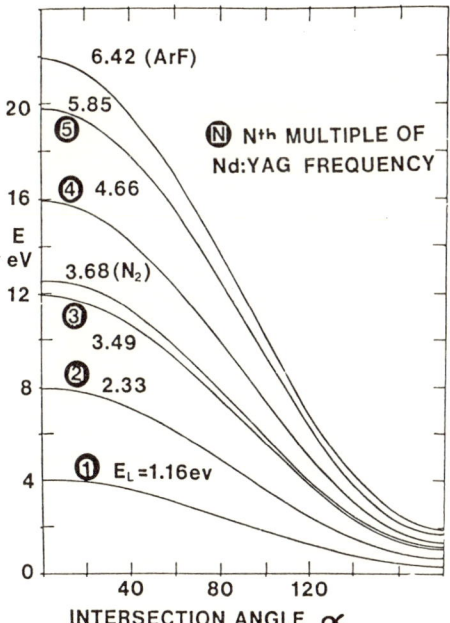

FIGURE 2. Center-of-mass photon energies E (in eV) produced by Doppler-shifting various laser (Nd: YAG and N_2) lines at LAMPF ($\beta = 0.842$).

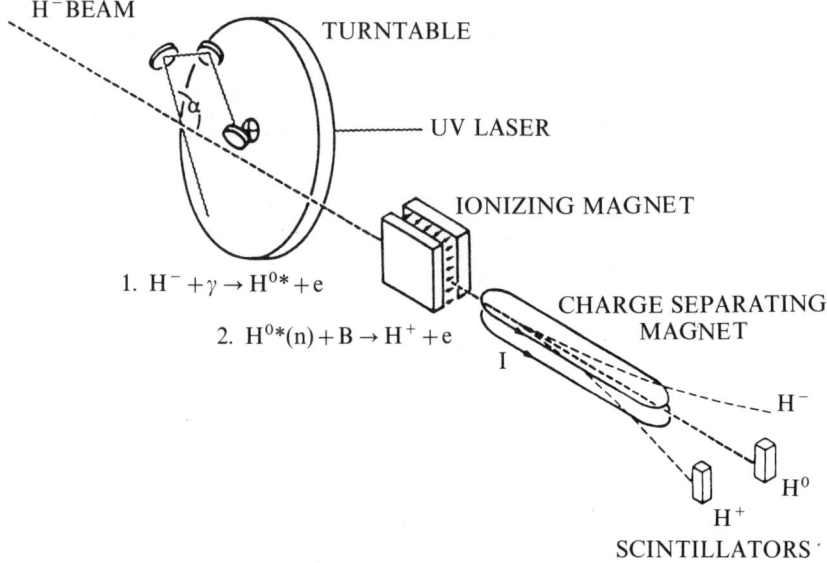

FIGURE 3. Typical experimental setup for high-lying resonance searches at LAMPF.[43]

Figure 3 is a schematic diagram of a typical experimental setup at the LAMPF high-resolution atomic beam facility (HIRAB). The laser beam, whose angle of intersection with the ion beam is varied, detaches electrons from the H^- ions by the reaction

$$H^- + h\nu \rightarrow H(N) + e^- \qquad (4.3.16)$$

or

$$H^- + h\nu \rightarrow p + 2e^- \qquad (4.3.17)$$

where $h\nu = E_{CM}$ and $H(N)$ is the neutral hydrogen atom with principal quantum number N. Because of the high laboratory velocity and the small kinetic energy in the center of mass frame, the detached eletrons (as well as the neutral atoms and protons) travel along with the H^- beam until they are deflected by applied fields.

The fragment $H(N)$ atom may be field-ionized by the laboratory motional field from a strong magnet, with the resulting protons detected downstream, so that decay to particular $H(N)$ channels can be monitored. Alternatively, if $N \geqslant 10$, the motional electric field of an electron spectrometer may be used to ionize, producing as the signal of interest the electron from the hydrogen atom. These electrons are easily distinguishable by their trajectories from the photodetached electrons from H^-. For total cross-section measurements, either neutral atoms or electrons may be detected.

Another successful method for counting Hs with a particular N is to excite the $H(N)$ to a higher state N' (for example $N' = 10$) using a second laser beam in another scattering chamber downstream of the first interaction. The electron spectrometer is then able to ionize the electron from the N' state. This is a cleaner method than field-ionizing the first state because background essentially vanishes.

The absorption of single photons by H^- means that only $^1P^0$ continuum states may be produced, according to electric dipole selection rules. Electron scattering experiments, where electrons are scattered from hydrogen atoms, are capable of exciting other angular momentum states of H^-, but are limited by fairly low energy resolution compared with that of photon beams. The precision of the interactions of laser beams with relativistic H^- beams is discussed next.

3.6. Energy Resolution

An appreciable energy spread in the photon energy of a laser beam can arise in the CM frame of the atomic system from three sources: spread in the angle α, spread in the velocity ratio β, and the usually negligible linewidth of the laser.

3.6.1. Spread in α. We can write

$$\delta E_{CM} = \left[\frac{\partial E_{CM}}{\partial \alpha}\right]_\beta \delta\alpha + \frac{1}{2}\left[\frac{\partial^2 E_{CM}}{\partial \alpha^2}\right]_\beta \delta\alpha^2 + \cdots \quad (4.3.18)$$

and, since normally the first term would dominate, we have

$$\overline{\delta E_{CM}} = \gamma\beta E_L \sin\alpha\, \overline{\delta\alpha} \quad (4.3.19)$$

where the bar indicates rms average. Typically, for the measurements described below, the energy spread from this source is about 4 meV for an E_{CM} of 11 eV. For example, using a nitrogen laser ($E_L = 3.68$ eV) to produce an 11-eV CM photon in the H^- system requires an angle $\alpha = 43.35°$. Here

$$\overline{\delta E_{CM}} = 3.94[\text{meV/mr}]\overline{\delta\alpha} \quad (4.3.20)$$

Since the spread $\overline{\delta\alpha}$ arises principally from the laser divergence as delivered (about 1 mr), with appropriate expansion optics much better resolution could be achieved.

3.6.2. Spread in β. A spread in β can be expressed as a spread in γ arising from a beam energy spread. From relativistic kinematics the kinetic energy T of a particle of mass M is given by $T = (\gamma - 1)Mc^2$, so that $\overline{\delta\gamma} = \overline{\delta T}/Mc^2$. From Eq. (4.3.15),

$$\frac{\partial E_{CM}}{\partial \gamma} = \frac{E_L}{\beta}(\beta + \cos\alpha) \tag{4.3.21}$$

For the example above we would have

$$\partial E_{CM} = \frac{7.3 \text{ meV}}{\text{MeV}} \partial T \tag{4.3.22}$$

Since the LAMPF beam energy spread is typically about 0.4 MeV at 800 MeV, this source of experimental uncertainty is comparable to that from the angular spread. It is interesting to note that when $\cos\alpha = -\beta$ (the "magic" angle), $\partial E_{CM}/\partial \gamma = 0$, and to first order there would be no contribution from this source. There would be a second-order term, however, which we leave for the reader to compute.

3.6.3. Laser Linewidth. The linewidth of the laser will also contribute to the CM energy resolution. For pulsed lasers this comes mostly from the uncertainty principle. In our case, the spreading from this source has been negligible, since the transit time of the ion through the laser beam is much shorter than the pulse duration. For example, a laser pulse duration of 10 ns requires a bandwidth of 10^8 Hz. This is an energy spread of $4(10^{-15})$ eV s $\times$ 10^8 Hz $= 0.4$ μeV. Resolutions have so far been in the meV range, so this contribution is negligible.

3.7. Transit Time Broadening (an Illusion?)

Although the duration of the laser pulse may be tens of nanoseconds in the laboratory, the pulse may appear much shorter to a relativistic particle passing through the laser beam. For example, an 800-MeV H^- ion passing through a 1-mm-diameter laser beam at right angles to it would experience a pulse of duration

$$\Delta t = \frac{0.1 \text{ cm}}{0.842 \times 1.853 \times 3 \times 10^{10} \text{ cm/s}} = 2(10^{12}) \text{ s} = 2 \text{ ps} \tag{4.3.23}$$

Via the uncertainty principle, this would correspond to an energy spread in the atom's frame of 2 meV. Therefore, one might think that this effect should be included in the energy resolution "budget," but *it should not*! Transit time broadening is just the same effect as the broadening due to the angular spread in the laser beam,[15] and is therefore included in Eq. (4.3.18).

4. ONE-ELECTRON DETACHMENT THRESHOLDS

4.1. Ground State Detachment in Zero Field

In the case of the single-electron photodetachment threshold at 0.7542 eV, where the residual hydrogen atom is left in its ground state ($n = 1$), no long-range forces act on the departing electron.[16] This means that the Wigner threshold law, which gives the cross section depending on energy E as

$$\sigma \sim (\Delta E)^{l+1/2} \qquad (4.4.1)$$

where l is the angular momentum of the outgoing electron and $\Delta E = E - E_t$ is the energy above threshold, should be valid.* The incoming photon contributes one unit of angular momentum, which is carried away by the detached electron. Thus in an experiment where one photon is absorbed, one should observe the photodetachment cross section $\sigma \sim (\Delta E)^{3/2}$ in a small energy range above threshold.

In their data analysis, Lykke et al.[6] (see also Section 4.2.1) fit the threshold to

$$\sigma(E) = A + B(\Delta E)^{3/2} U(\Delta E) + W(\Delta E - S)^{3/2} U(\Delta E - S) \qquad (4.4.2)$$

where W is the relative strength of the two hyperfine levels in H, S is the hyperfine splitting, A is the background offset, and B is a normalization constant. As can be seen in Lykke,[6] the experimental cross section is an excellent fit to (4.4.2), providing a threshold energy $E_t = 0.7542$ eV. The function was found to be valid to *at least* 0.074 eV above threshold.

4.2. Ground State Detachment in dc Fields

The application of a static field affects the long-range propagation of the photoelectron, and the behavior of the cross section at the lowest threshold changes dramatically: wave-like structure or "ripples" can clearly

---

*The origin of this law can be easily seen in terms of the center-of-mass momentum k of the outgoing electron. We simply plug into Fermi's golden rule a phase-space factor, which goes as k, and the part of the matrix element squared that goes as k^{2l} for angular momentum l (see Friedrich[17]).

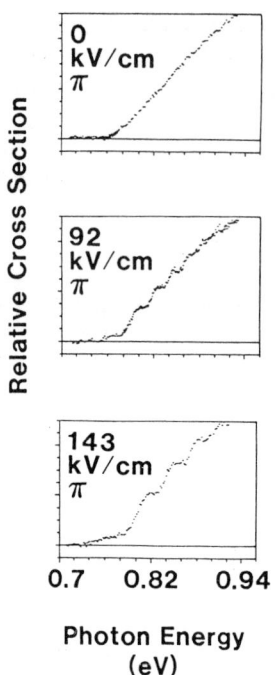

FIGURE 4. Relative cross sections for π polarization with external electric fields of 0, 92, and 143 kV/cm applied to the interaction region.[18]

be seen in the plot of cross section *vs.* energy for applied fields as small as 64 kV/cm.[18,19,20] Such effects were first predicted by Fabrikant.[21]

Measurements of the H^- photodetachment thresholds in the presence of static fields first exhibited the expected ripples in 1987.[18] Figure 4 shows the experimental results for applied fields of 0, 92, and 143 kV/cm. These data were taken in the presence of a static electric field produced by stainless steel plates positioned above and below the laser–ion beam interaction region. A typical experimental setup is shown schematically in Fig. 5. To attain higher electric field strengths in the frame of the ion, a laboratory magnetic field has been used, showing similar effects.[19] By taking advantage of the Lorentz transformation, electric fields of up to 10^6 V/cm in the ion's center of mass have been generated.

4.2.1. π vs. σ Polarization. An important consideration in these studies is the polarization of the incident light. Modulations on the cross section at threshold are only observed when the incident light is π-polarized (electric vector of laser light parallel to direction of external electric field). Why is this so?

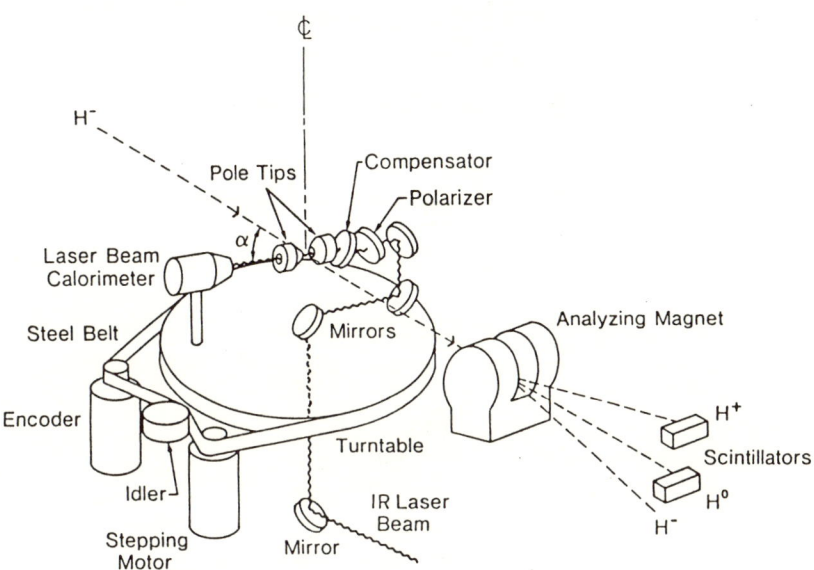

FIGURE 5. Schematic of LAMPF apparatus for studies of photodetachment in an electric field using an electromagnet.

A simple stationary-state picture gives one an intuitive feeling for the difference between π and σ (electric vector of light perpendicular to the external field) laser polarization influences. In photodetachment the electron is predominantly ejected along the electric vector of the light. Thus, when the light is π-polarized, most electrons are ejected parallel to the external field. The wave function can then reflect from the sloping barrier of the potential and interface with the outgoing part of the wave near the origin. In the case of σ-polarized incident light, most detached electrons travel initially perpendicular to the external field, so that the probability for reflection from the barrier is small and ripples are not observed (see Fig. 6).

4.2.2. Theory: Autocorrelation Approach. Since no resonances are known to exist near the lowest threshold, the field-induced modulations near threshold are caused simply by the outgoing wave function reflecting from the potential barrier arising from the applied field. The structure has been predicted and described in terms of a time-dependent autocorrelation in the outgoing wave function.[18,22] It agrees well with experiment up to about 100 meV above the threshold, where theory begins to diverge from the measurements (see Bryant et al.,[18] Fig. 2a). The energies of the ripples are spaced according to the extrema of the Airy function which solves the

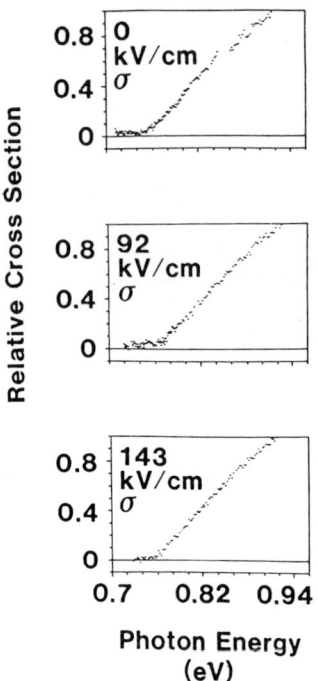

FIGURE 6. Relative cross sections for σ polarization with external electric fields of 0, 92, and 143 kV/cm applied to the laser–H$^-$ beam interaction region.

one-dimensional Schrödinger equation:

$$\left(\frac{\hbar^2}{2m}\right)\frac{d^2\psi}{dx^2} + (E - eFx)\psi = 0 \qquad (4.4.3)$$

predicting that the structure should scale as $F^{2/3}$. The data, however, show the experimental oscillations to be out of phase with those of the Airy function.

4.2.3. Theory: Semiclassical Closed Orbit Approach. A remarkably simple formula for the cross section at threshold in a field was formulated by Du and Delos[23] in a semiclassical closed orbit theory. They find

$$\sigma = 0.3604 \, F a_0^2 \frac{D(x)}{(E_b + E)^3} \qquad (4.4.4)$$

where E_b is the binding energy of H^-, $x = (2m)^{1/3}E/(eF)^{2/3}$, and for π-polarization

$$D(x) = \frac{1}{4\pi}[\tfrac{4}{3}x^{3/2} + \cos(\tfrac{4}{3}x^{3/2})] \tag{4.4.5}$$

which is strictly valid for $x \geqslant 4$, and a good approximation for all $x > 0$. The oscillations come from the cosine term, which may be written as $\cos\phi$, where ϕ is the Aharonov–Bohm[24] scalar potential phase shift:

$$\phi = \oint V(t)\,dt = \frac{4}{3}\frac{\sqrt{2m}E^{3/2}}{eF} \tag{4.4.6}$$

The minima of the ripples are found by setting $\phi = b\pi$, with b an odd integer.[19] Thus

$$E_{\min} = \left(\frac{1}{2m}\right)^{1/3}\left(\frac{3}{4}b\pi eF\right)^{2/3} \tag{4.4.7}$$

Table II compares these values with the experimentally extracted values of Stewart.[25]

4.2.4. Theory: Frame Transformation Approach. Following the frame transformation approach of Fano[26] and Harmin,[27] Rau and Wong[31] adapted the theory to H^- photodetachment in an electric field. Basically, the method treats the photoabsorption process in spherical coordinates, and the electron escape process in cylindrical coordinates. The justification for

TABLE II
Average Values for $E_{\min}^{3/2}/eF$ (Treated as a Parameter) under the Heading "Exp." for Each Order of Oscillation b in the Threshold.[19]

b	$E_{\min}^{3/2}/eF$	Exp.
1	1.77	1.75 ± 0.16
3	3.68	3.67 ± 0.13
5	5.17	5.15 ± 0.21
7	6.48	6.49 ± 0.14
9	7.66	7.66 ± 0.24
11	8.76	8.79 ± 0.17
13	9.79	9.81 ± 0.16

this "frame transformation" follows from the conclusion that the Stark potential is small when compared with the Coulomb potential for $r \ll F^{1/2}$ (a.u.). That is, the applied field F is completely negligible for $r \leqslant 100$ a.u. The transformation between these different coordinate bases or "frames" is facilitated through the use of a reaction matrix or "K-matrix,"[28] and the cross section is found to be the zero-field cross section times a modulating factor, describing the observed oscillations fairly well. The Rau and Wong[31] results also appear to be in excellent agreement with those of Du and Delos.[23]

Frame transformation has also been used by Fabrikant[29] to calculate the cross section including the effect of rescattering of the photodetached electrons from the electric field potential barrier. He finds that these effects should be observable for fields greater than 1 MV/cm when the photons are π-polarized.

4.3. $N = 2$ Threshold

The threshold energy dependence of the cross section for photodetachment of an electron from H^- leaving the final state H atom in the excited state with $N = 2$ is quite different from the threshold for leaving H(1) behind. The outgoing electron views the H(2) atom as a permanent dipole, meaning that it sees a *long-range* $1/r^3$ force. Thus the Wigner law does not hold as it does for the induced dipole force which falls much faster ($1/r^5$). The cross section at this particular threshold is further complicated by the appearance of the doubly-excited shape resonance immediately above threshold (Fig. 7). A shape resonance (described in detail in Section 5.4) is definitely known to affect an H^- threshold only for $N = 2$. No functional form is known that exactly describes this resonant cross section, but the five-parameter Fano function[32] [Eq. (4.5.6)] is in fairly close agreement with experimental data, as demonstrated in Section 5.4. This describes the shape resonance itself, but really does not tell us much about the threshold behavior.

When static electric fields are applied, the threshold behavior is not observed to change much in this region of the spectrum, although multichannel quantum defect (MQDT) calculations have predicted a small change in the rising slope depending on field strength.[33] An experimental resolution of 2 meV, attainable at LAMPF, but not yet applied to this region, should be able to show this effect.

4.4. Higher Thresholds in Zero Field

Characteristics of long-range interactions, specifically the effect of the dipole potential, can be examined in H^- photodetachment. For $N > 2$ the

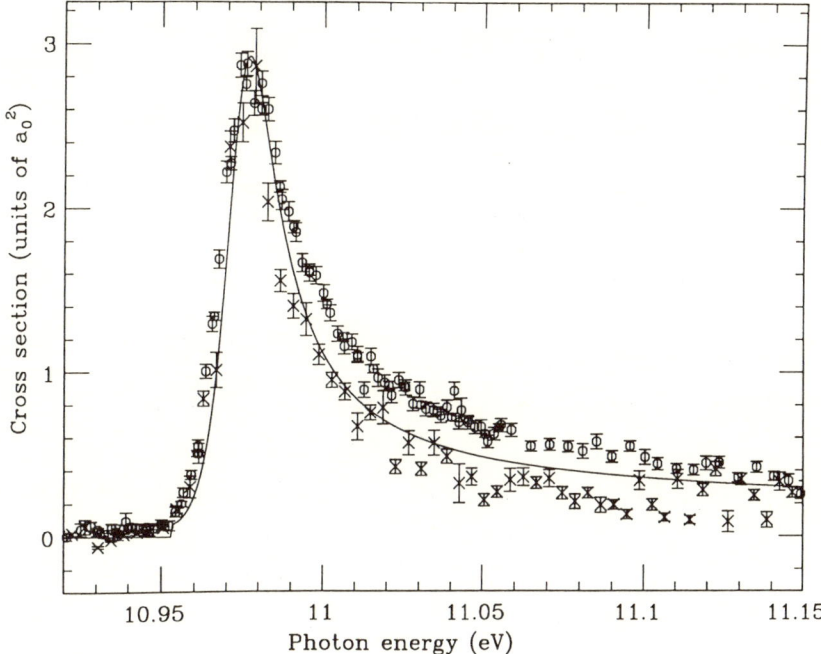

FIGURE 7. Partial cross-section *vs.* photon energy above the $n = 2$ threshold.[70] Experimental amplitudes have been normalized to theory. Error bars are statistical only. Open circles, 1983 data. Crosses, 1990 data. Solid line is the theoretical profile of Sadeghpour *et al.*[38]

exponent in the Wigner law becomes complex,[34] so that one expects oscillations in the cross section near threshold.[35] Greene and Rau,[36] however, in a calculation based on a Coulomb-dipole description of the ion, show that the amplitude of the oscillations should be extremely weak and unobservable by current experimental methods. Given the validity of this point of view, the leading term in the photodetachment amplitude dominates, and the cross section is constant, meaning that the threshold should look like a step function for $N > 2$.[37] Experimental data[42] seem to fit a step function well for the thresholds H(5) and H(6). The fits provide threshold energies of 13.808 (13.975) eV for the $N = 5(6)$ hydrogen production threshold. The H(4) threshold fit is interesting in that the confidence level (CL) for this threshold to be a step function is found to be only 0.1%, while the $N = 5$ and 6 thresholds both have a 95% CL. A small bump appearing in the cross section near the H(4) threshold is intriguing. A repetition of this measurement with better sensitivity is needed to show if this feature actually exists, and might prove or disprove the prediction[38] of a shape resonance above the $N = 4$ threshold.

4.5. Higher Thresholds in Applied dc Fields

For all thresholds, in the presence of an applied dc field, one might expect to see the nearly periodic field-induced modulations of the type observed in the lowest threshold data. After all, they are caused by the wave function of the outgoing electron reflecting from the linearly increasing potential barrier produced by the field. The physics is somewhat altered however when the H^0 neutral core is in an excited state. The outgoing electron's wave function is then no longer only weakly perturbed by a small residual atom in its ground state, but rather it is moving in the vicinity of the much more extended excited state whose contribution to the potential tends to wash out the oscillatory structure expected from barrier reflection.[39] In addition, the numerous resonant states just below threshold can cause phase changes that greatly increase the complexity of the problem. Rigorous theoretical calculations including the barrier reflection effect are needed for these higher thresholds.[40]

Some theoretical work on the effects of external fields on the $N = 4$ and 5 thresholds has been carried out by Zhou and Lin[41] using hyperspherical coordinates. This work was an attempt to interpret experimental observations[20,42] of threshold shifts. The measured thresholds* displayed unexpectedly large shifts under the influence of applied static electric fields ranging from 0 to 90 kV/cm. A schematic of the experimental setup can be seen in Fig. 8. For a comprehensive discussion of these experiments, see Harris[43] and Halka.[45]

The excited neutral hydrogen atoms $H(N > 1)$ formed by the photodetachment of H^- are field-stripped and detected as protons downstream in a scintillation counter. An ionizing magnet located between the interaction region and the scintillator allows the selection of the specific N state to be monitored. For example, at CM photon energies between 13.50 and 13.80 eV, only H atoms with principal quantum number $N \leqslant 4$ are produced.† The ionizing magnet is set to a field which strips $H(4)$, but leaves lower states unaffected. Thus in the range of photon energies between 13.5 and 13.8 eV, only the $H(4)$ final states are detected, so that the cross section for this particular channel can be cleanly studied. The resulting protons are deflected by a charge-separating magnet into a separate counter from the neutral atoms, and constitute the signature for photodetachment when observed in coincidence with a laser pulse. At higher photon energies, atoms with $N = 5$ and 6 were counted in a similar manner.

Figure 9 plots the amplitude of the threshold shift ΔE relative to the zero-field threshold for each field strength. These decreases in the onset

---

*The use of the term "threshold," which is not well-defined in the presence of fields, is here taken to mean the apparent threshold or onset of electron detachment.

†This is strictly true only for $F = 0$ (see Halka et al.[42]).

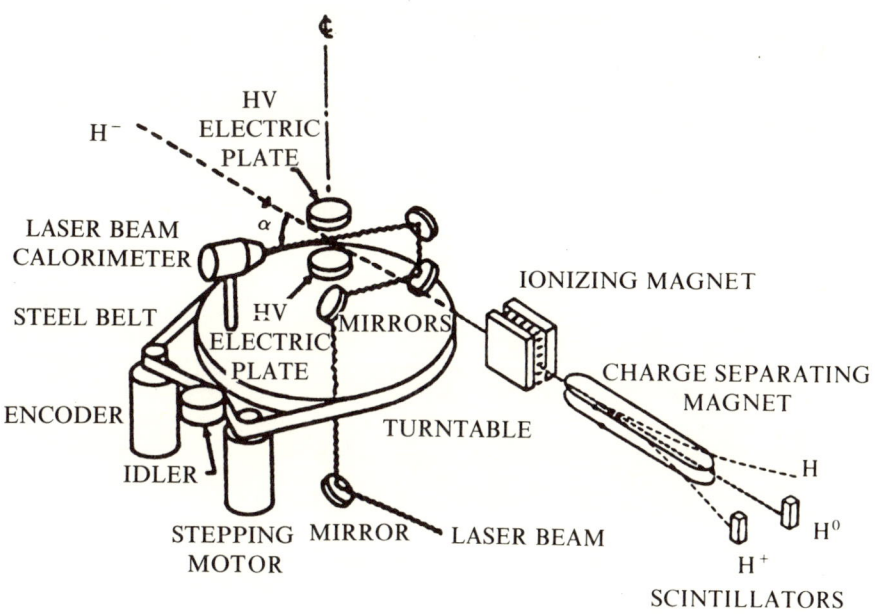

FIGURE 8. Schematic of LAMPF apparatus for studies of photodetachment in an electric field using a parallel plate capacitor.

energy of the particular channel are nearly an order of magnitude larger than what one would expect from Stark splitting of the H levels, as shown by the dashed lines. The solid lines are fits to a function which is proportional to $F^{2/3}$, first suggested to us by Fabrikant.[45] The shift should have this dependence if it results from field-lowering of the potential barrier seen by the outer electron. Classically

$$V(R) = \frac{-a}{2R^2} - eFR \qquad (4.4.8)$$

where $-a/2R^2$ is the dipole potential and $-eFR$ is the potential due to the external field F. The dipole parameter a is the strength of the effective dipole resulting from the Stark-mixing of the degenerate states associated with N. The classical threshold shift is found by taking the derivative with respect to R to find $R \equiv R'$ at which V is maximum. The shift is given approximately by

$$V(R') \equiv \Delta E = -\tfrac{3}{2}|a|^{1/3}F^{2/3} \qquad (4.4.9)$$

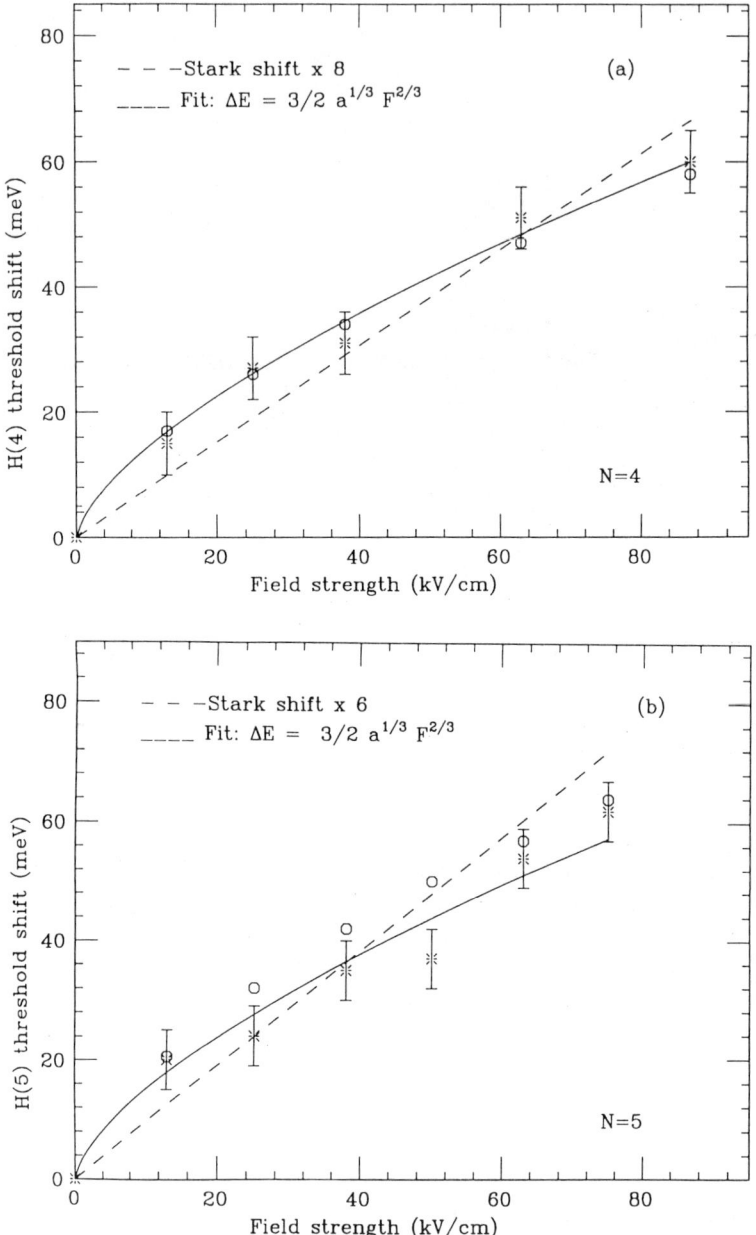

FIGURE 9. Threshold shift ΔE relative to the zero-field threshold vs. field strength.[42] Solid line is a fit to the function $\Delta E = \frac{3}{2}|a|^{1/3} F^{2/3}$ with the dipole moment a as a parameter. Open circles indicate values from Zhou and Lin.[41] (a) H(4) threshold shift values. Fit gives $a = 11.04 \pm 0.19$ a.u. Dashed line shows that the shifts are eight times larger than those expected from the first-order Stark effect in H. (b) H(5) threshold shift values. Fit gives $a = 13.03 \pm 0.17$ a.u. Dashed line shows that the shifts are six times larger than those expected from the first-order Stark effect in H.

Least-squares fits to the observed threshold shifts — keeping $|a|$ as a parameter — provide $|a| = 11.0 \pm 0.2$ (13.0 ± 0.2). The value of a depends on the nature of the correlation between the motions of the two electrons, and, as Zhou and Lin[41] have shown, this correlation changes with the applied electric field. Thus, values of $|a|$ (18.5 for $N = 4$ and 37.8 for $N = 5$) computed for the zero-field case are not those which should be entered into Eq. (4.4.9). Values for ΔE from their quantum mechanical hyperspherical calculations are plotted as open circles in Fig. 9. Excellent agreement with experiment is seen for the H(4) threshold, while theoretical values are somewhat high compared with the experimentally measured H(5) threshold shifts.

The H(6) yield was examined in only two field strengths, 13 and 25 kV/cm. The threshold shifts to lower energies as it does in the other channels, but no unusual structure is observed in the $N = 6$ continuum at these field strengths.

4.6. Two-Electron Detachment Threshold

The classic work on double detachment from negative ions is that of Wannier.[50] The key point in his theoretical analysis is the same as applied in the Wigner law for one-electron detachment: the energy dependence of the cross section at threshold is determined only by the final state. Consideration of the initial state and the transition operator is unnecessary, greatly simplifying the analysis. It is the long-range behavior, when the electrons are far from the proton ($R \gg Z/E$ in a.u.), that is important in the three-body breakup.

Wannier's analysis employs hyperspherical coordinates (see Section 5.3) as the practical coordinate basis to follow the evolution leading to breakup by varying R. Classical mechanics is applicable to this problem in the asymptotic region, and the cross-section dependence on energy is found to be

$$\sigma(E) \propto E^{\mu/2 - 1/4} \qquad (4.4.10)$$

where E is the energy above threshold, and

$$\mu = \frac{1}{2}\left(\frac{100Z - 9}{4Z - 1}\right)^{1/2} \qquad (4.4.11)$$

For H^-, $Z = 1$ gives $\sigma(E) = E^{1.127}$.

In a recent calculation, however, Kazansky and Ostrovsky[51] found the exponent to be strongly dependent on the reaction volume size R over which integration is performed. For small E their exponent approaches that of Wannier only for very large R. Their analysis also shows that for energies E more than 1 eV above threshold, the Wannier law breaks down completely.

According to Wannier, electron correlations are particularly strong right at the breakup threshold with $r_1 \approx r_2$ or $E_1 \approx E_2$. This means that the orbits which are concentrated near $\alpha \equiv \tan^{-1}(r_2/r_1) = \pi/4$ — on the Wannier ridge — are those which evolve to double escape, leading to configuration where the electrons are ejected back-to-back ($\theta_{12} = \pi$).

A different treatment of the double detachment threshold law, put forth by Temkin,[52] gives a modulated linear form to the cross section, stemming from his deduction of an asymmetric escape configuration. His model is called the "Coulomb-dipole," where the electron–electron interaction is approximated as the dipole term of its expansion. In this case, one electron would be excited much more than the other, and sequential, rather than simultaneous, escape is expected.

Greene and Rau,[36] however, maintain that while these oscillations should in principle exist *if* escape is asymmetric, for finite large α as given by Temkin[52] the oscillations would be negligibly small, since the oscillating term is modified by a factor $e^{-\pi\alpha}$. Greene and Rau find for this situation that α is infinitely large, and therefore the term containing oscillations should vanish.

It is not yet clear from experiment which of these analyses comes closest to the true behavior.[53] Evidence supports the Wannier theory within experimental error, but the resolution has not been good enough to rule out small modulations that would support Temkin's analysis (Fig. 10). This important question, which bears directly on a more complete understanding of electron correlations and chaotic behavior just below the double detachment threshold should be pursued in more precise experiments.[54]

5. DOUBLY-EXCITED STATES

5.1. Introduction

Theoretical work on quasi-bound states, in which both electrons are excited out of their ground states, was stimulated by experimental observation. Before 1957 it was generally understood that negative ions should have no structure in the detachment cross section above the binding energy of the outermost electron. An indication to the contrary was provided in an electron–helium scattering experiment when Schulz[55] observed a linear rise in the cross section with energy ΔE above the single-electron detachment threshold, whereas the Wigner theory[56] predicted a $(\Delta E)^{1/2}$

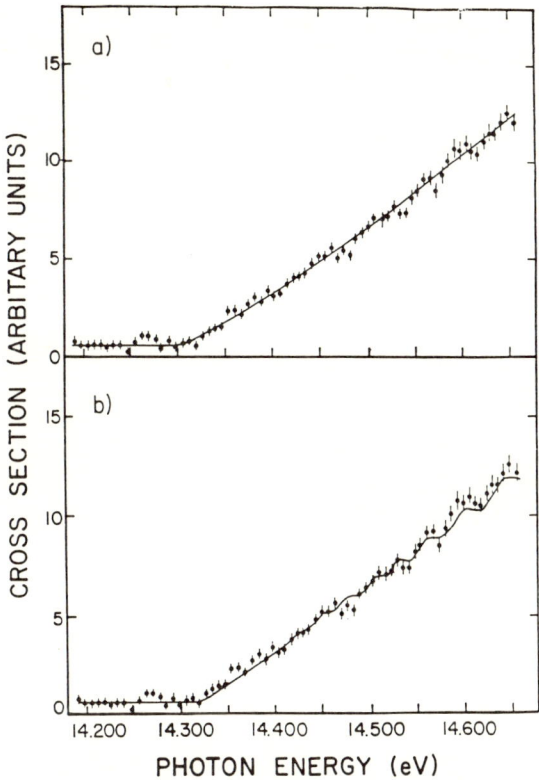

FIGURE 10. Cross section for the two-electron photodetachment process. (a) The curve is the result of a best fit by a power law. (b) Curve is the result of a best fit by a modulated linear law as suggested by Temkin.

dependence. This behavior was interpreted[57] as evidence for at least one Breit–Wigner type of resonance above threshold, a doubly-excited state embedded in the continuum of free electrons.

The first clear experimental evidence of such resonant states produced by photoabsorption appeared in 1959 in inelastic scattering of electrons by helium.[32,58] The data showed two asymmetric peaks which Fano[32] has attributed to the interference of the continuum with discrete, doubly-excited, autoionizing states. The inclusion of the continuum means that many different configurations can have the same energy, and ordinary perturbation theory is inadequate. Making use of the configuration interaction (CI) method, Fano presented a functional form which describes the profile in terms of five parameters: the central energy E_0, the width Γ, an asymmetry parameter q, and two parameters describing the strength of the cross section, σ_a and σ_b. Here we develop his expression for the cross section near a resonance by considering the time dependence of the amplitude of a decaying state:

$$\Psi_R(t) = A(r)e^{i\omega_0 t}e^{-t/2\tau} \tag{4.5.1}$$

where τ is the mean lifetime. Its amplitude in the frequency (energy) domain is

$$\Psi_R(\omega) = \int \Psi(t)e^{-i\omega t}\,dt = A[(\omega - \omega_0) + i/2\tau]^{-1} \tag{4.5.2}$$

or

$$\Psi_R(E) = A[(E - E_0) + \Gamma/2]^{-1} \tag{4.5.3}$$

where $E = \hbar\omega$ and $\Gamma = \hbar/\tau$. Then

$$\Psi_R^*\Psi_R = A^2[(E - E_0)^2 + \Gamma^2/4]^{-1} \tag{4.5.4}$$

a Lorentz-type profile. What happens if this resonance is embedded in a continuum? Say the amplitude of the continuum is $\Psi_C = B\exp(i\phi)$. Then the intensity is given by

$$\Psi^*\Psi = (\Psi_R + \Psi_C)^*(\Psi_R + \Psi_C) \tag{4.5.5}$$

After much algebra this leads to the Beutler–Fano profile or "Fano function"

$$\sigma(\varepsilon) = \sigma_b + \sigma_a \frac{(q + \varepsilon)^2}{1 + \varepsilon^2} \tag{4.5.6}$$

with $\varepsilon = 2(E - E_0)/\Gamma$. The parameters σ_a, σ_b, and q are related in a complicated way to A, B, and ϕ.

In 1962 resonance structure appeared below the $n = 2$ excitation threshold in close coupling (CC) calculations of Burke and Schey[59] for electron–hydrogen scattering. In 1963 Madden and Codling[60] used the 180-MeV electrons synchrotron at the National Bureau of Standards (now NIST) to photoionize helium. Many different photon energies were available, as the synchrotron has a range of wavelengths from the infrared down to 10 nm. A beautiful Rydberg sequence of eight resonances was observed converging to the $He^+(n = 2)$ threshold.

Cooper et al., in the same issue of *Physical Review Letters* as Madden and Codling's report, interpreted these resonances as belonging to the "+"

series of

$$\Psi(2n \pm) = \frac{1}{\sqrt{2}} [u(2snp) \pm u(2pns)] \qquad (4.5.7)$$

where the symmetrized independent-electron wave functions $u(2snp)$ and $u(2pns)$ have been replaced by symmetric and antisymmetric combinations to form the wave function $\Psi(2n \pm)$ of the new states, which reflects the interaction between the electrons. That is, there are two symmetrized independent wave functions that could describe the lowest double excitations. These authors supposed that they combined to give two series: "+" and "−". One can picture the two electrons as two balls or pendulums bouncing on opposite sides of a wall (see Fig. 11). The wall in our case is

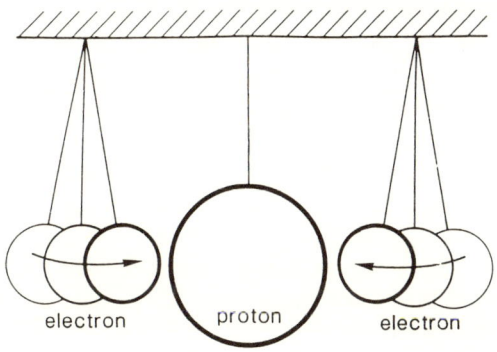

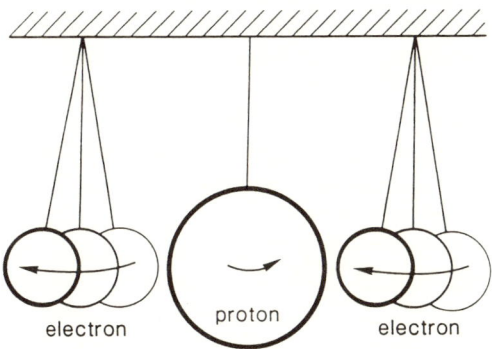

FIGURE 11. Heuristic idea of "+" and "−" states of H⁻. Top picture is the "+" mode. Bottom picture is the "−" mode.

the infinitely massive proton. In the "+" mode, the balls approach the wall and bounce off simultaneously, oscillating in phase with each other. This is analogous to the electrons in the real atom simultaneously overlapping the same orbital. In the "−"mode, one approaches the wall while the other moves away, so the oscillation is out of phase. The electrons are correlated in both modes, but more strongly interacting in the "+"mode. Since the electrons come closer to each other, the Coulomb force is stronger. In this simple mechanical picture, autodetachment from "−" states is quasi-forbidden, since there is no mechanism for energy transfer from one electron to the other as in the "+" mode. The weaker long-range dipole potential produced by the polarized "core" hydrogen atom controls the action.

5.2. Two Very Different Types of Resonances: Feshbach and Shape

An explanation of the resonant or "quasi-bound" states may be adopted from Bohr's[62] suggestion regarding nuclear resonances. The idea is that a projectile such as an electron approaching a hydrogen target can transfer most of its energy to the proton and the other electron, so that it cannot escape again until a fluctuation causes energy to be transferred back to it. The concept is the same for photoexcitation—a "half-scattering" process. Thus a doubly-excited state may be formed resembling an excited hydrogen atom with an extra electron also occupying an excited state. Such temporarily bound states, in which one electron can be imagined as bound to an excited atomic core, are now called Feshbach resonances because Herman Feshbach developed the quantum mechanical description of this "closed channel" resonance phenomenon.*[63] These resonances generally form a manifold which converges from below on each photodetachment threshold for production of $H(N)$, $N = 2 \to \infty$. Within each series the states become longer-lived as they approach the threshold. Feshbach resonances which approach the $H(n)$ threshold are classified as $H^{-**}(n)$ resonances. They decay by autoionization to a fragment hydrogen atom plus an electron. The hydrogen atom may have a principal quantum number ranging from 1 to $n - 1$ for the resonances discussed here. In this text the principal quantum number of the fragment H atom is denoted by N.

Resonances appearing above a threshold, but near to and associated with it, are denoted "shape" (or "open channel" or "core excited II") resonances, referring to the shape of the potential well in which they are temporarily bound. The potential is a combination dipole potential and centrifugal barrier, so that H^- shape resonances exist only for $l > 0$. In the H^- spectrum, only one $^1P^0$ shape resonance has been observed, that just

---

*In the past these states have also been labeled as core excited I, compound state, or hole–particle resonances.[116]

above the H($N = 2$) threshold (see Section 5.4). R-matrix theory[38] has predicted another just above the H(4) threshold, but this has yet to be experimentally verified. According to Taylor,[114] as l increases, the states should become broader, and cease to exist when l is high enough to produce centrifugal dissociation—when the potential is completely repulsive.

5.3. Convenient Coordinate Representations

The correlated behavior of the two excited electrons is reflected largely in the relative lengths of the radii $\mathbf{r}_1$ and $\mathbf{r}_2$ and the angle between them. The correlations are essentially invariant under rotations or scale changes. Hyperspherical coordinates (HSC) seem to offer an appropriate representation for this situation, allowing the segregation of one coordinate—the hyperradius R—which represents the size of the system. The evolution of correlations can be followed as R increases. For a more complete description of HSC, the reader is directed to Appendix A.

The Schrödinger equation in HSC incorporates what is called a "grand angular momentum" Λ (Eq. 4.A.5) which accounts for the pairwise structure of the total potential. It can be thought of as an orbital angular momentum operator for a single particle in six-dimensional space, and is analogous to the usual orbital angular momentum operator of a particle in three dimensions.

The Schrödinger equation becomes separable if channel couplings are neglected in the adiabatic approximation, where the motion in R is assumed to be much slower than the motion in the angles. This is reasonable because for small R, the centrifugal term dominates over the potential term. Diabatic "channel" functions and eigenpotentials, $U_\mu(R)$, are then obtained by integrating Schrödinger's equation with respect to the five hyperangles, while keeping R as a slowly varying parameter. The channel is denoted by the symbol μ, and depends on the correlations between the electrons. The $^1P^0$ dominant potential curves for H$^-$($n = 2$ to 11) are shown in Fig. 12.

Another less standard coordinate basis which obtains quasi-separability of the Schrödinger equation in an adiabatic approximation is the prolate spheroidal coordinate system. It also is useful for describing symmetries of doubly excited states, and is preferred by some because it more clearly reflects the nodal structure of the states. A description of these coordinates is given in Appendix B.

Calculations of the magnitudes of channel couplings have demonstrated the soundness of the adiabatic approximation. An analysis of weak couplings between different adiabatic channels has shown that the lowest "+" channels in each hydrogenic series wields the controlling influence on the H$^-$ resonant spectrum.[65] This has been borne out by experiment.[46]

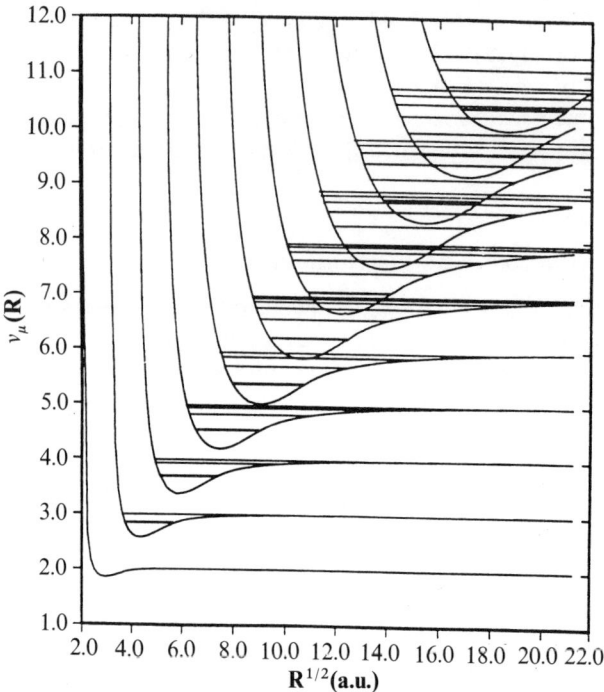

FIGURE 12. Adiabatic potential curves for $^1P^0$ H^- shown as effective quantum number vs. $\sqrt{R}$. Only the lowest "+" channels within each n manifold are plotted along with the level positions in each potential.[65]

5.4. H^- (n = 2) Resonances

5.4.1. Zero-Field Spectrum.
C. D. Lin[66] has worked out hyperspherical potential curves for many doubly-excited states. Since only $^1P^0$ states are dipole-allowed for photodetachment from the ground state, let us consider those eigenpotentials $U_\mu(R)$ associated with $n = 2$, shown in fig. 13. Asymptotically, $U_\mu(R)$ goes as $-3.71/R^2$ for the "−" curve, $2/R^2$ for the "+" curve, and $9.71/R^2$ for the pd curve. The channel labeled "pd" is completely repulsive (no radial correlations) and cannot support any state. It is sometimes called the "0" channel.

The hyperspherical "−" channel (or potential) has a shallow broad minimum and, neglecting fine structure and relativistic effects, can support an infinite number of Feshbach resonances, the first two of which are calculated to have the energies -0.25191 and -0.25006 Ry. (These are 0.00190×13.605 eV = 26 meV and 0.00006×13.605 eV = 0.8 meV below the $n = 2$ threshold, respectively.)

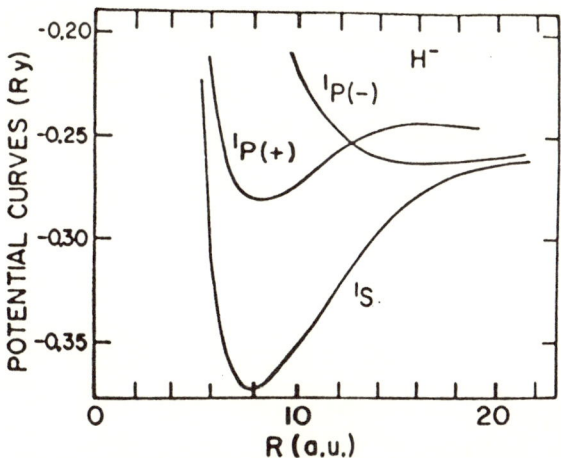

FIGURE 13. Field-free potential curves of H$^-$ converging to the H($n = 2$) limits. The "+" and "−" $^1P^0$ curves support the shape and Feshbach resonances, respectively (from Lin[64]).

The lowest ^1P Feshbach resonance was observed for the first time in photodetachment experiments at LAMPF in 1977.[67] Its energy of 10.9264(6) eV was measured in 1985, and agreed well with theoretical predictions.[68] This resonance is extremely narrow, predicted to be only about 30 μeV in width.[69] The best experimental resolutions have so far been around 2 meV, but there are schemes (see Section 3.2) which could yield a value for a width in the future. Figure 14 shows the experimental data of Halka et al.[70] with theoretical numerical data of Broad and Reinhardt for the lowest Feshbach resonance and the shape resonance.

The second lowest ^1P Feshbach resonance lies nearly on top of the $n = 2$ threshold, embedded in the low-energy shoulder of the $n = 2$ shape resonance, and is much narrower and smaller than the first. It has not yet been observed.

The "+" potential is more attractive at small R than the "−" curve, but it has a potential barrier height of 4.8×10^{-3} Ry (65.3 meV) at larger R. The potential curve thus presents a classical turning point to an incoming particle. If the particle were to tunnel through this barrier, it would find itself in a well bounded by two more classical turning points. Although this potential is not strong enough to support a resonance below the $n = 2$ threshold, it can support a shape resonance above the threshold between the two inside classical turning points.

The H$^-$ shape resonance (SR) was first predicted in 1967 by Taylor and Burke[71] for electron–hydrogen scattering, and by Macek for photoionization of H$^-$. The first observations of this resonance were reported in 1969

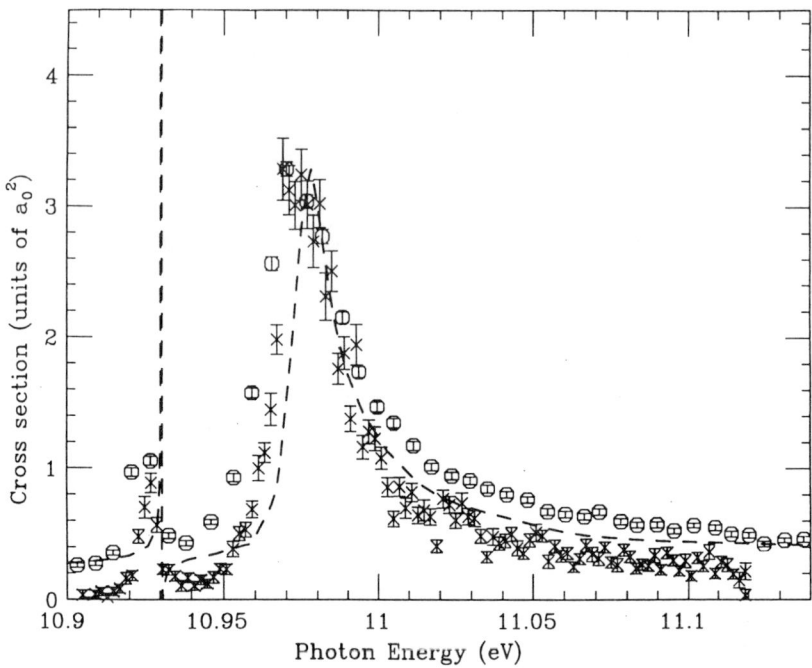

FIGURE 14. Photodetachment total cross section *vs.* photon energy near the $n = 2$ threshold.[70] The narrow Feshbach resonance below threshold is evident near 10.93 eV. Experimental peak amplitudes normalized to theory. Solid line is the theoretical profile of Broad and Reinhardt.[78]

by McGowan *et al.*[72] and later by Williams and Willis[73] in 1974. Both measurements involved electron scattering from neutral hydrogen atoms. A series of high-resolution photodetachment experiments[74] on H$^-$ at LAMPF pinned down the position and width of this feature and led to some understanding of its behavior when subjected to external electric fields. The shape resonance can be spotted with difficulty in the hydrogen-arc emission spectrum;[75] other early attempts at observing the shape resonance in the laboratory[76] and in stellar spectra[77] were less successful.

As the most prominent resonance in the H$^-$ photodetachment spectrum, the SR must also play a prominent role in understanding electron correlations. Unique in that its potential has three classical turning points as discussed above, it is well-known as the only resonance of its type yet observed in the H$^-$ photodetachment spectrum (or equivalently in electron-impact excitation of H atoms). The SR results from a centrifugal barrier potential as noted above, and appears above the threshold for excitation of the H($N = 2$) state. It can therefore autodetach to either the H(2) or H(1) continuum, whereas the lowest Feshbach resonance is energetically capable of decay only to the ground state of neutral hydrogen, the $N = 1$ channel. Since the SR is the only H$^-$ resonance which has been observed to decay to its parent state, its branching ratio $\sigma(2)/\sigma(\text{tot})$ is of interest, and has been extracted using relative measurements normalized to theory.[70] The maximum branching ratio (≈ 0.8) appears at an energy about 20 meV higher than the central energy of the resonance (10.974 eV) (see Fig. 15). The central energy was found by fitting to the Fano function (see Section 5.1) although the appropriateness of this function for fitting this particular resonance is debatable owing to its large width.[70]

Lin finds the central energy E_0 of the resonance 32 meV above threshold at 10.985 eV with a width Γ of 28 meV. (lifetime = 2.35×10^{-14} s). Other more accurate methods[38,78] give $E_0 \approx 10.971$ eV and $\Gamma \approx 15$ meV. The broad width of the resonance comes from the ease with which it can penetrate the barrier. Since the time the system spends between the two inside turning points is small, its energy uncertainty Γ is correspondingly large in compliance with the uncertainty principle.

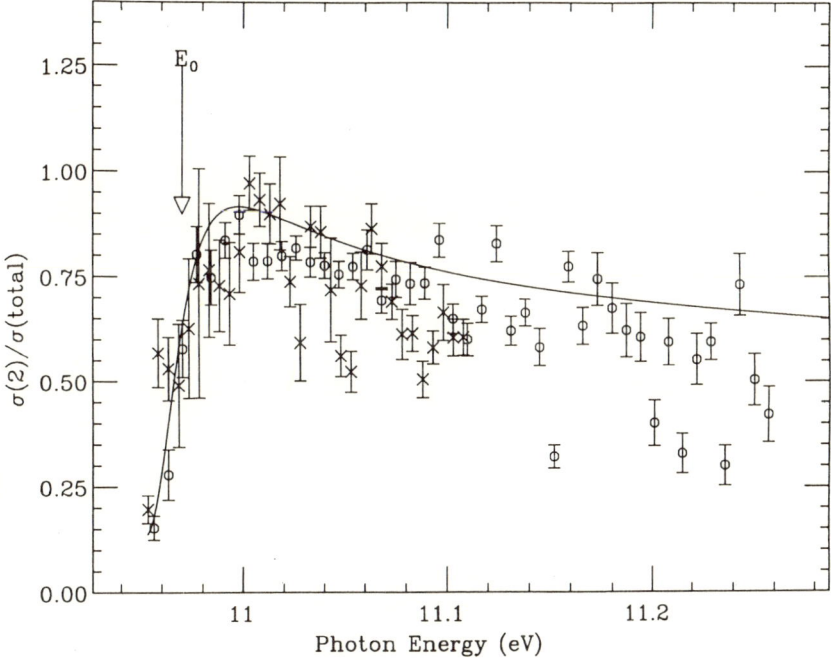

FIGURE 15. Branching ratio *vs.* photon energy for the H$^-$ ($n = 2$) shape resonance. The solid line is the branching ratio prediction of Sadeghpour.[38] The arrow points to the approximate central energy position of the shape resonance.

5.5. Field Effects on $H^-(n = 2)$ Resonances

The $H^-(2)$ lowest Feshbach resonance has been observed to split and quench when electric fields are applied to the photon–H^- interaction region.[67] The splitting into two LoSurdo–Stark resonances[79] corresponding to mixing of the $^1S^e$ state is evident for fields larger than 90 kV/cm (see Fig. 16). Controversy regarding whether the ^1P or ^1S state[80-82] was the lower-lying state seems to have been resolved by data comparison favoring

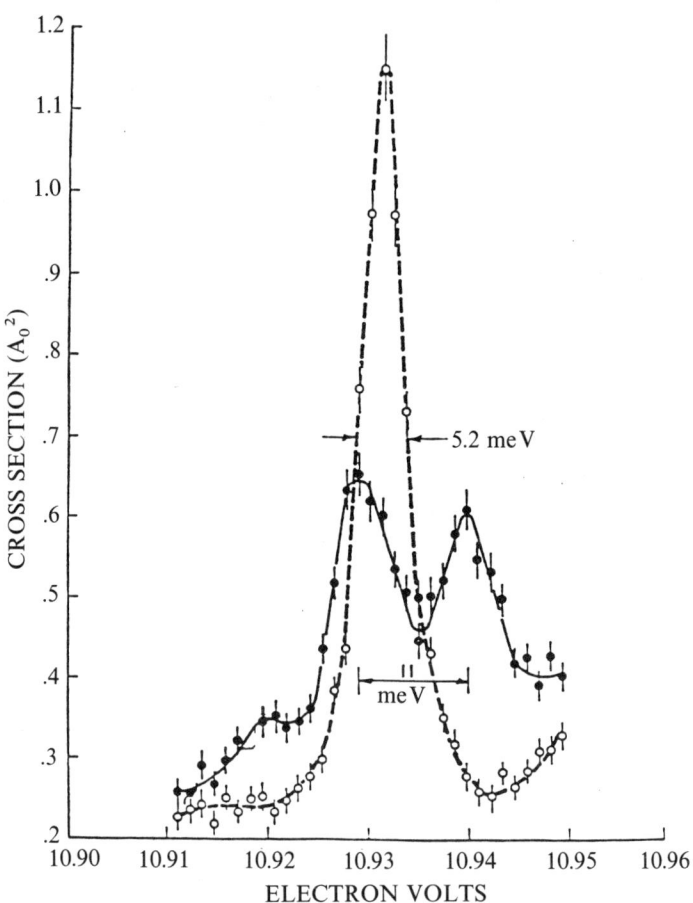

FIGURE 16. Superposition of cross sections of Gram et al.[67] in the $n = 2$ Feshbach region of H^- for two different applied field values. Open circles, zero field. Closed circles, 110 kV/cm in the center-of-mass frame.

the Callaway and Rau[80] and Bryant et al.[83] theoretical results, which place the ^1S above the ^1P state by 2.3 meV.

As the applied field is increased, making escape of the outer electron easier, the two resonances are quenched with the $^1P(^1S)$ branch disappearing when the field reaches about 1.4×10^5 (2.7×10^5) V/cm. For fields greater than 400 kV/cm the LoSurdo–Stark state corresponding to intermingling with the shape resonance of the ^1D state was also observed at 10.87 eV, lower in energy than the ^1P and ^1S admixtures, embedded in the low-energy shoulder of the ^1P shape resonance (see Bryant et al.[83] Fig. 11). The experimental position is in fair agreement with theoretical predictions. With increasing field the ^1D branch moves farther into the shoulder of the shape resonance, so it is impossible to ascertain the quenching field magnitude for this Stark state.

The ^1P shape resonance itself is extremely hardy and is not observed to diminish for fields as high as 2.5 MV/cm.[84,85] An interesting feature is that the width of the shape resonance depends on the strength of the electric field.[40] The trend appears to be a width increase with field strengths up to 2.5 MV/cm, but the evidence for this tendency is by no means conclusive. MQDT calculations of Slonim and Greene[33] predict that a plot of the width vs. the magnitude of the field should show oscillatory behavior. The experimental steps in the electric field strength should be smaller than those used in the cited experiment in order to show this behavior decisively, however — a point to keep in mind for the planning of future measurements. Another problem may lie in the procedure for analyzing the width of the shape resonance, as theories disagree on the best way to parametrize this resonance.[70,83]

The predicted oscillatory behavior of the SR width vs. field strength stems from the theoretical evidence that applied fields should cause the appearance of modulations in the high-energy shoulder of the resonant cross section itself. So far, experimental resolution has not been sufficient to see these modulations. A resolution of about 2 meV should be sufficient to prove or disprove this prediction.

5.6. Higher n Resonances

5.6.1. In Zero-Field (n = 3,4). The first two members of the ^1P Feshbach series converging from below to the hydrogenic $n = 3$ threshold were observed in 1979 at LAMPF[86] at photon energies of 12.650 ± 0.004 and 12.837 ± 0.004 eV (see Fig. 17). These energies were obtained by crossing an ultraviolet laser beam — the fourth harmonic of a Nd:YAG laser — with the LAMPF H$^-$ beam. These resonances have dip-to-peak structure resulting in a negative asymmetry parameter q, and, unlike the $n = 2$ shape resonance,

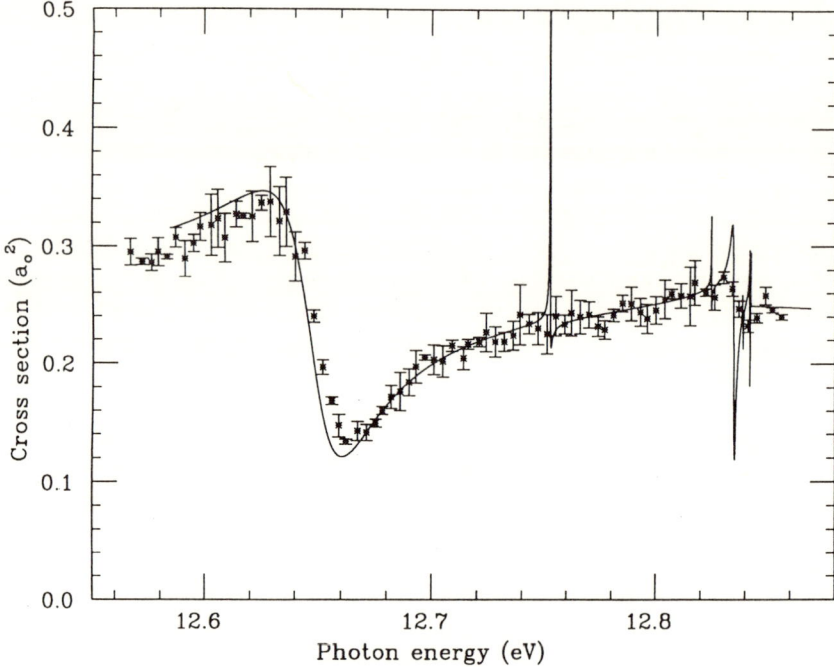

FIGURE 17. H⁻ photodetachment cross section just below the H(3) threshold.[86] Solid curve is the theory of Sadeghpour.[38]

fit a Fano function quite well.* These energies are in good agreement with most theoretical predictions as shown by Hamm et al.[86] The relative total cross sections show good agreement with J-matrix calculations of Broad and Reinhardt[78] and with R-matrix calculations of Sadeghpour et al. [38] See also the recent work of Tang et al.[9]

The H⁻(3) resonances are the highest-n states observed in *total* cross section measurements by the photodetachment experiments at LAMPF. All experiments detecting higher-n resonances measured partial cross sections.

The lowest H⁻(4) resonance was studied in a 1990 two-laser experiment by detecting neutral hydrogen in its $N = 2$ state.[87]† In contrast to the $n = 3$ resonances, the $n = 4$ has a positive asymmetry parameter q. That is, in the profile of the $n = 4$ resonance, the dip occurs at a lower energy than the peak, whereas the $n = 3$ profile is just the opposite. This is also apparent

---

*The Feshbach resonances are much narrower than the shape resonance, so the likelihood of any Fano parameter changing across the width of the Feshbach resonances is slim.

†This experiment also measured $N = 2$ partial decay of the lowest H⁻(3) resonance and confirmed the expected proclivity for this state to decay to its nearest lower continuum channel (see, e.g., Sadeghpour et al.[38] and Chrysos et al.[88]).

in calculations of Sadeghpour et al.[38] It simply indicates that the phase shift induced by continuum interference is different for the different n states. The energy position of the observed $n = 4$ resonance agrees well with theory, but the experimental width is somewhat smaller than theories predict (see Halka et al.[87]). We point out, however, that statistics were rather poor for this experiment.

5.6.2. *In Zero Field (n = 5–8).* In 1989 the LAMPF group[43,46,89] observed a distinctive series of window resonances converging on each of the continuum thresholds, H(n) + e with $n = 5, 6, 7, 8$. At least three resonances of each series could be discerned, associated with each limit. The observation of these resonances was made possible by the development of a new technique, magnetic-stripping of H(n), which allowed the experimenters to study the production of only one n-channel at a time for certain photon energy intervals.

These families of resonances occurred at photon energies $E(m, n)$, which can be described by a remarkably simple formula:

$$E(m, n) = E_t - \frac{Ry}{n^2} + 2\,Ry\, e^{-2\pi(m-n)/\alpha_n} \left[\frac{1}{2n^2} - \frac{0.696}{(n + 0.333)^2} \right] \quad (4.5.8)$$

Here n is the principal quantum number of the hydrogen plus free electron associated with the series, m is the quantum number of the second electron such that $m \geqslant n$, E_t is the energy at the threshold for the continuum, and α_n is given by $\alpha_n = (a_n - \frac{1}{4})^{1/2}$, where a_n is the dipole strength of the relevant photoionization channel. Ry is the Rydberg constant. The numerical values in the formula are the result of a fit of the experimental data with the theoretically-derived formula of Sadeghpour and Greene.[65]

This remarkable formula is reminiscent of the Balmer formula, which was the Rosetta stone for the hydrogen atom. The lowest resonance in each series ($m = n$) is in the form of a quantum-defect Rydberg series, dramatically highlighting the extremely strong correlation between the two electrons: it is as if the two electrons have formed a quasi-particle (with some internal energy given by the quantum defect) which then orbits the proton. The higher resonances within a given family can be viewed as an excited hydrogen atom having a large permanent dipole moment which in turn binds the more distant electron.[90]

5.6.3. *Field Effects.* The lowest $n = 3$ resonance at 12.65 eV was studied experimentally in fields ranging from 0 to 2.36 MV/cm by Cohen et al. in 1987 at LAMPF.[91] In this experiment a Lorentz-transformed barycentric field was provided by pulsed Helmholtz coils surrounding the laser–H⁻ interaction region.[84,85] The amplitude of the resonance decreases with increasing field, reaching about half its original amplitude in a field of about 1.2 MV/cm, and completely disappearing in 2.36 MV/cm. There was no evidence of a shift in the position or width of the resonance with changing field, but this finding may not be the final word. The parameters were found using a fit to the Fano function, which is a good fit for the zero-field case, but gives large errors in the parameters when the field is high. An alternate method for finding these parameters is so far not clear, but taking smaller steps in the field strength might show more clearly any change in position and width if statistics are high enough.

The $n = 5, 6, 7$, and 8 resonances all appear as "window" resonances, or dips in the photodetachment spectrum. Windows are wavelength regions for which a gas of H⁻ would be relatively transparent. In the 1989 LAMPF experiment, at least three resonances were clearly visible converging on each threshold (Fig. 18). The observed resonances are all of the "+" type, meaning that those associated with the lowest "+" diabatic channel in each n-manifold dominate in photodetachment of ^1P resonances. This is demonstrated theoretically by Sadeghpour and Greene[65] using hyperspherical treatment of adiabatic potential curves and calculated diabatic curve crossings. They develop a generalized two-electron formula which seems to describe the observed resonance positions quite accurately, although, interestingly, the energy of the second lowest resonance in each channel seems to be underestimated by 5 to 9 meV.

At higher photon energies the Feshbach resonances begin to approach and even overlap the region of the threshold *below* the one for which they are labeled. For example the lowest $N = 8$ resonance has been observed to modify the $n = 7$ threshold.[46] At even higher energies one should be able to observe the lowest $n = 9$ resonance *below* the $n = 8$ threshold. The number of resonances behaving in this way should proliferate near thresholds of higher n.

The resonances below the $n = 5, 6$, and 7 thresholds also diminish in the presence of electric fields[46] up to 87 kV/cm. The maximum field strengths F_Q needed to quench the mth resonance as observed in the $H(n - 1)$ channel are shown in Table III, where $m = n$ indicates the lowest (in energy) resonance in a series.

The maximum field strength used in this experiment was 87 kV/cm for the $n = 5$ and 6 resonance spectra and 25 kV/cm for the $n = 7$ resonance spectrum. The lowest field strength (greater than zero) was 13 kV/cm — the fourth $n = 6$ resonance would probably quench in a lower field. The higher energy resonances are more easily quenched because the outer electron is more weakly bound in the higher energy states, and the field assists in detaching it. Field-induced tunneling, however, also contributes to the apparent "quenching" in these partial cross-section measurements, effectively changing the branching ratios.

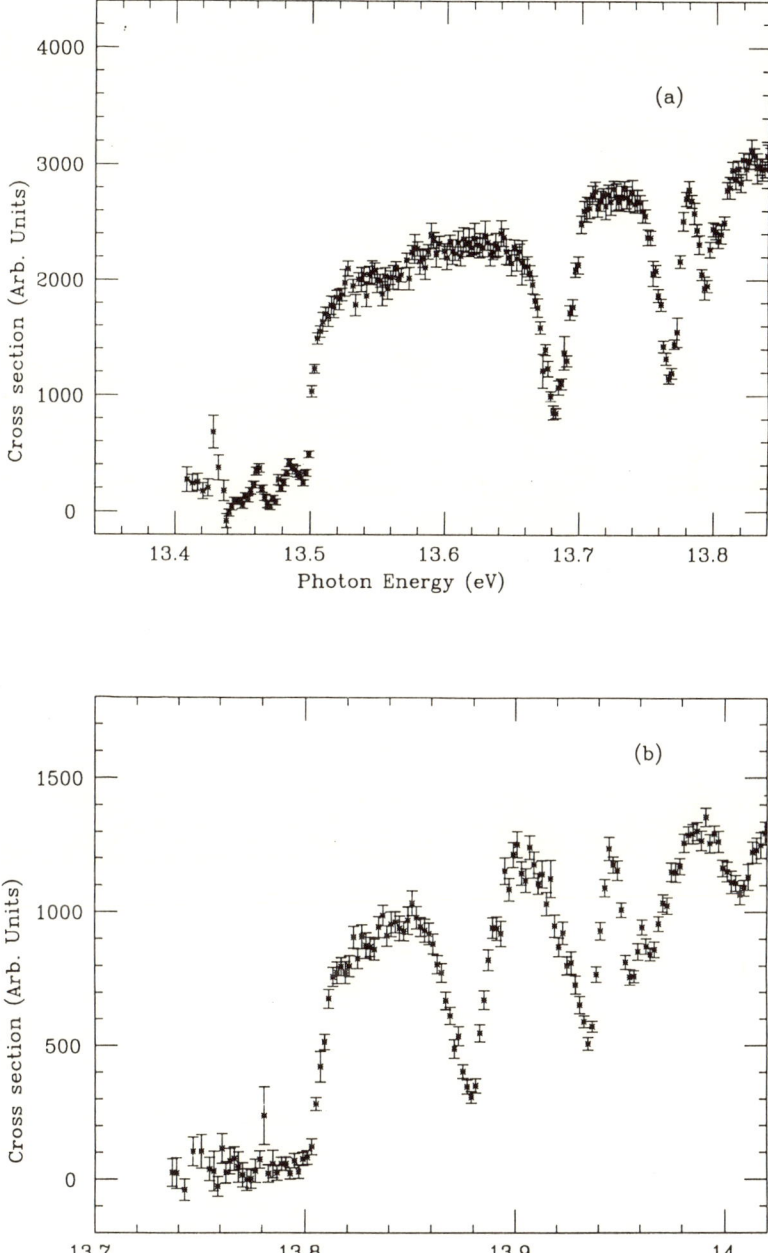

FIGURE 18. H⁻ partial photodetachment cross section. (a) H(4) production by photodetachment below H(5) threshold. (b) H(5) production below H(6) threshold. (c) H(6) production below H(7) threshold.

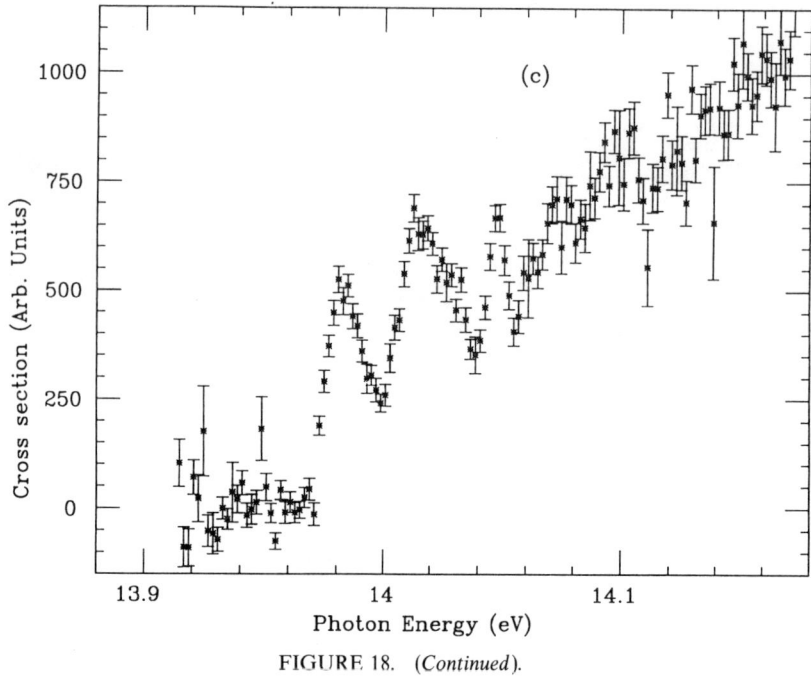

FIGURE 18. (Continued).

The lowest resonance in each series is quite stable as compared with the others, and its central energy and width can be monitored as a function of field strength. A recent analysis[42] suggests that the widths and positions of these resonances may in fact be oscillating with field strength. That the positions may oscillate with the field may be conceptually understood in terms of spectral repulsion by Stark states caused by mixing in of the ^1S or

TABLE III
Maximum Field Strengths F_Q Needed to Quench the mth Resonance as Observed in the $H(n-1)$ Channel.[43,44]

n	m	F_Q (kV/cm)
5	7	13
	6	87
6	9	13
	8	13
	7	50
	6	87
7	9	25

^1D for example. An intuitive understanding of the possible width oscillation is not easy, but this behavior has been predicted in MQDT calculations of Slonim and Greene.[33]

5.6.4. The Induction of New Structures by the Applied Field. With a field applied, a new structure is observed in the region below the zero-field threshold. This may be partly attributable to field-assisted tunneling of $H^{-**}(n)$ doubly-excited autodetaching resonances which converge from below to each H(n) threshold. (A description of these states can be found in Harris et al.[46]) In zero field these are observed only in the $H(N \leq n-1)$ continuum because the inner electron must exchange energy with the outer one if autodetachment is to occur. As suggested by Lin,[47] however, the field may supply the outer electron with enough energy to allow autodetachment without the participation of the inner electron. It would therefore remain in the n level, and the resonance would be observable in the $H(N = n)$ channel. Order-of-magnitude calculations show that fields used in this experiment are of sufficient strength for this process to occur. More detailed calculations of Zhou and Lin[41] explain this behavior in terms of the changing shape of the potential curves in a field.

A particularly intriguing change in the $H(N = 4)$ cross section appears in the $F = 87$ kV/cm data, where a dip develops which is not seen in lower field strengths. A fit to the Fano function (see Section 5.1) places this feature at 13.513 ± 0.001 eV — 10 meV higher in energy than the zero-field threshold. The width from the fit is 15 ± 6 meV. It has been suggested that this feature is the result of the modification of the $^1P^0$ "+" potential curve where an effective centrifugal potential barrier is induced when a field is applied.[41] This new potential does not modify the positions of the first two resonances associated with the "+" channel, but the third resonance is lifted to a position above the zero-field threshold. Without cross-section calculations, it is unknown whether this feature should emerge as a peak or a dip, but the energy of the observed resonance compares favorably with that calculated by Zhou and Lin using the WKB approximation.

6. ANGULAR DISTRIBUTION OF PHOTOELECTRONS

The differential cross section for ejection of a single photoelectron (in a single-electron atom) in the direction (θ, ϕ) is[4]

$$\frac{d\sigma(\theta, \phi)}{d\Omega} = 4\pi^2 e^2 \frac{\omega}{c} |\Sigma_{lm} Y_{lm}(\theta, \phi) \langle \varepsilon lm|z|0\rangle|^2 \quad (4.6.1)$$

where $\langle \varepsilon lm|z|0\rangle$ is the dipole amplitude, $\varepsilon = k^2/2$ is the total energy of the electron, and $|0\rangle$ is the initial (ground) state of the atom. For detachment from the ground state ($J = M = 0$) by linearly polarized light, one finds

$$\frac{d\sigma(\theta, \phi)}{d\Omega} = \frac{\sigma_{tot}}{4\pi}[1 + \beta P_2(\cos \theta)] \qquad (4.6.2)$$

Here θ is the angle between the direction in which the electron is ejected and the light polarization vector $\hat{\varepsilon}$, and $P_2(\cos \theta) = (3 \cos 2\theta + 1)/2$. For unpolarized light, θ is the angle between ejection and light propagation, and the asymmetry parameter of the angular distribution β is replaced by $-\beta/2$ in Eq. (4.6.2).

In a theoretical analysis of the relationship of relative orientations of angular momentum and radial vectors with the approximate angular correlation quantum number T (described in Appendix A), Greene[92,93] shows that $\beta \approx -1$ for double photodetachment of H^- in its ground state, meaning that very near the detachment threshold energy the observed electron escape should be predominantly orthogonal to $\hat{\varepsilon}$ (sideways photoejection) owing to torque exerted by the unobserved excited electron.

The above predictions of Greene apply only to nonresonant photodetachment differential cross sections, away from any autodetaching doubly-excited state. For the case of resonant cross sections just below the $n = 3$ threshold, R-matrix calculations of Burkov et al.[94] find a large variation in the asymmetry parameter β ranging from about 0 to 2 across the width of the lowest resonance in this channel.

Calculations of β-parameters for higher n states have not been carried out, but it is not yet really necessary, as experiment is behind theory in this area. The experimental limitation lies in the relativistic nature of the $n > 2$ experiments. In these experiments, the ejected photoelectron travels with the velocity of the incident beam in the lab frame, making angular ejection measurement virtually impossible.

A recent, very interesting, calculation of Berakdar and Klar[95] predicts circular dichroism in double photoionization of H^-. This means that, in an experiment detecting the two escaping electrons in coincidence, a difference in angular correlation should be observed depending on whether the incident light is right- or left-circularly polarized. This hypothesis also has yet to be checked in the laboratory.

7. MULTIPHOTON STUDIES

When an atomic system is bathed in light of sufficient density, transitions involving the simultaneous absorption of several photons may be observed. In particular, the photodetachment process in H^-, perhaps the simplest, almost archetypic, example of the photoelectric effect, can be accomplished by the absorption of several photons, and as we shall describe below, has been observed to occur with photon multiplicities from one to six. Einstein, and Millikan too, would have been surprised; Maria Goeppert-Mayer,[96] the first to study multiphoton processes in atoms, would have been pleased!

At what photon intensity is multiphoton activity important? We can estimate the intensity required using the following reasonable assumptions: In order that several photons be absorbed together, they must impact on the ion in a time comparable to the period of the light wave. Thus, for n photons to be absorbed, if N is the number of photons per unit area per unit time in the incident beam, σ is the cross section for absorption, and $\tau = 2\pi/\omega$ is the period:

$$(N\sigma\tau)^n \approx 1 \qquad (4.7.1)$$

Then

$$N \approx 1/\sigma\tau \qquad (4.7.2)$$

Putting $\tau \approx 10^{-14}$ s and $\sigma \approx 10^{-17}$ cm^2, we get $N \approx 10^{31}$ photons/s/cm^2. If the energy per photon for H$^-$ detachment is around 10^{-19} J, we find that the intensity (irradiance) would be of the order of $I \approx 10^{12}$ W/cm^2. As we shall see below, multiphoton effects in electron detachment from H$^-$ become observable in the LAMPF experiments at somewhat lower intensities, which are consistent with the above considerations.

The H$^-$ ion is a good "laboratory" for the exploration of multiphoton effects because of its extreme simplicity. It has no intermediate states between the ground state and the single-electron continuum to serve as "stepping stones" that could turn a multiphoton process into an upward cascade. With the relativistic beams technique, one can have tunable, intense lasers and very little background.

Before describing the results of such studies, a further discussion of the important concepts of the multiphoton process is in order: when an electron is ejected from an atom in the multiphoton regime, it finds itself undergoing oscillatory motion in a dense sea of coherent radiation. Take as an example the 10.6 μm beam of a CO_2 laser at an irradiance of 10 GW/cm^2. The period of this light is $2\pi/\omega = 35.3$ fs. It is very simple to solve Newton's second law for the electron in such a field if one neglects, as is reasonable and customary, the magnetic field effects. One finds that, in this laser field of 2.7 MV/cm, the electron will swing back and forth in space with an amplitude of 4.4 Å. Its average kinetic energy $\bar{W}_p$ (105 meV in this case) is known as the "ponderomotive energy," or using popular jargon in this field,

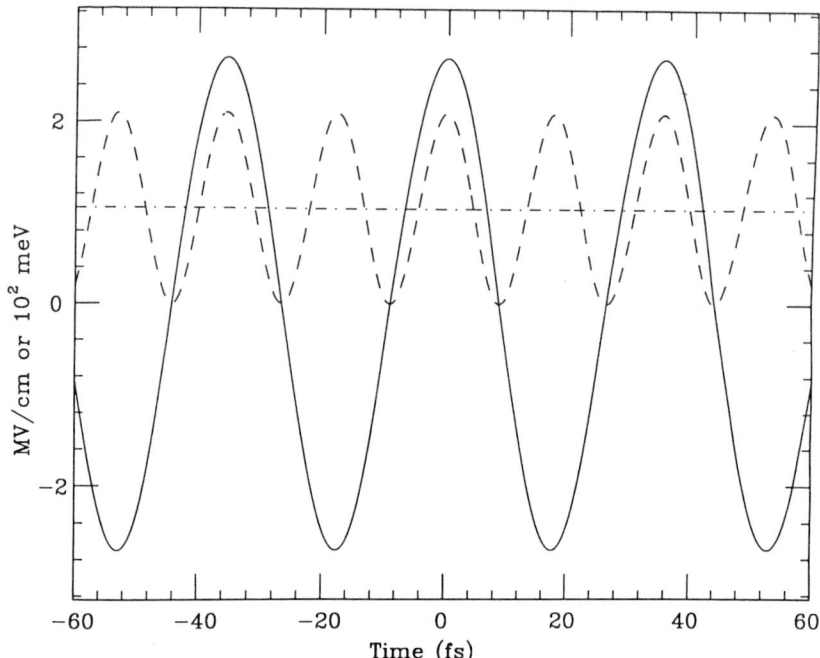

FIGURE 19. The solid line plots the electric field strength (MV/cm) *vs.* time for the 10.6μ line of a CO_2 laser in an irradiance of $10\,GW/cm^2$. An electron in this field would experience a quiver motion, the kinetic energy of which is given by the dashed curve as a function of time. The dot-dash line shows the average value of this quiver energy, 105 meV. Effects of the magnetic part of the laser beam have been ignored.

"quiver energy." Figure 19 illustrates the ambience in which the electron quivers.

In the spirit of quantum mechanics, the electron transits from its home in the ground state of the H^- into its quivering state in the radiation field. Therefore, the required energy for this transition is the field-free binding energy of the electron — 754 meV for H^- — plus the ponderomotive energy of its final state. In our example, therefore, the system would have to absorb a minimum of 859 meV for electron detachment to occur.

It is interesting to note that the ponderomotive energy is a *relativistic invariant*. This is particularly relevant for the measurements we shall describe. The invariance can be understood by examining the expression for the average ponderomotive or quiver energy (in S.I.):

$$\bar{W}_p = \frac{e^2 I}{2\varepsilon_0 mc\omega^2} \tag{4.7.3}$$

where m is the electron rest mass, c is the speed of light, e is the electron charge, and ε_0 is the permittivity of free space; ω transforms according to the Doppler shift and I transforms as the Doppler shift squared [Eq. (4.3.8)]. Thus I/ω^2 is the same in all frames.

This ponderomotive or quiver energy is also sometimes known as the "ponderomotive potential" because, if the laser field is not uniform in space, the gradient of the quiver energy constitutes a force. For example, if the electron were "born" into a focused laser beam it would pick up translational kinetic energy at the expense of its quiver energy as it moves to less intense regions of the pulse. In fact, if the laser pulse persists long enough for the electron to "roll" free of it, its kinetic energy would be the same as if it had been ejected from an atom in no field. On the other hand, if the laser pulse diminishes so rapidly in time that the electron has little chance of converting its quiver energy into kinetic energy, the effective binding energy is unchanged.[97]

Now let us discuss some recently explored aspects of the multiphoton detachment of H^-. Our purpose is not to provide a critique or even a thorough presentation of the theory (the interested reader is referred to Tang[98]), but rather to motivate the experimental measurements. For this reason we shall highlight the theoretical work of Becker et al.,[99] which displays the essential features of the measurements so far performed. Since it is based on a very simplistic model, it will certainly be superseded by more detailed and realistic treatments.

In the spirit of Armstrong,[3] Becker et al.[99] treat the H^- ion as an electron bound to a zero-range potential, in the regularized form,

$$V_0(\mathbf{r}) = \frac{2\pi\hbar^2}{m\varepsilon^2} \delta(\mathbf{r}) \frac{\partial}{\partial r} r \qquad (4.7.4)$$

with symbols defined as in Section 2.1. The initial ground state of the ion is given by Eq. (4.2.2). The solution to the Schrödinger equation for this potential in the presence of an intense monochromatic laser field is taken to have the Floquet form (see Tang[98]):

$$\Phi(\mathbf{r}, t) = \exp\left(\frac{-iE_F t}{\hbar}\right) \Phi_F(\mathbf{r}, t) \qquad (4.7.5)$$

where E_F is the complex Floquet eigenvalue,

$$E_F = -E_B + \Delta - i\hbar\Gamma/2 \qquad (4.7.6)$$

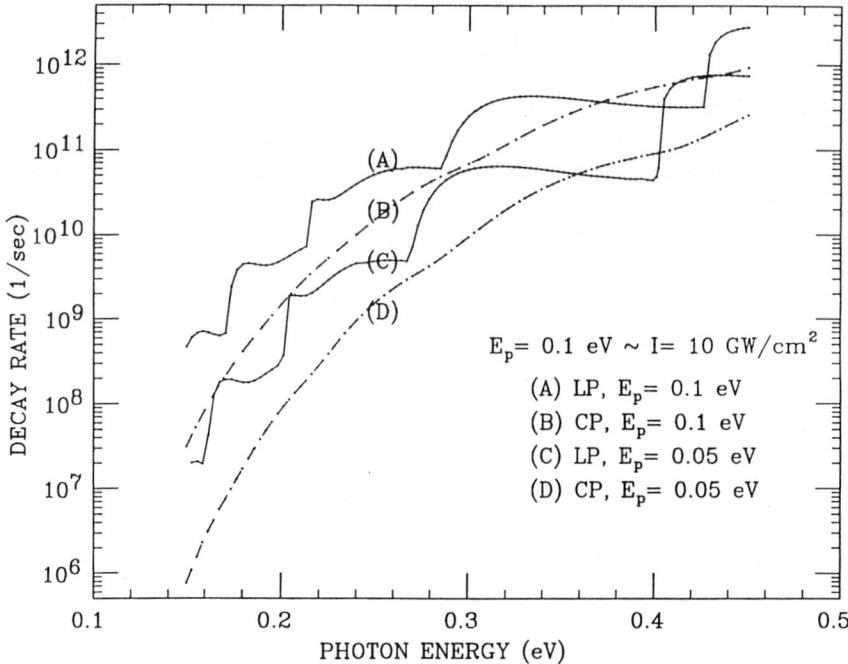

FIGURE 20. Dependence of multiphoton detachment rates on laser frequency, intensity, and polarization. $E_p \equiv$ ponderomotive energy, CP $\equiv$ circular polarization, LP $\equiv$ linear polarization (Becker et al.[99]).

and E_B is the binding energy, Δ is the ponderomotive shift, and Γ is the detachment rate. The results of the Becker et al.[99] method for laboratory CO_2 laser intensities of 5 and 10 GW/cm² are shown in Fig. 20, in which the detachment rate Γ in the H^- rest frame is plotted against the Doppler-shifted photon energy. Note that the intensities in the ion's frame vary with photon energy (going as the square of the Doppler shift), whereas the ponderomotive energy is constant for each curve—50 and 100 meV, respectively, for the lower and upper pairs. The solid lines are for linearly polarized light, and the dot-dashed lines are for circularly polarized light.

Turning our attention to the results for linear polarization first, we note pronounced thresholds corresponding to the boundaries between n and $n + 1$ photon-driven detachment as we decrease the CM laser frequency, with appropriate ponderomotive energy shifts. Thus the transition between three- and two-photon detachment should occur at low intensities at photon energies of 377 meV. For 5 GW/cm², we must add 25 meV and for 10 GW/cm², 50 meV. Because of this shift we see that the detachment rates between these two thresholds are *higher* for the 5 GW/cm² than for the 10 GW/cm² case!

The reader may also notice that the threshold behavior alternates in character. At the 3–2 transition, i.e., where the photon energy has dropped to the point where it takes three photons rather than two to produce the same transition, the yield is rising sharply, more smoothly at the 4–3 transition, again sharply at the 5–4 transition, and so on. How can we explain this behavior? The answer is that we are dealing with the Wigner threshold law (discussed in Section 4.1) involving final states of different l's. For two photons, linearly polarized, the final l could be 2 or 0. In the Wigner law, the $l = 0$ curve clearly dominates at threshold, giving a rate that goes as $\Delta E^{1/2}$. For three photons, $l = 1$ dominates so that the rate rises as $\Delta E^{3/2}$. For four photons we are back to $l = 0$ again, and so on.

We can also explain the circularly polarized light results. For two photons $l = 2$, for three $l = 3$, etc. The Wigner law predicts ever more slowly

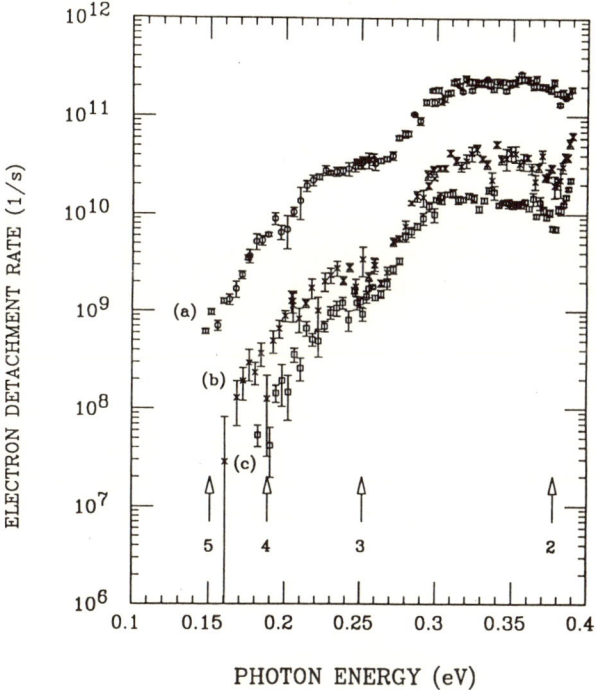

FIGURE 21. Intensity-averaged multiphoton detachment rate *vs.* photon energy as the peak intensity is set at (a) 12, (b) 6, and (c) 4 GW/cm². Arrows indicate the 5-, 4-, 3-, and 2-photon detachment thresholds.

rising cross sections as the multiplicity increases, almost completely smoothing out the threshold behaviors. A study is presently underway of the two-photon threshold region comparing the cross section resulting from linearly-polarized light with that of circularly-polarized light.[100]

Figure 21 shows results of recent measurements. The application of multiphoton processes to the study of the H^- ion opens up yet another dimension — intensity — in the spectroscopy of the H^- ion.

8. FOIL TRANSMISSION STUDIES

A unique experiment was performed in 1989 to determine the relative yield of neutral hydrogen excited states when H^- is sent through amorphous carbon foils.[101] The foil thicknesses ranged from 15 to 300 $\mu g/cm^2$ with H^- beam energies of 226, 500, 581, 716, and 800 MeV.

The interaction time of the 800-MeV H^- ions with a carbon foil in the thickness range considered is of the order of a femtosecond. Since this is the approximate orbital period of loosely bound electrons, complicated energy transfer and chaotic processes may occur. The fields in the foils are in excess of those needed to ionize H atoms and H^- ions, yet ions and atoms exit in abundance. This apparent increase in magnitude of the necessary ionizing field may also be evidence of stabilization in a broad-band, high-frequency electromagnetic wave, which, for the relativistic H^- ion, a foil may effectively be.

In traversing the amorphous foil, a projectile ion sees a superposition of different periodic fields, owing to its different impact parameter with each carbon atom. Thus, frequencies may be encountered that could directly excite and ionize the H^- ions and hydrogen atoms.

The principal results of this experiment are the relative yields of $H(n)$ states vs. beam thickness for $n = 1-5$, 10, and 11 — some at varying beam energies. Of great interest are the plots of foil thickness of optimum $H(n)$ transmission vs. n, shown in Fig. 22. From these data, one can deduce that the production mechanisms for the $n = 1$ and 2 states are similar to each other, but that different or additional mechanisms are probably involved for states with $n \geqslant 3$.

Additionally, the effect of H^- beam energy on the optimum thickness for production of the $n = 2$ and 3 states was studied. Figure 23 indicates that at lower energy, processes other than Lorentz contraction (see Section 3.1) of the foil thickness are important. Interest in the processes involved in the stripping of high-energy H^- ions by thin foils is currently quite intense as there are important practical implications for the injection of beams in proton accelerators and in storage rings. Further experimental measure-

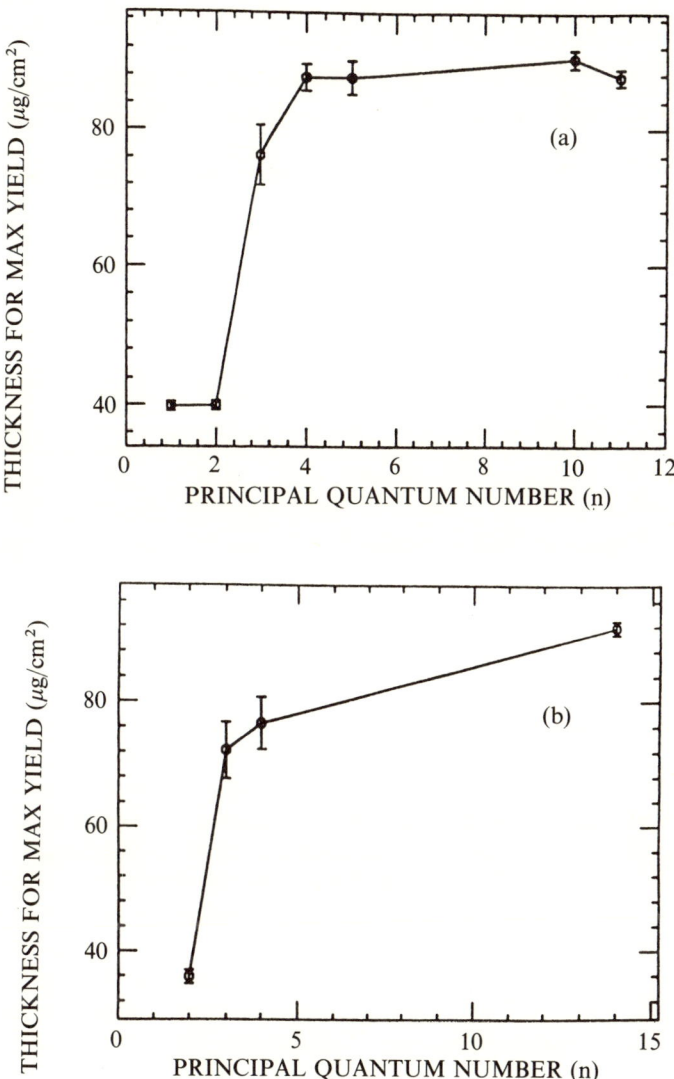

FIGURE 22. Optimum carbon foil thickness *vs.* n at (a) 800 MeV and (b) 581 MeV H⁻ beam energy.[101]

ments in which individual LoSurdo–Stark states can be resolved are underway.[102]

Rigorous theoretical treatment of these phenomena is not yet available, although some promising ideas, including a theory based on classical stochastic dynamics,[103] have been put forth.

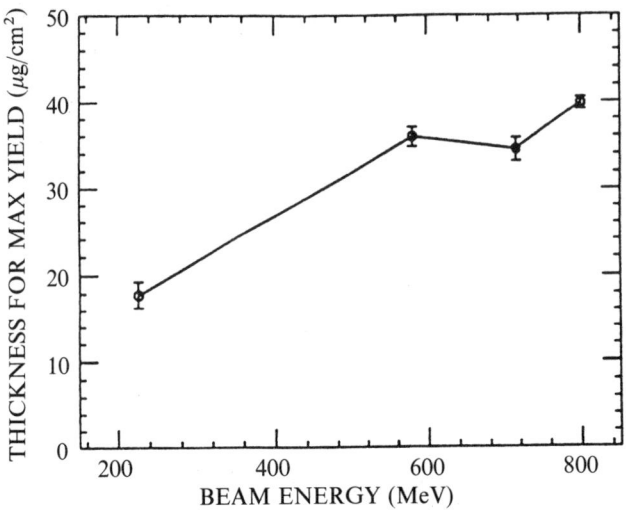

FIGURE 23. Optimum carbon foil thickness for $n = 2$ production as a function of H⁻ beam energy.[101]

9. SUMMARY AND OUTLOOK

Experiments in photodetachment of the negative ion of hydrogen have strongly supported the importance of electron correlations in the quantum-mechanical three-body Coulomb problem. The method of crossing laser beams with a relativistic H⁻ beam provides a uniquely powerful technique for probing H⁻ in the vacuum ultraviolet and in the high-intensity regime, and has delivered many new, even unpredicted, results.

Particularly important have been recent studies of high-n thresholds and doubly-excited resonances with and without applied dc fields,[48] multiphoton experiments,[104] and foil transmission work.[101] Understanding of this ion and its electron correlations is far from complete, however. A spectroscopist's work is never done. For example:

- The energy of the lowest-lying $^1P^0$ Feshbach resonance at a photon energy of 10.9264 eV has been observed with a precision of 1 meV.[68] Theory, however, predicts a width of the order of 30 μeV.[69] A measurement of this width would be useful as a gauge to theorists making very high-precision calculations. If this resolution can be achieved, the second-lowest $n = 2$ Feshbach resonance might also be observed.
- The large ^1P shape resonance just above the $n = 2$ threshold has been studied in static electric fields,[105] but a more systematic experiment,

using smaller steps in the electric field strength, should most certainly be performed to look for predicted oscillations on the high-energy shoulder[33] of this unique resonance.
- The lowest electron-detachment threshold at 0.7542 eV should be studied in applied field strengths greater than 1 MV/cm. Ths is the region in which Fabrikant[29] predicts the observability of rescattering of the electron wave.
- Using the two-laser method described in Section 3.5, branching ratios for the excited states of H⁻ to decay into the various excited states of neutral hydrogen could be measured.
- A most interesting experimental study would be to continue the measurement of resonances to energies above the $n = 8$ threshold. Doubly-excited states converging to thresholds with $n \geqslant 9$ should overlap the next lower threshold, and interfere with those converging there, possibly exhibiting chaotic behavior.[106]
- Cross-section measurements to date have only provided relative values, owing to the difficulties in determining the overlap volume between the laser and ion beams. Large differences among various theoretical predictions for photodetachment cross sections[70] indicate however that absolute cross-section measurements should be attempted in the region of the H($n = 2$) threshold at least.
- A remeasurement of the two-electron detachment threshold is desirable. An experimental resolution of 2 meV should be sufficient to observe the presence or absence of modulations on the threshold, such as have been predicted in a Coulomb-dipole model[52] and in a semiclassical orbit calculation.[107]
- The use of multiphoton processes for spectroscopy is still in its infancy, but with the relativistic beam technique it could be an extremely useful probe of doubly-excited levels that cannot be accessed directly from the ground state. For example, by two-photon absorption from the ^1S ground state one can excite states in the ^1S and ^1D continua. By triggering on the excited hydrogen decay channel one could strip off the overlay of single-photon detachments.

APPENDIX A. HYPERSPHERICAL COORDINATES

In hyperspherical coordinates the two electrons are treated as a quasi-particle mapped in six dimensions—five angles and one "radius." The hyperradius is defined as $R = \sqrt{(r_1^2 + r_2^2)}$, where r_1 and r_2 are the radial distances to the two electrons. The hyperangles are $\alpha = \arctan(r_2/r_1)$ with $0 \leqslant \alpha \leqslant \pi/2$ and $\theta_{12} = \cos^{-1}(\hat{r}_1 \cdot \hat{r}_2)$, plus three Euler angles (ϕ, θ, ψ) locating the triangular figure in space. The set of six coordinates (R, α, θ_{12}, ϕ, θ,

ψ) is equivalent to $(\mathbf{r}_1, \mathbf{r}_2)$. In this coordinate system the potential energy

$$V(r_1, r_2) = -\frac{Ze^2}{r_1} - \frac{Ze^2}{r_2} + \frac{e^2}{r_{12}} \qquad (4.\text{A}.1)$$

with $r_{12} = |r_1 - r_2|$, becomes[108]

$$V(R, \alpha, \theta_{12}) = \frac{e^2}{R} C(\alpha, \theta_{12}) \qquad (4.\text{A}.2)$$

where

$$C(\alpha, \theta_{12}) = -\frac{Z}{\cos\alpha} - \frac{Z}{\sin\alpha} + \frac{1}{\sqrt{[1 - \sin(2\alpha)\cos\theta_{12}]}} \qquad (4.\text{A}.3)$$

is the correlation potential for two electrons in the field of a proton.

Neglecting the spin–orbit coupling, the Schrödinger equation in hyperspherical coordinates for the electron pair is then

$$\left[\frac{1}{2}\left(-\frac{\partial^2}{\partial R^2} + \frac{\Lambda^2}{R^2}\right) + V(R, \alpha, \theta_{12}) - E\right]\Psi(R, \alpha, \theta_{12}) = 0 \qquad (4.\text{A}.4)$$

where

$$\Lambda^2 = -\frac{\partial^2}{\partial \alpha^2} - \frac{1}{4} + \frac{l_1^2}{\cos^2\alpha} + \frac{l_2^2}{\sin^2\alpha} \qquad (4.\text{A}.5)$$

The quantity $-(\Lambda/R)^2$ describes a generalized angular momentum barrier, and Λ is designated the "grand angular momentum."

Correlation Quantum Numbers: The highly degenerate eigenvalues of Λ^2 are compatible with a new set of approximate quantum numbers $(K, T)^4$. These have reasonably been called "correlation" quantum numbers, and correspond to different subgroups of rotations in six dimensions. The specification of a channel μ is given in terms of these quantum numbers.

K and T describe angular correlations, and effectively replace the independent particle angular quantum numbers l_1 and l_2. They are related to the total angular momentum and the principal quantum number n by[109]

$$T = 0, 1, 2, \ldots, \min(L, n - 1) \qquad (4.\text{A}.6)$$

$$K = n - 1 - T, n - 3 - T, \ldots, -n + 3 + T, -n + 1 + T \qquad (4.\text{A}.7)$$

Nonzero T values express the effect of the torque exerted on the inner electron by by outer one, while $T = 0$ indicates that the electrons move in the same plane.

The quantum number A describes radial correlations, but is not independent of K and T,

$$A = \begin{cases} \pi(-1)^{T+S} & \text{for } K \geq L - n + 1 \\ 0 & \text{for } K < L - n + 1 \end{cases} \quad (4.A.8)$$

so that $A = \pm 1$ or 0. Here π is the parity, L is the total angular momentum quantum number, and S is the total spin. An eigenchannel with $A = +1$ is often called a "+" channel, or a "−" channel if $A = -1$. The tendency of "+" channels to produce the strongest resonances has been substantiated by several authors.[65,109,110]

APPENDIX B. PROLATE SPHEROIDAL COORDINATES

The six independent coordinates in this classification are R, θ, ψ, λ, μ, and ϕ where $\lambda = (r_1 + r_2)/R$ and $\mu = (r_1 - r_2)/R$ are the prolate spheroidal coordinates, and ϕ is the azimuthal angle of r, the position of the nucleus with respect to the CM of the two electrons.

Adapted from molecular theories, the formulation allows for approximate quantum numbers (n_μ, n_λ, m) which can be related to the hyperspherical quantum numbers. An excellent review on the subject is found in Rost and Briggs,[111] where it is shown that

$$K = (n_\mu/2) - n_\lambda \quad (4.B.1)$$

$$T = m \quad (4.B.2)$$

$$A = (-1)^{n_\mu} \quad (4.B.3)$$

Here $(n_\mu/2)$ is the integer value of $n_\mu/2$ and m is the familiar magnetic quantum number; n_μ designates the number of nodes in the hyperspherical coordinate α, which coincides with one of the prolate spheroidal directions, and n_λ gives the number of elliptical molecular orbital nodes.

It is interesting to note that n_λ, n_μ, and m can be extracted from consideration of the potential curves at any R, while the hyperspherical quantum numbers K, T, and A were developed in the asymptotic limit of the escaping electron ($R \to \infty$).

APPENDIX C. PROPENSITY RULES

Propensity rules are, as the name implies, not exact selection rules. Rather they tell us what transitions are most likely when the atom or ion in question passes from one state to another. Not surprisingly, they have been formulated in both hyperspherical and prolate spheroidal coordinates for electron-scattering and for photodetachment.

In one of the earliest and most comprehensive studies of the application of HSC to atomic systems, Watanabe and Lin[112] introduce "systematics" of autoionization widths relating to the overlap between the doubly-excited resonance wave function with a continuum wave function. When $\alpha = \pi/4$, $\theta_{12} = 0$, the overlap is expected to be at a maximum, especially when channels have similar characteristics. These authors therefore introduced "rules of thumb" (propensity rules) for autoionization which give

$$\Delta n = -1, \quad \Delta K = -1, \quad \text{and } \Delta T = 0 \qquad (4.C.1)$$

with A unchanged. A measure of angular excitation, the number of nodes in θ_{12} is found to be $(n - 1 - K - T)/2$,* but this is equivalent to the PSC quantum number n_λ. Using Eq. (4.C.1), we find $\Delta n_\lambda = 0$. The number of nodes in the hyperspherical pseudoangle α is $n_\mu = n - 1 - T + K$.[114] Again using Eq. (4.C.1), one finds $\Delta n_\mu = -2$.

For the relation to united atom and separated atom classifications, see Rost and Briggs.[111]

REFERENCES

1. H. A. Bethe and E. E. Salpeter, *Quantum Mechanics of One- and Two-Electron Atoms* (Academic Press, New York, 1957).
2. D. M. Bishop and A. M. Frolov, *J. Phys. B* **25**, 3049 (1992); A. M. Frolov, *J. Phys. B* **25**, 3059 (1992).
3. B. H. Armstrong, *Phys. Rev.* **131**, 1132 (1963).
4. U. Fano and A. R. P. Rau, *Atomic Collisions and Spectra* (Academic Press, New York, 1986).
5. S. Chandrasekhar, *Astrophys. J.* **100**, 176 (1944).
6. K. R. Lykke, K. K. Murray, and W. C. Lineberger, *Phys. Rev. A* **43**, 6104 (1991) and references therein.
7. B. H. Bransden and C. J. Joachain, *The Physics of Atoms and Molecules* (Longman, New York, 1983).

---

*This quantity has been called $_n v^A$ in Sadeghpour[113] and others, but is not to be confused with v in Watanabe and Lin,[112] where $v = n - K - 1$. In Watanabe and Lin,[112] our n is called N, and $_n v^A$ is called n.

8. C. L. PEKERIS, *Phys. Rev.* **112**, 1649 (1958); **126**, 1470 (1962).
9. J. Z. TANG, Y. WAKABAYASHI, M. MATSUZAWA, S. WATANABE, AND I. SHIMAMURA, *Phys. Rev. A*, to be published.
10. R. N. HILL, *Phys. Rev. Lett.* **38**, 643 (1977).
11. D. W. MACARTHUR, K. B. BUTTERFIELD, D. A. CLARK, J. B. DONAHUE, P. A. M. GRAM, H. C. BRYANT, C. J. HARVEY, W. W. SMITH, G. COMTET, *Phys. Rev. Lett.* **56**, 282 (1986).
12. L. D. LANDAU AND E. M. LIFSHITZ, *Quantum Mechanics: Non-Relativistic Theory* (Addison-Wesley, 1965).
13. W. RINDLER, *Essential Relativity* (Springer-Verlag, Berlin, 1977).
14. A. B. STEWART, *Scientific American*, March, 100 (1964).
15. C. R. QUICK AND H. C. BRYANT, *J. Opt. Soc. Am. B* **7**, 708 (1990).
16. M. H. SHARIFIAN-ATTAR, Ph.D. Dissertation, University of New Mexico, 1977.
17. H. FRIEDRICH, *Theoretical Atomic Physics* (Springer-Verlag, Berlin, 1990).
18. H. C. BRYANT, A. MOHAGHEGHI, J. E. STEWART, J. B. DONAHUE, C. R. QUICK, R. A. REEDER, V. YUAN, C. R. HUMMER, W. W. SMITH, S. COHEN, W. P. REINHARDT, AND L. OVERMAN, *Phys. Rev. Lett.* **58**, 2412 (1987).
19. J. E. STEWART, H. C. BRYANT, P. G. HARRIS, A. H. MOHAGHEGHI, J. B. DONAHUE, C. R. QUICK, R. A. REEDER, V. YUAN, C. R. HUMMER, W. W. SMITH, AND S. COHEN, *Phys. Rev. A* **38**, 5628 (1988).
20. P. G. HARRIS, H. C. BRYANT, A. H. MOHAGHEGHI, C. TANG, J. B. DONAHUE, C. R. QUICK, R. A. REEDER, S. COHEN, W. W. SMITH, J. E. STEWART, AND C. JOHNSTONE, *Phys. Rev. A* **41**, 5968 (1990) and references therein.
21. I. I. FABRIKANT, *Zh. Eksp. Teor. Fiz.* **79**, 2070 (1980); [*Sov. Phys.-JEPT* **52**, 1045 (1980)].
22. W. P. REINHARDT, *J. Phys. B* **16**, L635 (1983).
23. M. L. DU AND J. B. DELOS, *Phys. Rev. A* **38**, 5609 (1988), and *Phys. Lett.* **A134**, 476 (1989).
24. Y. AHARONOV AND D. BOHM, *Phys. Rev.* **115**, 485 (1959).
25. J. E. STEWART, Ph.D. Dissertation, University of New Mexico, 1987 (Los Alamos Report No. LA-11152-T, UC-34).
26. U. FANO, *Phys. Rev. A* **24**, 619 (1981).
27. D. A. HARMIN, *Phys. Rev. Lett.* **49**, 128 (1982).
28. H. Y. WONG, A. R. P. RAU, AND C. H. GREENE, *Phys. Rev. A* **37**, 2393 (1988).
29. I. I. FABRIKANT, *Phys. Rev. A* **40**, 2373 (1989).
30. E. LUC-KOENIG AND A. BACHELIER, *Phys. Rev. Lett.* **43**, 921 (1979).
31. A. R. P. RAU AND H. Y. WONG, *Phys. Rev. A* **37**, 632 (1988).
32. U. FANO, *Phys. Rev.* **124**, 1866 (1961).
33. V. Z. SLONIM AND C. H. GREENE, *Rad. Eff. Def. Sol.* **122–123**, 679 (1991).
34. A. R. P. RAU in *Comm. At. Mol. Phys.* **14**, 285 (1984).
35. M. GAILITIS AND R. DAMBURG *Proc. Phys. Soc.* **82**, 192 (1963).
36. C. H. GREENE AND A. R. P. RAU, *Phys. Rev. A* **32**, 1352 (1985).
37. C. H. GREENE AND J. MACEK, private communications.
38. H. R. SADEGHPOUR, C. H. GREENE, AND M. CAVAGNERO, *Phys. Rev. A* **45**, 1587 (1992).
39. J. B. DELOS, private communication.
40. N. Y. DU, I. I. FABRIKANT, AND A. F. STARACE, submitted to *Phys. Rev. A*, June 10, 1993.
41. B. ZHOU AND C. D. LIN, *Phys. Rev. Lett.* **69**, 3294 (1992).
42. M. HALKA, P. G. HARRIS, A. H. MOHAGHEGHI, R. A. REEDER, C. Y. TANG, H. C. BRYANT, J. B. DONAHUE, AND C. R. QUICK, *Phys. Rev. A* **48**, 419 (1993).
43. P. G. HARRIS, Ph.D. Dissertation, University of New Mexico, 1990 (Los Alamos Report No. LA-11843-T, UC-413).
44. M. H. HALKA, Ph.D. Dissertation, University of New Mexico, 1993 (Los Alamos Report No. LA-12533-T, UC-410).
45. I. FABRIKANT (private communication, 1991).

46. P. G. Harris, H. C. Bryant, A. H. Mohagheghi, R. A. Reeder, C. Y. Tang, J. B. Donahue, and C. R. Quick, *Phys. Rev. A* **42**, 6443 (1990).
47. C. D. Lin (private communication, 1991).
48. Y. K. Ho and J. Callaway, *Phys. Rev. A* **34**, 130 (1986).
49. A. Pathak, A. E. Kingston, and K. A. Berrington, *J. Phys. B* **21**, 2939 (1988).
50. G. H. Wannier, *Phys. Rev.* **90**, 817 (1953).
51. A. K. Kazansky and V. N. Ostrovsky, *J. Phys. B* **25**, 2121 (1992).
52. A. Temkin, *J. Phys. B* **7**, L450 (1974).
53. C. A. Frost, Ph.D. Dissertation, University of New Mexico, 1984 (Los Alamos Report No. LA-8796-T, UC-34).
54. J. R. Friedman, X. Q. Guo, M. S. Lubell, and M. R. Frankel, *Phys. Rev. A* **46**, 652 (1992).
55. G. J. Schulz and R. E. Fox, *Phys. Rev.* **106**, 1179 (1957).
56. E. P. Wigner, *Phys. Rev.* **73**, 1002 (1948).
57. E. Baranger and E. Gerjuoy, *Phys. Rev.* **106**, 1182 (1957).
58. S. Silverman and E. N. Lassettre, Rept. 9, R. F. Project 464, Ohio State University Research Foundation, Columbus, Ohio (unpublished).
59. P. G. Burke and H. M. Schey, *Phys. Rev.* **126**, 147 (1962).
60. R. P. Madden and K. Codling, *Phys. Rev. Lett.* **10**, 516 (1963).
61. J. W. Cooper, U. Fano, and F. Prat, *Phys. Rev. Lett.* **10**, 518 (1963).
62. N. Bohr, *Nature* **137**, 351 (1936).
63. H. Feshbach, *Ann. Phys.* **5**, 357 (1958); **19**, 287 (1962).
64. C. D. Lin, *Phys. Rev. A* **28**, 1876 (1983).
65. H. R. Sadeghpour and C. H. Greene, *Phys. Rev. Lett.* **65**, 313 (1990).
66. C. D. Lin, *Phys. Rev. A* **10**, 1986 (1974).
67. P. A. M. Gram, J. C. Pratt, M. A. Yates-Williams, H. C. Bryant, J. Donahue, H. Sharifian, and H. Tootoonchi, *Phys. Rev. Lett.* **40**, 107 (1978).
68. D. W. MacArthur, K. B. Butterfield, D. A. Clark, J. B. Donahue, P. A. M. Gram, H. C. Bryant, C. J. Harvey, W. W. Smith, and G. Comtet, *Phys. Rev. A* **32**, 1921 (1985).
69. L. Lipsky and J. Conneely, *Phys. Rev. A* **14**, 2193 (1976).
70. M. Halka, H. C. Bryant, C. J. Harvey, B. Marchini, E. P. MacKerrow, W. Miller, A. H. Mohagheghi, C. Y. Tang, K. B. Butterfield, D. A. Clark, S. Cohen, J. B. Donahue, P. A. M. Gram, R. W. Hamm, A. Hsu, D. W. MacArthur, C. R. Quick, J. Tiee, and K. Rozsa, *Phys. Rev. A* **46**, 6942 (1992).
71. A. J. Taylor and P. G. Burke, *Proc. Phys. Soc.* **92**, 336 (1967).
72. J. W. McGowan, J. F. Williams, and F. K. Carley, *Phys. Rev.* **180**, 132 (1969).
73. J. F. Williams and B. A. Willis, *J. Phys. B* **7**, L61 (1974).
74. H. Tootoonchi, Ph.D. Dissertation, University of New Mexico, 1977; H. C. Bryant, D. A. Clark, K. B. Butterfield, C. A. Frost, H. Sharifian, H. Tootoonchi, J. B. Donahue, P. A. M. Gram, M. E. Hamm, R. W. Hamm, J. C. Pratt, M. A. Yates, and W. W. Smith, *Phys. Rev. A* **27**, 2889 (1983); H. C. Bryant in *Electronic and Atomic Collisions* (N. Oda and K. Takayanagi, eds.) (North-Holland, Amsterdam, 1980); G. Comtet, C. J. Harvey, J. E. Stewart, H. C. Bryant, K. B. Butterfield, D. A. Clark, J. B. Donahue, P. A. M. Gram, D. W. MacArthur, V. Yuan, W. W. Smith, and S. Cohen, *Phys. Rev. A* **35**, 1547 (1987); P. A. M. Gram, J. C. Pratt, M. A. Yates-Williams, H. C. Bryant, J. Donahue, H. Sharifian, and H. Tootoonchi, *Phys. Rev. Lett.* **40**, 107 (1978).
75. K. Behringer and P. Thoma, *Phys. Rev. A* **17**, 1408 (1978).
76. W. R. Ott, J. Slater, J. Cooper, and G. Gieres, *Phys. Rev. A* **12**, 2009 (1975).
77. T. P. Snow, *Astrophys. J.* **198**, 361 (1975).
78. J. T. Broad and W. P. Reinhardt, *Phys. Rev. A* **14**, 2159 (1976).
79. H. White, *Introduction to Atomic Spectra* (1934) (See J. P. Connerade's summary of the NATO ASI conference.)

80. J. CALLAWAY AND A. R. P. RAU, *J. Phys. B* **11**, L289 (1978).
81. J. CALLAWAY, *Phys. Lett.* **68A**, 315 (1978).
82. L. LIPSKY, R. ANANIA, AND M. J. CONNEELY, *Atomic and Nuclear Data Tables* **20**, 127 (1977).
83. H. C. BRYANT, D. A. CLARK, K. B. BUTTERFIELD, C. A. FROST, H. SHARIFIAN, H. TOOTOONCHI, J. B. DONAHUE, P. A. M. GRAM, M. E. HAMM, R. W. HAMM, J. C. PRATT, M. A. YATES, AND W. W. SMITH, *Phys. Rev. A* **27**, 2889 (1983).
84. K. B. BUTTERFIELD, C. J. HARVEY, H. C. BRYANT, D. A. CLARK, J. B. DONAHUE, P. A. M. GRAM, D. W. MACARTHUR, AND W. W. SMITH, in *Symposium on Atomic and Surface Physics* (F. Howorka, W. Lindinger, and T. D. Märk, eds.) (STUDIA Studienförderungsgesellschaft m. b. H., Innsbruck, 1984).
85. K. B. BUTTERFIELD, Ph.D. Dissertation, University of New Mexico, 1984 (Los Alamos Report No. La-10149-T, UC-34).
86. M. E. HAMM, R. W. HAMM, J. DONAHUE, P. A. M. GRAM, J. C. PRATT, M. A. YATES, R. D. BOLTON, D. A. CLARK, H. C. BRYANT, C. A. FROST, AND W. W. SMITH, *Phys. Rev. Lett.* **43**, 1715 (1979).
87. M. HALKA, H. C. BRYANT, E. P. MACKERROW, W. MILLER, A. H. MOHAGHEGHI, C. Y. TANG, S. COHEN, J. B. DONAHUE, A. HSU, C. R. QUICK, J. TIEE, AND K. ROZSA, *Phys. Rev. A* **44**, 6127 (1991).
88. M. CHRYSOS, Y. KOMNINUS, AND C. A. NICOLAIDES, *J. Phys. B* **25**, 1977 (1992).
89. P. G. HARRIS, H. C. BRYANT, A. H. MOHAGHEGHI, C. Y. TANG, H. TOOTOONCHI, H. SHARIFIAN, J. B. DONAHUE, C. R. QUICK, R. A. REEDER, E. RISLOVE, W. W. SMITH, AND J. E. STEWART, *Phys. Rev. Lett.* **65** (1990).
90. H. C. BRYANT AND C. H. GREENE, in *Physics News in 1991* (Philip F. Schewe, ed.), (American Institute of Physics, New York, 1991).
91. S. COHEN, H. C. BRYANT, C. J. HARVEY, J. E. STEWART, K. B. BUTTERFIELD, D. A. CLARK, J. B. DONAHUE, D. W. MACARTHUR, G. COMTET, AND W. W. SMITH, *Phys. Rev. A* **36**, 4728 (1987); S. COHEN, Ph.D. Dissertation, University of New Mexico, 1986 (Los Alamos Report No. LA-10726-T, UC-34A); J. E. STEWART, Ph.D. Dissertation, University of New Mexico, 1987 (Los Alamos Report No. LA-11152-T, UC-34).
92. C. H. GREENE, *Phys. Rev. Lett.* **44**, 869 (1980).
93. C. H. GREENE, *J. Phys. B* **20**, L357 (1987).
94. S. M. BURKOV, N. A. LETYAEV, AND S. I. STRAKHOVA, *Phys. Lett. A* **150**, 31 (1990).
95. J. BERAKDAR AND H. KLAR, *Phys. Rev. Lett.* **69** (1992).
96. M. GOEPPERT-MAYER, *Ann. Phys. (Leipzig)* **9**, 273 (1931).
97. L. D. VAN WOERKOM, R. R. FREEMAN, S. C. DAVEY, T. J. MCILRATH, AND W. E. COOKE, OSA Topical Conference on Applications of Subpicosecond High Intensity Lasers, Snowbird, CO, September, 1989.
98. C. Y. TANG, Ph.D. Dissertation, University of New Mexico, 1992 (Los Alamos Report No. LA-12254-T, UC-414).
99. W. BECKER, S. LONG, AND J. K. MCIVER, *Phys. Rev. A* **42**, L753 (1990).
100. E. P. MACKERROW *et al.* Sixth International Symposium on Resonance Ionization Spectroscopy and Its Applications, Sante Fe, NM, May 24–29, 1992.
101. A. H. MOHAGHEGHI, H. C. BRYANT, P. G. HARRIS, R. A. REEDER, H. SHARIFIAN, C. Y. TANG, H. TOOTOONCHI, C. R. QUICK, S. COHEN, W. W. SMITH, AND J. E. STEWART, *Phys. Rev. A* **43**, 1345 (1991); A. H. MOHAGHEGHI, Ph.D. Dissertation, University of New Mexico, 1990 (Los Alamos Report No. LA-11925-T, UC-411).
102. E. P. MACKERROW, P. KEATING, M. GULLEY, M. HALKA, W. MILLER, H. C. BRYANT, J. DONAHUE, D. CLARK, S. COHEN, D. FITZGERALD, S. FRANKLE, R. HUTSON, R. MACEK, C. QUICK, O. VANDYCK, AND C. WILKINSON, Postdeadline poster WE115, American Physical Society (DAMOP) Meeting, Reno, NV, May 16–19, 1993.

103. J. Burgdörfer and C. Bottcher, *Phys. Rev. Lett.* **61**, 2917 (1988).
104. C. Y. Tang, P. G. Harris, A. H. Mohagheghi, H. C. Bryant, C. R. Quick, J. B. Donahue, R. A. Reeder, S. Cohen, W. W. Smith, and J. E. Stewart, *Phys. Rev. A* **39**, 6068 (1989).
105. G. Comtet, C. J. Harvey, J. E. Stewart, H. C. Bryant, K. B. Butterfield, D. A. Clark, J. B. Donahue, P. A. M. Gram, D. W. MacAthur, V. Yuan, W. W. Smith, and S. Cohen, *Phys. Rev. A* **35**, 1547 (1987).
106. W. Reinhardt, New Developments in Two-Electron Atoms and Ions, Workshop, Boulder, CO, July 22–25, 1992.
107. J.-M. Rost (private communication, 1992).
108. J. H. Macek, *J. Phys. B* **1**, 831 (1968).
109. D. R. Herrick, *Phys. Rev. A* **12**, 413 (1975).
110. C. D. Lin, *Phys. Rev. Lett.* **35**, 1150 (1975).
111. J. M. Rost and J. S. Briggs, *J. Phys. B* **24**, 4293 (1991).
112. S. Watanabe and C. D. Lin, *Phys. Rev. A* **34**, 823 (1986).
113. H. R. Sadeghpour, *Phys. Rev. A* **43**, 5821 (1991).
114. H. S. Taylor, *Adv. Chem. Phys.* **18**, 91 (1970).

CHAPTER 5

COULOMB FORCES IN THREE-PARTICLE ATOMIC AND MOLECULAR SYSTEMS

J. S. BRIGGS

1. INTRODUCTION

In the solution of atomic and molecular few-body problems one is in the fortunate position that the force operating, the Coulomb force, is known exactly and has a simple analytic form both in configuration and momentum space. For systems involving nuclei of low charge the situation is even simpler since relativistic and quantum-electrodynamic effects can also be neglected in most cases. Then it appears remarkable that even the simplest few-body Coulomb problem, that involving three particles, is still not solved completely. The reason is equally simple. The infinite range of the Coulomb force and the singular behavior at the origin, both in configuration and momentum space, result in mathematical difficulties not present for short-range potentials. For example, traditional formal scattering theory is not applicable and must be modified appropriately. Similarly in the formation of resonances in two-electron atoms and in the motion in states of the three-body continuum, the correlation between the particles extends to infinite separation; the motion is never free. Of course this feature is well understood in the two-body Coulomb problem, which luckily is soluble in closed form in both the quantum-mechanical and classical cases. The infinite-range force results in an infinite number of bound states converging to the two-body breakup threshold.

In the three-body Coulomb problem two of the particles must have the same charge sign. Hence, only two of the three two-body subsystems have

---

J. S. BRIGGS • Fakultät für Physik, Albert-Ludwigs-Universität, D-79104 Freiburg, Germany.
Coulomb Interactions in Nuclear and Atomic Few-Body Collisions, edited by Frank S. Levin and David A. Micha. Plenum Press, New York, 1996.

a bound-state spectrum, the remaining one having only a continuum. This has the important result that the spectrum of the three-body system has an infinity of two-fragment thresholds (two particles bound in some state, the other free) converging to the three-body breakup threshold. Near to each threshold, resonant states of the three-body system occur where the third particle can be viewed as bound temporarily in the potential provided by the other two, so long as these two do not have a residual charge of the same sign as the third particle, i.e., the long-range force between the two fragments is attractive. This results in an enormously complicated spectrum of interacting states converging to the three-body breakup threshold. Only comparatively recently has an unraveling of this spectrum begun to emerge.

This complexity of the three-body Coulomb spectrum extrapolates through the complete breakup threshold with the result that Wigner threshold laws, applicable to short-range potentials are invalid for Coulomb potentials.[1] This is another manifestation of the infinite range of the Coulomb force. When two particles of like charge recede from one of opposite charge there is always the possibility, no matter what the spatial separation of the like charges, that one will proceed to a bound state around the particle of opposite charge and the other will move off to infinity. That this possibility exists out to infinite separation is essentially the cause of the departure from the Wigner threshold law for three-body breakup.

The intricacy of the spectrum below the breakup threshold is mirrored by the mutual angular and energy distribution of the three charged particles above threshold. Since no free three-body collision experiments have been made to date, experimental interest is confined to those situations where the initial state is bound (e.g., double photoionization of the helium atom) or where two particles are bound and the other free (e.g., electron-impact ionization of the hydrogen atom). For low total energies, where the particles move slowly, there is a strong correlation of the particles in the final continuum state. Only in the case of total energies very large compared to the initial binding energy and where all particles move quickly relative to one another is the correlation less noticeable. However, even in this situation there are measurable deviations from free motion.

Classically the three-body Coulomb problem in atomic physics is closely similar to the three-body gravitational problem in celestial mechanics. The gravitational force suffers from the same infinite range and singular nature as the Coulomb force and after centuries of effort the gravitational three-body problem is still not completely solved. However the fact that in the gravitational case all pair forces are attractive, whereas in the Coulomb case only two are, leads to significant differences in the classical phase space structure in the two cases.

The successful semiclassical quantization of the regular closed orbits of the classical hydrogen atom launched the "old" quantum theory. The logical next step was the quantization of the helium atom. This was notably unsuccessful.[2-4] Indeed the problem of the quantization of the helium atom was the rock upon which the ship of quantum state theory foundered for some years, before being refloated by wave mechanics. Subsequently this fundamental problem of semiclassical physics was largely forgotten until interest revived only some years ago.[5] The reason for the lack of success lay partly in the unavailability of computers with which to solve the three-body classical problem but more importantly in the lack of a suitable semi-classical theory. The existing Einstein–Brillouin–Kramers (EBK) theory sufficed only to quantize regular classical motion on phase space tori. It is now known that much of the classical phase space of the helium atom is irregular (chaotic). Only comparatively recently have the necessary semiclassical theories been developed.[6,7] This has opened the way, after almost 80 years, to the first successful semiclassical quantization of the helium atom.[8]

The three-body Coulomb problem has played a central role in atomic and molecular physics since the inception of quantum mechanics. The simplest nontrivial atomic problem is the He atom, consisting of an α-particle and two electrons. The simplest stable negative ion is H^-, consisting of a proton and two electrons. The simplest molecular positive ion is H_2^+, consisting of one electron and two protons. All of these elementary three-body systems have their analogs in which one or more of the constituents is replaced by another particle of like charge. For example Ps^-, consisting of a positron and two electrons, is obviously similar to H^-. The molecular ion ($dt\mu^-$), consisting of deuteron, a triton and a negative muon, is in some respects similar to H_2^+ and so on.

The three-body Coulomb system has assumed a corresponding importance in the development of scattering theory. The first application of the Born approximation in atomic physics was to the excitation and ionization of the hydrogen atom by incident electrons and protons.[9] This can be viewed as the theory of the continuum states of H^- and H_2^+, respectively. Despite this fundamental importance, the nonrelativistic three-body Coulomb problem is still not completely solved. Nevertheless considerable progress has been made on several fronts in the past few years and this article reports selected aspects of this progress.

The problem separates naturally into two regimes. For total energies negative with respect to the three-body fragmentation threshold, one is concerned predominantly with the theory of bound states, resonant states dissociating into two-fragment channels and scattering states involving only two-fragment channels. For total energies above this three-body threshold one is concerned with the momentum distribution of three correlated particles arising from excitation out of an initial state where all three particles, or only two of them, are bound. Although progress has been made in the theory of both of these regimes separately, the transition threshold region remains largely not well understood.

In the following the theory of atomic structure below the three-body threshold will be considered first and then new results in the theory of ionization of the hydrogen atom by charged-particle impact presented. In both cases methods which include all three two-body interactions on an equal footing will be discussed. This is to be contrasted with more traditional methods in which the repulsive interaction between particles of like charge is often included as a perturbation. The theoretical description of the transition region, as stated above, requires further development and will be discussed only briefly.

2. BELOW THE THREE-BODY BREAKUP THRESHOLD

In this section the fundamental atomic problem of one nucleus and two electrons and the fundamental molecular problem of one electron and two nuclei will be emphasized. Important examples of the atomic problem are the helium atom and the hydrogen negative ion H^-. The hydrogen molecular ion is exemplary of the molecular systems. Traditionally the atomic and molecular problems have been approached using quite different methods. In both cases, however, the nuclei have been taken in zeroth order as fixed centers of attraction for the electrons (He) or electron (H_2^+).

In the atomic case the fixed nucleus as the center of attraction leads naturally to an independent electron picture in which the repulsive electron–electron interaction is included explicitly at a second stage. This is the basis of configuration interaction methods which have found wide application in atomic physics since they can be readily extended to many-electron atoms. For two-electron atoms it was recognized from the outset that explicit correlated wave functions in which all two-body interactions are treated equally[10] are essential to achieve high accuracy. Indeed for the H^- ion such methods are necessary to obtain binding at all. Nowadays these methods can be pushed to digital computer accuracy. However, they give scant information as to the nature of the three-body motion involved.

In the molecular case the consideration of *two* nuclei as fixed centers of attraction leads to the adiabatic or Born–Oppenheimer (BO) approximation. The attractive (bound state) motion of the electron in the field of the two nuclei is diagonalized first and then the repulsion between the two nuclei is considered in solving for their motion in the field of the electron. This procedure automatically gives wave functions whose analog in the atomic case are explicitly correlated wave functions. Despite this similarity adiabatic "molecular" methods were not applied to atoms until the 1960s.[11] This was probably due to the popular, but mistaken, belief that the validity of the BO approximation relies on the small ratio of electron to proton mass. More correctly it relies on the smallness of the interprotonic velocity to the electron velocity. Hence in cases where the interelectronic velocity is small compared with the velocity of the nucleus relative to the electronic center-of-mass the adiabatic approximation will also be valid in atomic systems. Recently it has been shown that certain classes of doubly-excited resonant states of He and H^- fulfill this criterion to a good approximation.[12,13]

One further adiabatic method also allows a unified approach to the atomic and molecular three-body problems. This is the use of hyperspherical coordinates in which the hyperspherical radius, the root mean square of the six cartesian coordinates required to specify the internal motion of the three-body system, is used as an adiabatic coordinate. Although this method has been very successful, particularly in the interpretation of resonant atomic states,[14] it isolates only one coordinate so that a diagonalization in the remaining five dimensions must be performed. In addition no symmetries arise naturally in the method other than the total symmetries of spin, orbital angular momentum, and parity. By contrast the molecular adiabatic method considers the reduced motion in three dimensions initially. Furthermore, it also leads quite naturally to new "internal" approximate quantum numbers in addition to the aforementioned good total quantum numbers.[13] The method also allows a connection to be made to the periodic motions underlying the semiclassical quantization procedures.[15]

2.1. The Adiabatic Approximation

In the following the restriction will be made that two of the particles are identical with mass $m_1 = m_2 \equiv \mu$ and charge $Z_1 = Z_2 \equiv Z$. This covers the cases of H_2^+, Ps^-, H^-, and He, important in atomic and molecular physics. Without loss of generality we assume for the third particle of mass m_3 a unit charge $Z_3 = 1$ and use charge scaled atomic units (cau)

$$[\text{energy(cau)}] = Z^{-2}[\text{energy(au)}]$$
$$[\text{length(cau)}]d = Z[\text{length(au)}] \tag{5.2.1}$$

The vector **R** between the two identical particles will be taken as a body-fixed quantization axis. The remaining internal coordinate is **r** the position of the third particle with respect to the center-of-mass (ECM) of the two identical particles. If the body-fixed $\hat{z}$- axis is defined along $\hat{R}$, then six independent coordinates are $R, \Theta, \Psi, \mathbf{r}$ where Θ, Ψ and φ, the azimuthal angle of **r**, are Euler angles relating the body-fixed and space-fixed frames. The remaining two body-fixed coordinates of **r** will be taken to be the prolate spheroidal coordinates

$$\lambda = (r_1 + r_2)/R, \qquad \mu = (r_1 - r_2)/R \qquad (5.2.2)$$

where $\mathbf{r}_1 = \mathbf{r} - \hat{z}R/2$, $\mathbf{r}_2 = \mathbf{r} + \hat{z}R/2$ are the separations between particles 1 and 3 and particles 2 and 3, respectively.

If spin–orbit effects are first neglected the three-body wave function can be written in terms of uncoupled products $\Psi_{LM}^{S,t}(\mathbf{r}, \mathbf{R})\chi_{SM_S}\xi_{sM_s}$ where Ψ_{LM} is a spatial wave function for total orbital angular momentum L, space-fixed projection M and total spin S. The spin wave function for the identical pair is χ_{SM_S}, and ξ_{sM_s} is that for the remaining particle 3. In the cases of H_2^+, H^-, or He considered here $S = 0$ or 1 corresponding to singlet or triplet configurations, respectively. The spatial wave functions can be decomposed[12,13] in body-fixed components $\psi_{Lm}^t(\mathbf{r}, R)$

$$\Psi_{LM}^{S,t}(\mathbf{r}, \mathbf{R}) = \sum_{0 \leq m \leq L} [D_{Mm}^{L*}(\Psi, \Theta, \Phi) + (-1)^{S+t+L+m} D_{M-m}^{L*}(\Psi, \Theta, \Phi)]\psi_{Lm}^t(\mathbf{r}, R)$$

$$(5.2.3)$$

These functions are eigenfunctions of the permutation operator P_{12} for the identical particles

$$P_{12}: \mathbf{R} \to -\mathbf{R}, \qquad \mathbf{r} \to \mathbf{r} \qquad (5.2.4)$$

with eigenvalue $(-1)^S$. They are simultaneously eigenfunctions of the parity operator P

$$P: \mathbf{R} \to -\mathbf{R}, \qquad \mathbf{r} \to -\mathbf{r} \qquad (5.2.5)$$

with eigenvalue π and therefore of the product operator PP_{12} with eigenvalue $(-1)^t = (-1)^S \pi$. The remaining quantum number m is the eigenvalue of the angular momentum operator l_z, where $\mathbf{l} = -i\mathbf{r} \times \nabla_r$, and also of the body-fixed component L_z of the total angular momentum $\mathbf{L}$. Note that $\mathbf{L} = \mathbf{l} + \mathbf{L}_R$, where $\mathbf{L}_R = -i\mathbf{R} \times \nabla_R$.

Following Feagin and Briggs[12] the adiabatic representation of the body-fixed wave function ψ_{Lm}^t is written as

$$\psi_{Lm}^t(\mathbf{r}, R) = \sum_i \frac{f_{im}^L(R)}{R} \Phi_{im}^t(\mathbf{r}, R) \qquad (5.2.6)$$

where Φ_{im}^t is an internal moleular orbital (MO) wave function [to be defined in Eq. (5.2.12) below] with the azimuthal quantum number m and $t = 0, 1$ for gerade/ungerade MO. The remaining index i in (5.2.6) labels the alternative MO internal states. The molecular decomposition in terms of

MO, $\Phi^t_{im}(\mathbf{r}, R)$, vibrational, $f^L_{im}(R)$, and rotational, $D^L_{Mm}(\Psi, \Theta, 0)$, wave functions is evident in the adiabatic representation (5.2.3) and (5.2.6) of the total spatial wave function. The total spin-independent Hamiltonian in charge scaled coordinates (5.2.1) is

$$H = -\frac{1}{\mu}\nabla^2_R + \frac{1}{ZR} + h_v \qquad (5.2.7)$$

where

$$h_v = -\frac{1}{2v}\nabla^2_r - \frac{1}{r_1} - \frac{1}{r_2} \qquad (5.2.8)$$

is the two-center Coulomb Hamiltonian with $v = 2\mu m_3/(2\mu + m_3)$. Then the solution of the eigenvalue equation $(H - E)\Psi^{s,t}_{LM} = 0$ reduces, after integration over $\mathbf{r}$ and the Euler angles Ψ and Θ, to the following set of coupled equations for the vibrational amplitudes:

$$-\left\{\frac{\partial^2}{\partial R^2} - \mu\left[U^L_{im}(R) + \frac{1}{ZR} - E\right]\right\} f^L_{im}(R)$$

$$= \sum_{j \neq i} (\langle \Phi_{im}|\partial^2/\partial R^2 + \mathbf{l}^2/R^2|\Phi_{jm}\rangle + \langle \Phi_{im}|2\partial/\partial R|\Phi_{jm}\rangle \partial/\partial R) f^L_{jm}(R)$$

$$+ \Lambda_+ \langle \Phi_{im}|l_-/R^2|\Phi_{jm+1}\rangle f^L_{jm+1}(R) + \Lambda_- \langle \Phi_{im}|l_+/R^2|\Phi_{jm-1}\rangle f^L_{jm-1}(R)$$

$$(5.2.9)$$

where $\Lambda_\pm = [(L \mp m)(L \pm m + 1)]^{1/2}$.

The adiabatic channel energy $U^L_{im}(R)$ in (5.2.9) is defined by the diagonal matrix element

$$U^L_{im}(R) = \langle \Phi_{im}|h_v + \frac{1}{\mu}\left[-\frac{\partial^2}{\partial R^2} + \frac{\mathbf{l}^2 + L(L+1) - 2m^2)}{R^2}\right]|\Phi_{im}\rangle \qquad (5.2.10)$$

and the single-channel adiabatic approximation corresponds to the neglect of all nondiagonal coupling elements on the r.h.s. of (5.2.9).

In the case of H_2^+, where μ is equal to the proton mass and $v \approx 1$, the adiabatic channel energy can be approximated by the BO energy

$$\mathscr{E}^1_{im}(R) = \langle \Phi_{im}|h_1|\Phi_{im}\rangle \qquad (5.2.11)$$

independent of L, where h_1 is the two-center electronic Hamiltonian (5.2.8) for fixed R and reduced mass $v = 1$. The MO basis is then chosen as the set of eigenstates of h_1, i.e.,

$$h_1|\Phi_{im}\rangle = \mathscr{E}^1_{im}(R)|\Phi_{im}\rangle \qquad (5.2.12)$$

The eigenvalues $\mathscr{E}^1_{im}(R)$ define the correlation diagram (Fig. 1) of the H_2^+ problem. The full adiabatic energy $U^L_{im}(R)$ is then formed by calculating the matrix element (5.2.10) in the BO basis (5.2.12). If v is not small compared to μ (as, e.g., in atomic systems with $\mu = 1$ and $v \approx 2$) the BO approximation is inadequate. Nevertheless, adiabatic energies with the correct asymptotic separated-atom (SA) limit can be calculated from matrix elements constructed with an alternative basis to (5.2.12), designed to correct the inadequacy of the BO basis.

If particle 2 (or 1, which makes no difference in the symmetrical case) is removed in the limit $R \to \infty$, the two-center Hamiltonian reduces to the hydrogenic operator

$$h_v(R \to \infty) \to -\frac{1}{2v}\nabla_r^2 - \frac{1}{r} \qquad (5.2.13)$$

with the eigenvalues $\mathscr{E}_v(R \to \infty) \to -v/2N^2$, corresponding to principal quantum number N. The correct SA limit is given by the hydrogenic energy $-m_{13}/2N^2$ of the subsystem where particle 1 (or 2) and 3 are bound ($m_{13} = m_{23}$). A basis with correct SA energies when one of the identical particles 1 or 2 has been removed can therefore be constructed from the eigenvalues of the two-center Hamiltonian with the reduced mass $v = m_{13}$,

$$h_{m_{13}}|\Phi_{im}\rangle = \mathscr{E}^{m_{13}}_{im}(R)|\Phi_{im}\rangle \qquad (5.2.14)$$

The adiabatic potential energies can now be calculated in the MO basis (5.2.14) as

$$U^L_{im}(R) = \mathscr{E}^{m_{13}}_{im} + \frac{1}{\mu}\langle\Phi_{im}|D(R)|\Phi_{im}\rangle \qquad (5.2.15)$$

where

$$D = -\frac{\partial^2}{\partial R^2} + \frac{1}{4}\nabla_r^2 + \frac{L(L+1) - 2m^2 + l^2}{R^2} \qquad (5.2.16)$$

The kinetic energy $\frac{1}{4}\nabla_r^2$ is the residual part of the complete two-center Hamiltonian (5.2.8) not diagonalized in (5.2.14).

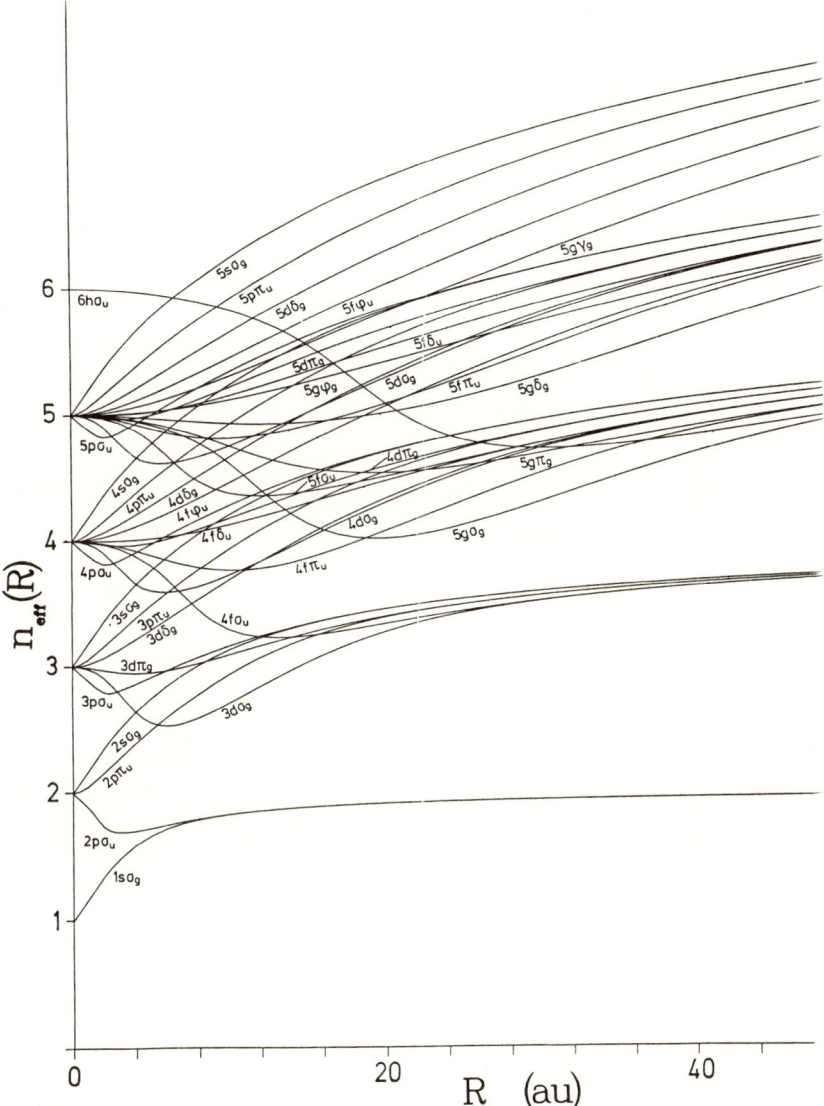

FIGURE 1. The correlation diagram of the two-center Coulomb problem (Helfrich[16]). The energy is in units of the effective quantum number $n_{\text{eff}} = [-2/\varepsilon_{im}(R)]^{1/2}$.

As anticipated from the construction, in the SA limit $R \to \infty$, i.e., $r_2 \to \infty$ with r_1 finite,

$$\langle \Phi_{im} | D | \Phi_{im} \rangle \to 0, \qquad \mathscr{E}_{im}^{m_{13}}(R) \to -\frac{m_{13}}{2N^2} \qquad (5.2.17)$$

and therefore

$$U_{im}^L(R) \to -\frac{m_{13}}{2N^2} \quad (5.2.18)$$

with the correct SA hydrogenic one-electron binding energy corresponding to a shell with principal quantum number N. The $R = 0$ united atom (UA) limit is also of simple analytic form with the channel wave functions of (5.2.14). The leading behavior of $U_{im}(R \to 0)$ is decided by the centrifugal barrier of the operator $D(R)$, which is diagonal for $R \to 0$:

$$U_{nlm}(R \to 0) \approx \frac{1}{\mu} \langle \Phi_{nlm} | D(R) | \Phi_{nlm} \rangle \approx \frac{L(L+1) - 2m^2 + l(l+1)}{\mu R^2} \quad (5.2.19)$$

Here the MO index im has been resolved into the UA quantum numbers nlm, where the good quantum number l is the key element in (5.2.19). A detailed discussion of the quantum numbers will be given in the next section. The atomic case, with the approximation of an infinitely heavy nucleus such that $m_{13} = 1$, has the important consequence that in the charge scaled units (5.2.1) the eigenfunctions Φ_{im} and the energies $\mathscr{E}_{im}^1(R)$ of (5.2.14) are exactly the H_2^+ ones of (5.2.12).

To summarize it has been shown that a common set of homonuclear BO basis MO wave functions and energies can be used to construct the single-channel adiabatic energies for H_2^+ and for two-electron atoms with nucleus of charge Z. The adiabatic energies are asymptotic to the correct one-electron limit as $R \to \infty$ in all cases. The correlation diagram of the set of energies $\mathscr{E}_{im}^1(R)$ plotted against the length R is shown in Fig. 1, taken from Helfrich.[16]

2.2. The Correlation Diagram

As is well-documented the two-center Hamiltonian (5.2.8) is separable in the λ, μ, φ coordinates and correspondingly MO states have quantum numbers n_λ, n_μ, m. These MO assignments are in one-to-one correspondence with parabolic SA ($n_1 n_2 m$) and spherical UA (nlm) quantum numbers as $R \to \infty$ and $R \to 0$, respectively. The connections are given in Rost and Briggs.[13] One notes the additional relation $(-1)^t = (-1)^l$, where l is the UA orbital angular momentum, arising from the parity of the MO wave function. In the SA limit one-electron parabolic states are produced and the additional designation $A = (-1)^{n_\mu}$ is necessary to specify the linear combination (LCAO), $\Phi_1 + A\Phi_2$, which forms an MO with correct nodal structure. In the chemical literature it is more usual to designate the MO by the

UA quantum numbers nlm and $(-1)^l$ (although this latter g, u label is redundant), e.g., $1s\sigma_g$, $2p\sigma_u$ etc.... One notes in the correlation diagram (Fig. 1) that some MO are depressed smoothly in energy as R decreases from infinity, whereas others are first depressed and then rise in energy so that the MO energy possesses a minimum. This is the phenomenon of promotion (increase of principal quantum number between SA and UA) known to be of fundamental importance for the formation of molecules and in ion–atom collisions.[17] This feature also plays a major role in the classification of MO for the two-electron atomic problem as described below.

The evenness or oddness of n_μ expressed by $A = (-1)^{n_\mu}$ is fundamental to the behavior of MO energies. For this reason in Fig. 2a and b the separate correlation diagrams for $A = +1$ and $A = -1$, respectively, are presented. As discussed extensively in Rost and Briggs,[13] $A = -1$ states have a node on the saddle $(r_1 \approx r_2)$ of the two-center potential of Eq. (5.2.8), whereas $A = +1$ MO have an antinode there. Accordingly $A = +1$ MO have more "molecular" character and this is seen qualitatively in Fig. 2 by the fact that the bundle of $A = +1$ MO emanating from a given SA hydrogenic N level is more strongly split in energy at finite R than its $A = -1$ counterpart. The latter are always promoted, i.e., $n > N$ while only $A = +1$ MO can correlate SA principal shells N with UA principal shells $n = N$.

The qualitative appearance of the correlation diagram (CD), that of lower levels of a given bundle crossing the remaining members (for $A = +1$ several higher bundles may be crossed) as they are promoted, is readily understood and is crucial for the formation both of molecular states and resonant electronic states of atoms. For large R, before molecule formation takes place, the shift of atomic levels is due to the Stark effect. Levels with the minimum value of the "electric" quantum number $k = n_1 - n_2$ for a given m are depressed most. These MO are the ones in which the two particles of like charge are roughly at 180° with the particle of unlike charge between them. Hence the two electric dipoles are maximally oppositely directed to give minimum energy. By contrast states with maximum k have particles of like charge on the same side of the third particle, leading to maximum energy configuration of interacting dipoles. The promotion condition is

$$n - N = [\tfrac{1}{2}(l - m + 1)] \equiv [\tfrac{1}{2} n_\mu] \equiv \begin{cases} n_2, & A = +1 \quad (5.2.20a) \\ n_2 + 1, & A = -1 \quad (5.2.20b) \end{cases}$$

where $[x]$ is the integer value of x. From (5.2.20) it is readily seen that MO maximally depressed by the Stark effect at large R have n_2 maximum and

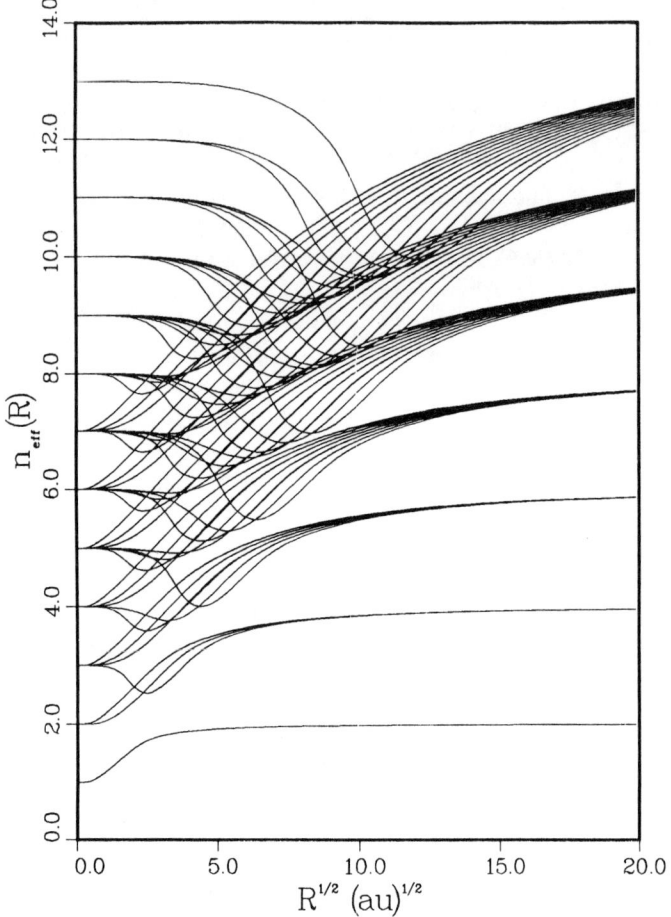

FIGURE 2a. The CD for $A = +1$ MO for $m \leq 2$. The effective quantum number n_{eff} is defined as in Fig. 1.

$n_1 = 0$. The MO with these quantum numbers are those most strongly promoted at small R. It is also clear from (5.2.20) that all $A = -1$ MO are promoted and additionally that promotion is directly proportional to the number of "angular" nodal surfaces n_μ. One also notes that promotion involves a minimum change in binding energy between UA and SA limits but a maximal change in the character of the MO. Such MO will be called "elastic" and are described by $n_2 \neq 0$. By contrast all MO with a *minimum* of angular excitation $n_2 = 0$ will be called "stiff." To this class belong all nonpromoted $A = +1$ MO with the minimum value of $n_\mu = 0$ and the $A = -1$ states with the corresponding minimum value $n_\mu = 1$ in accordance

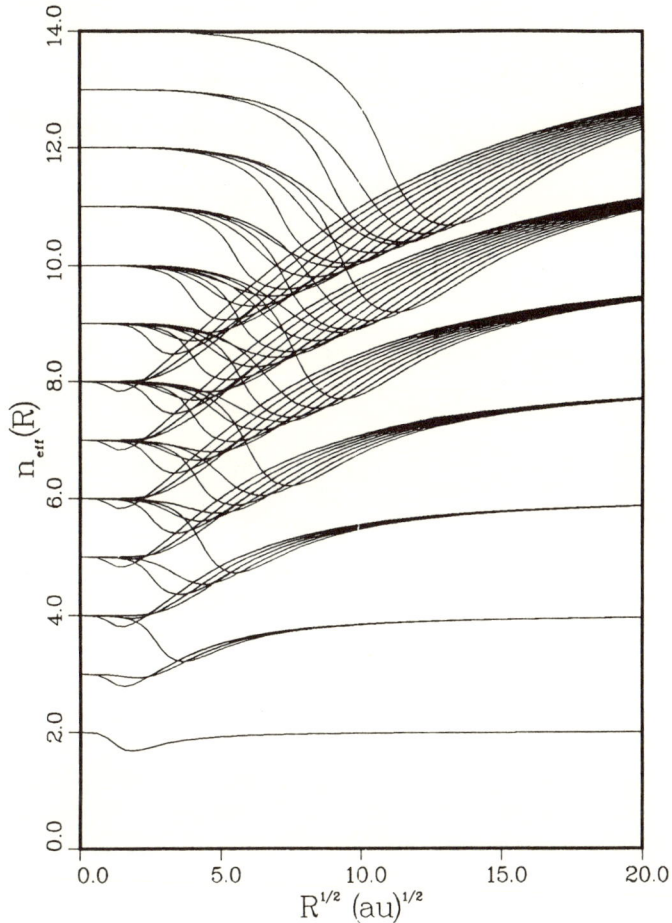

FIGURE 2b. The CD for $A = -1$ MO for $m \leq 2$. The effective quantum number n_{eff} is defined as in Fig. 1.

with Eq. (5.2.20b). For a given total number of nodal surfaces, the greater n_μ the more elastic the MO.

The $A = +1$ CD are now discussed in more detail. Three features will be stressed which describe the saddle behavior of the MO, i.e., their behavior in the neighborhood of the potential saddle point shown in Fig. 3. The first has already been discussed in connection with the definition of diabatic MO curves asymptotic to the $R \to \infty$ three-body breakup threshold. Each MO with a positive energy gradient near $R = 0$ has fixed quantum numbers $n_\lambda, 0, m$ and is the precursor of a complete sequence of MO whose other members are all promoted MO with quantum numbers $n_\lambda, n_\mu = $ even, m.

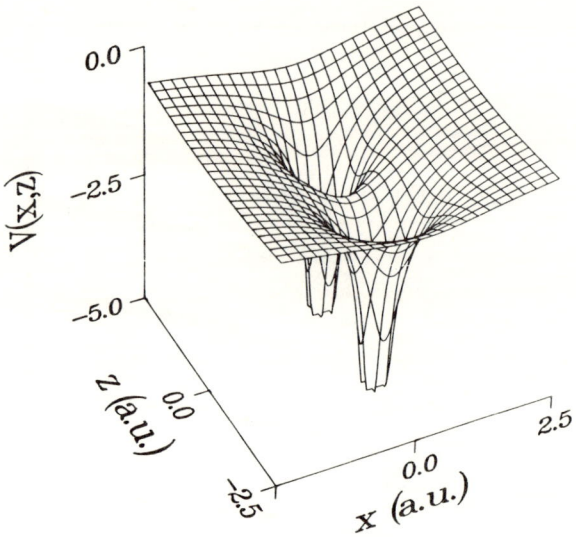

FIGURE 3. The two-center potential for interelectronic distance $R = 2$ a.u. The saddle point is at the origin of coordinates x, z and corresponds to the point $\lambda = 1$, $\mu = 0$ in spheroidal coordinates.

The promoted MO show avoided crossing with other members of the sequence. Examples of the $1s\sigma_g$ and $2p\pi_u$ sequences are shown in Fig. 4. The diabatic connection of all avoided crossings, i.e., potential curves drawn though the avoided crossings, connects the precursor MO near $R = 0$ with the $R \to \infty$ three-body breakup threshold. Note that the precursor MO are those with a positive gradient near $R = 0$ in Fig. 2a and are therefore easily recognized.

The second characteristic of MO distinguishing their behavior in the region of the saddle point in the two-body potential at $\mathbf{r} = 0$ (see Fig. 4) concerns the λ, φ part of the $\mathbf{r}$ motion in the plane perpendicular to $\hat{R}$. To describe saddle motion it is convenient to adopt cylindrical coordinates $\mathbf{r} = \rho, z, \varphi$. Then if we approximate in the region of the saddle point $\lambda \approx 1$, $\mu \approx 0$ we have

$$\rho = \frac{R}{2}[(\lambda^2 - 1)(1 - \mu^2)]^{1/2} \to R\left[\frac{(\lambda - 1)}{2}\right]^{1/2} \quad (5.2.21a)$$

$$z = \frac{R}{2}\lambda\mu \to \frac{R}{2}\mu \quad (5.2.21b)$$

$$\varphi = \varphi \quad (5.2.21c)$$

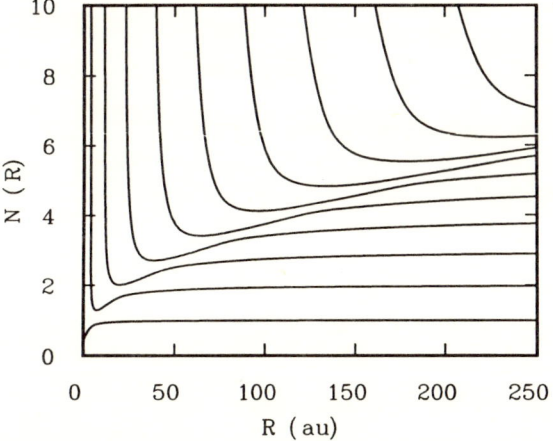

FIGURE 4a. The $1s\sigma_g$ saddle sequence of MO with $(n_1, n_2, m) = (0, N - 1, 0)$ for $N = 1, 2 \ldots$. In order of increasing energy they are the $1s\sigma_g$, $3s\sigma_g$, $5f\sigma_g$, $7h\sigma_g$, etc. MO.

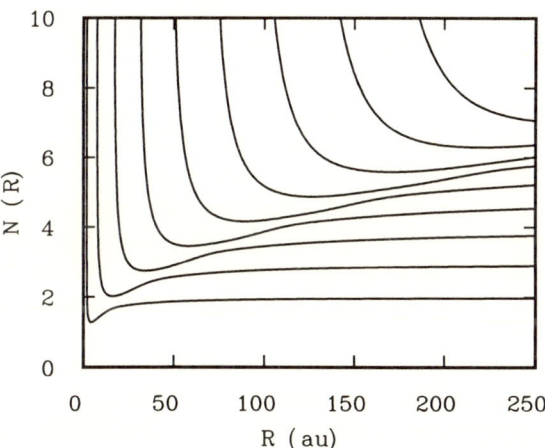

FIGURE 4b. The $2p\pi_u$ saddle sequence of MO with $(n_1, n_2, m) = (0, N - 1, 1)$ for $N = 1, 2 \ldots$. In order of increasing energy they are the $2p\pi_u$, $4f\pi_u$, $6h\pi_u$, $8i\pi_u$, etc. MO.

In contrast to the motion parallel to $\hat{R}$ (described by μ or z), the motion in the plane perpendicular to the $\hat{R}$-axis [described by (λ, φ) or (ρ, φ)] is bound near $\mathbf{r} = 0$. As may be readily shown from Eq. (5.2.8) the motion is that of an isotropic two-dimensional harmonic oscillator with Hamiltonian

$$H_\perp = -\frac{1}{2}\left(\frac{1}{\rho}\frac{d}{d\rho}\rho\frac{d}{d\rho} + \frac{1}{\rho^2}\frac{d^2}{d\varphi^2}\right) + \frac{1}{2}\omega^2\rho^2 \qquad (5.2.22)$$

where $\omega = 4R^{-3/2}$. The energies are given by $\hbar\omega(v_2 + 1)$ with

$$v_2 = 2n_\rho + m \tag{5.2.23}$$

According to (Eq. 5.2.21) the saddle nodes (n_ρ, m) go over to MO nodes (n_λ, m) and hence the degeneracy of the two-dimensional isotropic oscillator is expressed as the MO saddle degeneracy condition

$$v_2 = 2n_\lambda + m \tag{5.2.24}$$

for various combinations of n_λ and m.

2.3. Symmetries of Doubly-Excited States

In a series of papers in the 1970s and early 1980s, initiated by Wulfman and Sukeyuki[18] and by Sinanoglu and Herrick[19] and concluded by the review of Herrick,[20] new schemes for the classification of resonant states of two-electron species were proposed on the basis of various reductions of the $O_4 \times O_4$ product group of noninteracting electrons. In the case of one of these reductions, a classification of resonances in terms of new quantum numbers K, T, whose significance is explained below, was introduced. Lin[21] was able to correlate the empirical nodal structure of hyperspherical wave functions with the K, T quantum numbers. Lin also emphasized another Herrick quantum number, which he called A, having fundamental significance for the structure of two-electron resonances. Since then empirical labeling of resonances with K, T, A quantum numbers has become standard practice particularly with reference to hyperspherical calculations and to full quantum mechanical calculations, even though it is recognized that these quantum numbers cannot be related directly to the nodal structure of the states. In alternative group theory reductions, Herrick and Kellman[22] identified sequences of levels grouped according to rovibrational structure. This led also to many papers suggesting rotational and vibrational motion, as in a triatomic molecule, as the underlying dynamics of sequences of resonant states. These treatments were based on approximate three-body Hamiltonians or on model Hamiltonians and hence remain semiempirical.

All of this work hinted at, but did not identify, some approximate separability of the two-electron problem in terms of a set of collective variables. Here it is emphasized that the approximate separability and the corresponding collective cooordinates have been identified in the MO approach to resonant electronic structure. The collective coordinates are those of the relative vector **R** of the two electrons and the vector **r** of the ECM with respect of the nucleus. A description of the exact two-electron Hamiltonian in terms of a rotation to the body-fixed (**r, R**) frame provides automatically the molecular rotational structure. Similarly the treatment of the BF coordinate R as an adiabatic coordinate as in molecules provides vibrational structure in the adiabatic potentials obtained by averaging over the r ECM motion. Finally, and most importantly, the r motion itself is separable in MO prolate spheroidal λ, μ, φ coordinates. It has already been shown that the associated good quantum numbers n_λ, n_μ, m corresponding to this separability contain the K, T, A Herrick quantum numbers (Feagin and Briggs[12]). Recent work has indicated that a "vibrational" quantum number suggested by Herrick and Kellman and used by others is more exactly identified[23] as $2n_\lambda + m$. Furthermore this recent work has also shown that large basis set numerical wave functions exhibit the nodal structure expected from an approximate separability in MO coordinates.

The form (Eq. 5.2.3) of the total wave function limits the states of total LSt symmetry that can be built upon a given MO internal state i with fixed $n_\lambda n_\mu m$ quantum numbers and $t = 0, 1$. The rules can readily be derived from (5.2.3) since for certain quantum numbers the linear combination of rotational wave functions is zero. Hence not all states with given $LS\pi$ quantum numbers can be associated with particular MO. For example all σ_g MO ($m = 0, t = 0$) give total states with $(-1)^{S+L} = +1$. Therefore the series $^1S^e$, $^3P^o$, $^1D^e$,... is possible for increasing total angular momentum L. The superscript e (even) or o (odd) denotes total parity $\pi = +1$ or -1, respectively. Similarly for σ_u MO ($m = 0, t = 1$) we have the complementary spin series $^3S^e$, $^1P^o$, $^3D^e$.... For MO with $m \neq 0$ the situation is more complicated. Each member of a series consists of a pair of states with the same angular momentum $L \geq m$ but with opposite parity and spin. For example, every π_g MO ($m = 1, t = 0$) gives series $(^1P^e, ^3P^o)$, $(^1D^e, ^3D^o)$.... Similarly each π_u MO gives rise to series of pairs of states $(^3P^e, ^1P^o)$, $(^3D^e, ^1D^o)$.... Hence one sees already how the MO picture automatically leads to rotor series whose existence was postulated empirically by Herrick.

One aspect of the MO ordering of two-electron states should be mentioned explicitly. On the basis of ordering states according to group-theoretical considerations Herrick [20] noted that in zeroth-order, pairs of states, e.g., $(^1P^e, ^3P^o)$ or $(^1D^e, ^3D^o)$ are degenerate and he explained this in analogy to the familiar Λ-doubling found in diatomic molecules[16] (Λ corresponds to our m quantum number). Since the MO treatment of two-electron atoms is identical to that of diatomic molecules, the occurrence of degenerate pairs $(^1P^e, ^3P^o)$ or $(^1D^e, ^3D^o)$ is *exactly* the Λ-doubling phenomenon. As in molecules the degeneracy is lifted when rotational coupling to other MO of different body-fixed m is taken into account.

In a single-channel adiabatic approximation each state $(LS\pi)$ is associated with a potential curve in which vibrational states corresponding to increasing degrees of excitation in R may be defined. However, the vibrational structure of two-electron states introduced by Herrick[20] on the basis of a triatomic molecular model involves three normal modes and only the "breathing" v_1 longitudinal mode (symmetric stretch) corresponds to R. The other two modes appear when the exact λ, μ internal motion is approximated in the region of the saddle point where $\lambda \sim 1$, $\mu \sim 0$ [see Eqs. (5.2.21) and (5.2.22) and Fig. 4]. The corresponding modes are the asymmetric stretch (v_3, MO coordinate μ or z) and bending (v_2, MO coordinate λ or ρ).

Herrick introduced five quantum numbers K, T, A (called by him v) on the basis of SO_4 group theory and v_2 and d on the basis of a triatomic molecular model of two-electron atoms. All of these five quantum numbers are contained in the three quantum numbers $n_\lambda n_\mu m$ unique to each MO, as will now be shown.

For a one-electron atom the higher SO_4 symmetry of the pure Coulomb problem leads to the possibility of separating the Hamiltonian not only in spherical coordinates but also in parabolic coordinates $n_1 n_2 m$ or equivalently Nkm, where $k = n_1 - n_2$ is the electric quantum number and m the usual projection quantum number of angular momentum along the z-axis. Herrick considered two electrons with principal quantum numbers $n_>$ and N and recognized that in the limit $n_> \to \infty$ the inner electron has parabolic quantum numbers Nkm, with the quantization axis along $\hat{r}_2$. He repeatedly pointed out the analogy and similarity between the one-electron quantum numbers k and m and the two-electron quantum numbers K and T. Remarkably perhaps, he failed to recognize that if the MO axis $\mathbf{r}_1 - \mathbf{r}_2$ is taken as the quantization axis, the limit $R \to \infty$ gives the parabolic one-electron limit for all $n_>$, including the intrashell case $n_> = N$. Then the sets K, T and $-k$, m are not only analogous, they are *identical*.

The one-electron SO_4 group is reducible[18] as $SU_2 \times SU_2$. The SU_2 generators are pseudospins $\frac{1}{2}(\mathbf{l} + \mathbf{b}) = \mathbf{j}_\pm$. Then $\mathbf{l} = \mathbf{j}_+ + \mathbf{j}_-$ and $\mathbf{b} = \mathbf{j}_+ - \mathbf{j}_-$ are the one-electron angular momentum and Runge–Lenz vector, respectively. Each irreducible representation of SO_4 is labeled by two indices

$$p = j_+ + j_-, \qquad q = j_+ - j_- \qquad (5.2.25)$$

with states in each representation labeled $|pqlm\rangle$. For the pure Coulomb case $\mathbf{l} \cdot \mathbf{b} = 0$ and therefore $q = 0$ or $j_+ = j_-$. For two electrons there are four vector generators $\mathbf{l}_1$, $\mathbf{l}_2$, $\mathbf{b}_1$, $\mathbf{b}_2$, which can be coupled in a variety of ways. It turns out[20] that the meaningful way for two-electron atoms is to consider the Lie algebra SO_4 generators $\mathbf{L} = \mathbf{l}_1 + \mathbf{l}_2$ and $\mathbf{B} = \mathbf{b}_1 - \mathbf{b}_2$. Then coupled states are labeled

$$|PQLM\rangle = \sum_{l_1 l_2} |p_1 q_1 p_2 q_2 (l_1 l_2) LM\rangle (-1)^{l_2} \langle p_1 q_1 l_1, p_2 - q_2 l_2 | PQL\rangle \qquad (5.2.26)$$

where the coefficient is proportional to a 6-j symbol,[20] and P and Q are defined by Eq. (5.2.31) below.

Herrick recognized that the one-electron Stark basis diagonalizes $\mathbf{b}_2$ and that within a manifold N one has

$$\mathbf{r} \to \frac{3N}{Z} \mathbf{b} \qquad (5.2.27)$$

From the form of the leading dipole interaction term for $r_2 \to \infty$, r_1 finite and using (5.2.27) Herrick showed that the dipolar interaction leads to an effective "angular momentum" for the outer electron given by the channel invariant

$$\mathscr{A} = \mathbf{l}_2^2 + \frac{3N}{Z}(\mathbf{b}_1 \cdot \hat{r}_2) \qquad (5.2.28)$$

He also identified another channel invariant

$$W = \mathbf{L} \cdot \left(\frac{3N}{Z} \hat{r}_2 - 2\mathbf{b}_1 \right) \qquad (5.2.29)$$

and then argued that electron correlation is most pronounced when $Z \to 0$ and therefore chose the representation (5.2.26) to diagonalize the dominant parts of (5.2.28) and (5.2.29), i.e.,

$$\mathbf{l}_1 \cdot \hat{r}_2 = -Q \qquad (5.2.30a)$$

and

$$\mathbf{b}_1 \cdot \hat{r}_2 = -K \qquad (5.2.30b)$$

with two-electron quantum numbers K and Q. These quantum numbers were then related to the quantum numbers P, Q of (5.2.26), corresponding to invariants

$$\mathbf{L}^2 + \mathbf{B}^2 = P(P+2) + Q^2 \qquad (5.2.31a)$$

$$\mathbf{L} \cdot \mathbf{B} = |Q|(P+1) \qquad (5.2.31b)$$

by $K = P - N + 1$, although Herrick only stated that K and Q defined by (5.2.30) are "essentially the same" as those arising from the dipole-channel operators in the limit $Z \to 0$. He also stated that to form states of definite parity it is necessary to consider quantum number $|Q| = T$ rather than Q.

The MO explanation of the origin of Herrick's quantum numbers is much simpler. The question arises, why are quantum numbers based on SA wave functions and symmetries ($R \to \infty$ limit) valid for classifying resonant states involving intermediate values of R? The answer lies in the fact that the MO quantum numbers $n_\lambda n_\mu m$ are valid for all R. In particular for $R \to \infty$ they can be expressed uniquely in terms of the parabolic quantum numbers $n_1 n_2 m$ or Nkm of the SA state to which each MO is asymptotic. The "extra" quantum number $A = (-1)^{n_\mu}$ is necessary to specify the correlation (this is the v quantum number of Herrick). It also specifies the LCAO form of the MO in terms of SA states in which electron 1 or electron 2 is bound, i.e.,

$$|n_\lambda n_\mu m\rangle \approx |n_1 n_2 m\rangle_1 + A|n_1 n_2 m\rangle_2 \quad \text{for} \quad R \to \infty \quad (5.2.32)$$

The origin of the asymptotic channel invariants is also simple. The two-center problem is separable and this separability arises from the presence of an extra constant of the motion[24]

$$\Omega = \tfrac{1}{4}(\mathbf{l}_1 + \mathbf{l}_2)^2 - \tfrac{3}{4}(\mathbf{l}_1 - \mathbf{l}_2)^2 + \frac{N\mathbf{R}(\mathbf{b}_2 - \mathbf{b}_1)}{2} \quad (5.2.33)$$

and $\mathbf{l}_j = -i\mathbf{r}_j \times \nabla_j$. Considering the limit $\mathbf{r}_2 \to \infty$ with $\mathbf{r}_1$ finite, i.e., $\mathbf{R} = \mathbf{r}_1 - \mathbf{r}_2 \to -\mathbf{r}_2$, and using (5.2.29) one finds

$$\frac{\Omega}{R^2} \to \text{constant} + \frac{1}{R} \mathbf{b}_1 \cdot \hat{\mathbf{r}}_2. \quad (5.2.34)$$

Hence for fixed R one sees that this constant of the motion contains Herrick's channel invariant (5.2.30b) corresponding to the electric quantum number $k = n_1 - n_2 = -K$.

The projection quantum number $\mathbf{l} \cdot \hat{\mathbf{R}} = m$ is also a good quantum number for all R. In the limit $\mathbf{r}_1$, finite, $\mathbf{r}_2 \to \infty$ this assumes the form $-\mathbf{l}_1 \cdot \hat{\mathbf{r}}_2$, which is precisely Herrick's other channel invariant (5.2.30a) so that $Q \equiv m$. Furthermore, the fact that MO are classified by $|m|$ only accounts for the introduction of $T = |Q|$.

Herrick's other two quantum numbers, d and I ($= L - m$), were derived on the basis of an ad hoc model of a two-electron atom as a triatomic molecule. In particular he was able to organize states into I supermultiplets where those states corresponding to a given value of v_2 are almost degenerate. It has been established that the v_2 quantum number, corresponding to a bending vibration is nothing more than the quantum number $2n_\lambda + m$ describing degenerate saddle-point bending vibrations. The degeneracies for fixed v_2 in the supermultiplet classification have their origin in the underlying degeneracy of MO for fixed $2n_\lambda + m$. In addition the classification according to $I = L - T$ with $L - T \geqslant 0$ arises naturally since $T = m$, and since $\mathbf{L} \cdot \hat{R} = m$ then $L - m \geqslant 0$. In the MO picture $I = L - m$ describes the difference of total angular momentum L and MO angular momentum m.

There remains the quantum number d. It will emerge that classification according to d, for fixed N is *exactly* the same as classification according to n_λ denoting the number of elliptical MO nodes. The quantum number d arises out of the reduction $SO_4 \times SO_4 = SU_2 \times SU_2 \times SU_2 \times SU_2$ according to irreducible representations in which the operator $\mathbf{J}_>^2$ is invariant. Here $\mathbf{J}_>^2$ is the angular momentum operator with the eigenvalue $d = \max(J_+, J_-)$, where $\tfrac{1}{2}(\mathbf{L}+\mathbf{B})^2 = J_+(J_+ + 1)$ and $\tfrac{1}{2}(\mathbf{L}-\mathbf{B})^2 = J_-(J_- - 1)$. This representation is connected to that of (5.2.31) by $d = \tfrac{1}{2}(P + T) = \tfrac{1}{2}(K + N + T - 1)$. The MO interpretation follows from $K = n_2 - n_1$, $T = m$, to give

$$d = n_2 + m \tag{5.2.35}$$

The real meaning for intrashell resonance, however, relies on the fact that for fixed N, d gives the number of λ nodes, i.e., $n_\lambda = n_1 = N - n_2 - m - 1 = N - d - 1$.

To summarize, it has been shown that all of Herrick's *five* quantum numbers, arrived at by various physical arguments supported by the SO_4 group classification express different aspects of the MO character through the *three* unique MO numbers $n_\lambda n_\mu m$. The connection, emerging from the above arguments and discussed in detail in Feagin and Briggs,[12] Rost and Briggs,[13] and Vollweiter et al.,[23] is

$$\begin{aligned} T &= m \\ A &= (-1)^{n_\mu} \\ K &= n_2 - n_1 = \tfrac{1}{2}n_\mu - n_\lambda \\ v_2 &= 2n_1 + m = 2n_\lambda + m \\ d &= n_2 + m = \tfrac{1}{2}n_\mu + m \end{aligned} \tag{5.2.36}$$

2.4. Nodal Structure of Doubly-Excited States

The accuracy of the MO model can be tested in two ways. The obvious one is quantitative: do the rotation–vibration eigenenergies in the single-channel adiabatic motion agree with known resonance energy positions of He and H^-? The second test is more qualitatiave: do exact wave functions how the λ, μ nodel structure implied by the separation of ECM motion in spheroidal coordinates? Both questions have been answered largely in the affirmative. However, since experimental data are of limited accuracy the test of the MO method lies in comparison with large basis set diagonalizations of the Hamiltonian (5.2.7). Nowadays such calculations can be performed with machine accuracy for states up to $N \approx 10$. A detailed comparison is made by Rost and Briggs.[13] The general trend is that for He the accuracies are moderate (a few percent) but for H^- better than 1 percent. More importantly, the accuracy increases with increasing principal quantum number N. This is perhaps surprising since the adiabatic approximation is usually considered to become less accurate as the level of excitation in the internal r coordinate increases. That the opposite is true suggests a "stiffening" of the interelectronic axis, i.e., slow rotation and vibration as the size of the atom increases.

The second question, concerning nodal structure, has also thrown light on the dynamical motion underlying resonance formation both from the extent to which the MO nodal structure is conserved and the extent to which it is not. The dominant part of the probability density of two-electron wave functions is located in a restricted region of R. Within this region, accurate wave functions calculated by direct diagonalization exhibit, for fixed R, the nodal structure of the underlying MO.[15] An example is shown in Fig. 5a, where the quantity

$$P(R, \lambda, \mu) = |\Psi|^2 dV(R, \lambda, \mu) \qquad (5.2.37)$$

is plotted for the $^1S^e(N = 3)$ symmetric resonance in He for $R = 8$ au. This resonance has quantum numbers $n_\lambda = 1, n_\mu = 2, m = 0$ corresponding to the $4d\sigma_g$ MO. From Fig. 5a one sees that the number of nodal lines in the accurate wave function is one in the λ coordinate and two in the μ coordinate. To a good approximation they represent lines of constant λ and μ. The similarity to those of the $4d\sigma_g$ MO in the H_2^+ molecule (shown in Fig. 5b) is striking.

The λ motion corresponds to bending vibrations perpendicular to $\hat{\mathbf{R}}$ and therefore parallel to the stable direction of the saddle potential of Fig. 4. Hence the λ nodes are conserved for all R. By constrast the μ motion is in the internal coordinate parallel $\hat{\mathbf{R}}$, i.e., parallel to the unstable direction of the saddle point. Hence μ nodes are fragile and change with R, i.e., there is a mixing of R and μ motion.

The precise nature of the R, μ motion can be examined by fixing λ and plotting the probability density in the (R, μ)-plane. This plot is shown in Fig. 6 for the lowest $^1S^e(N = 4)$ symmetric resonance of helium. As expected the nodal lines, although well-defined, are not parallel either to the R- or the μ-axis. In addition one notes a strong clustering of probability density in a localized region of the (R, μ)-plane. It will be shown that this density clusters along the trajectory of a particular collinear periodic orbit of the classical

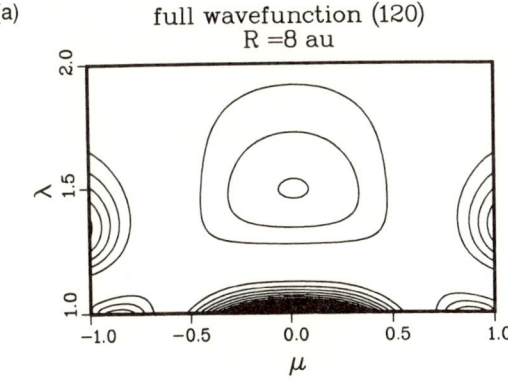

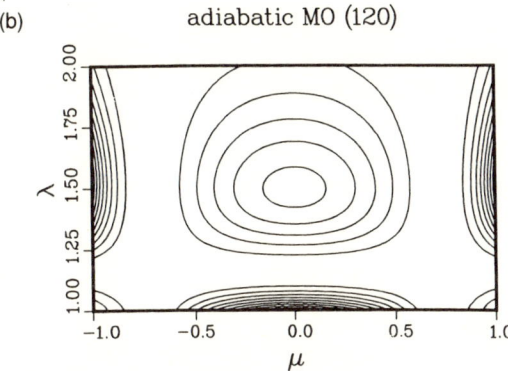

FIGURE 5. (a) The probability density of the $^1S^e(N = 3)$ resonance of helium for $R = 8$ a.u. (b) The corresponding density for the $4d\sigma_g$ MO of the H_2^+ molecular ion for $R = 8$ a.u.

helium atom.[8] The clustering of the accurate wave function along this orbit provided the first clue to the understanding of the classical motion underlying the formation of symmetric intrashell doubly-excited resonances in the helium atom, as will now be explained.

2.5. The Semiclassical Helium Atom

Recently there has been a renewed interest in the classical mechanics of the helium atom, beginning in 1980 with the work of Percival and Leopold.[5] As mentioned in the introduction a landmark was achieved by Wintgen and co-workers,[8] who accomplished the semiclassical quantization of the motion responsible for various classes of resonant states. In the solution of classical mechanics it has been shown that, for motion in which the electrons are predominantly on opposite sides of the nucleus, the

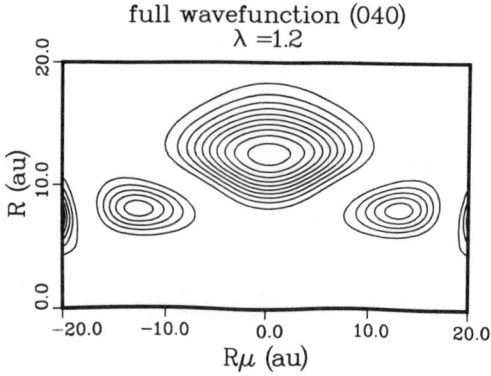

FIGURE 6. The probability density of the $^1S_e(N = 4)$ resonance of helium with $(n_\lambda n_\mu m) = (0, 6, 0)$ plotted in the $(R\mu)$-plane for fixed $\lambda = 1.01$.

bending vibration, first investigated by Langmuir[2] long ago, is stable but as indicated above, the "stretching" motion parallel to the $\hat{\mathbf{R}}$ direction is essentially unstable and leads to a chaotic phase space structure in which are embedded unstable periodic orbits (PO). In this case it is simplest to visualize the colliner configuration, where the angle Θ_{12} between the two electrons is restricted to the value π. The equipotential surface consists of two valleys $r_1, r_2 \to \infty$ and a broad plateau (Fig. 7) roughly in the region of the line $r_1 = r_2 \to \infty$. This is also indicated in Fig. 9. In Fig. 7 is shown

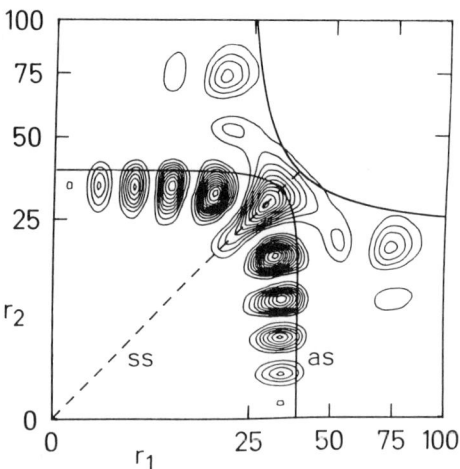

FIGURE 7. The probability density of the $^1S^e(N = 6)$ helium resonance $(n_\lambda n_\mu m) = (0, 10, 0)$ for $\hat{\mathbf{r}}_1 = -\hat{\mathbf{r}}_2$. Also shown are the fundamental asymmetric (as) and symmetric (ss) periodic orbits.

the probability density of an exact wave function of the lowest $^1S^e$ state of the (0, 10, 0) MO. Again there is a remarkable clustering or localization of the probability density. Also shown in these r_1, r_2 coordinates is the fundamental periodic orbit labeled 'as' for asymmetric stretch. It is strikingly clear that the quantum-mechanical resonance is built along this periodic orbit and not along the line $r_1 = r_2$ (labeled 'ss') which corresponds to a pure symmetric stretch motion. However, that the fundamental PO is not purely, although predominantly, of asymmetric character, is seen in Fig. 8, which shows the separate variation of $r(t)$ and $R(t)$. A pure asymmetric stretch has $R(t)$ = constant. From Fig. (8) it is clear that this is not the case. Again this corresponds to the mixing of R (symmetric stretch) and μ (asymmetric stretch) nodes.

The formalism for the semiclassical quantization of the helium atom is based on the PO theory of Gutzwiller.[6] This theory is itself based on the

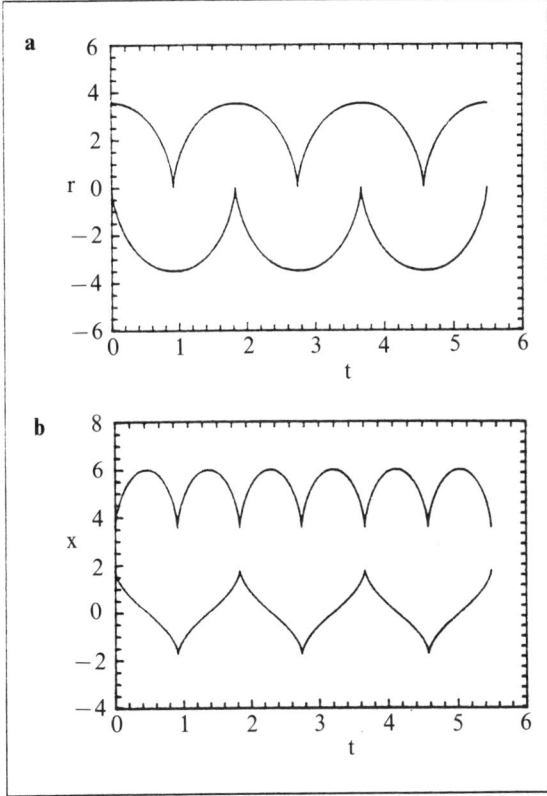

FIGURE 8. The time dependence of molecular coordinates in the asymmetric stretch orbit. (a) $r_1(t)$ and $r_2(t)$. (b) The molecular coordinates; upper curve $R(t)$, lower curve $r(t)$.

Feynman path integral formulation of quantum mechanics evaluated to leading order in $\hbar$, and can be applied both to stable (leading to EBK quantization) and unstable or chaotic motion. It was the absence of a suitable theory for the quantization of chaotic motion that hindered the early application of the "old" quantum theory to three-body Coulomb problems.

Details of the application of the Gutzwiller formalism are given in Wintgen et al.[8] The formalism is not without its problems when applied to concrete cases. The quantum density of states is represented as a sum over the contribution of all periodic orbits. However, there is no guarantee that the sum will converge. Much recent research in the field of semiclassical methods has been devoted to this problem and strategies devised to accelerate this convergence.[25] The results of one such strategy, "the cycle expansion," applied to a summation over 22 periodic orbits in the (r_1, r_2)-plane, where the stable motion perpendicular to $\hat{R}$ is taken into account in linear approximation, is shown in Table I. A representative PO is shown in Fig. 9. In the first column of Table I are the MO quantum numbers of selected states of $^1S^e$ symmetry. The column labeled WKB contains the energies, represented by the effective quantum number N, obtained from a WKB quantization of the fundamental symmetric stretch PO shown alone in Fig. 7. Remarkably, even this naive quantization gives reasonable energies. The column labeled PO contains the results of a "cycle expansion" summation over the 22 different PO. The last column contains numerically exact full quantum-mechanical calculations. One sees that, even down to the

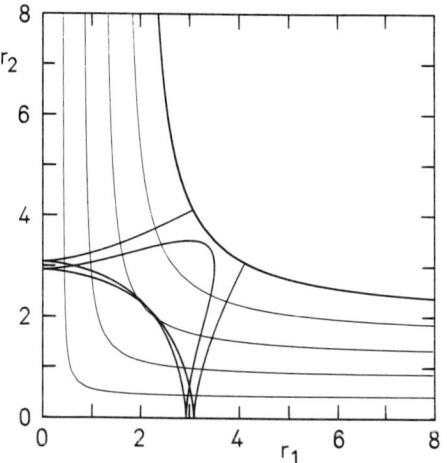

FIGURE 9. The equipotential surfaces of the collinear helium atom showing a typical classical trajectory included in the PO sum (from K. Richter, Thesis, University of Freiburg 1991, unpublished).

TABLE I
The Calculated Energies of the $(n^2)^1 S^e$ Series of Resonances in Helium[a]

N	$(n_\lambda n_\mu m)$	Energy (au)		
		QM	WKB	Cycle
1	(0, 0, 0)	2.904	3.097	2.932
2	(0, 2, 0)	0.778	0.804	0.777
3	(0, 4, 0)	0.354	0.362	0.353
4	(0, 6, 0)	0.201	0.205	0.199
5	(0, 8, 0)	0.129	0.131	0.129
6	(0, 10, 0)	0.0902	0.0917	0.0895
7	(0, 12, 0)	0.0663	0.0675	0.0657
8	(0, 14, 0)	0.0514	0.0517	0.0510

[a] The columns denote: QM, accurate quantum-mechanical calculation; WKB, quantization of the asymmetric stretch orbit alone; cycle, results of a cycle expansion. The data are taken from Wintgen.[8]

ground state, the PO calculations, which have only *classical* information as input give quite accurate values for the energies of quantum states of the helium atom.

3. THE THREE-BODY THRESHOLD—AND BEYOND

Much effort, probably a disproportionate effort, has been concentrated on the departure from the Wigner threshold law of the energy dependence of the cross section for three-body breakup. For short-range potentials the law is fulfilled and predicts a cross section increasing linearly with the excess E required to raise the system from an initial state to a three-body continuum state.[26] It should be emphasized that the linearity arises essentially from the normalization of the final state alone and is independent of the initial state and of the mechanism of transition. The precise normalization of three-body Coulomb continuum states is unknown. However, Wannier[1] recognized that the infinite-range force invalidates the Wigner threshold law and proceeded to give a classical argument to predict a threshold law of $E^{1.127}$ for the (e⁻ + H) system. In arriving at this prediction Wannier supposed that to achieve three-body breakup, against the competition of two-body channels, it is essential that the electrons move so that $\theta_{12} \approx \pi$. It was also assumed that the probability of double ionization is independent of the precise sharing of the total excess energy E between the two electrons.

Subsequently, Peterkop[27] and Rau[28] gave semiclassical derivations of the Wannier threshold law. Other authors[29,30] have developed this further, generalizing to different spin and angular momentum states, to arbitrary masses and charges and to more than three particles. More interesting than the index of the energy dependence alone, however, is the precise structure of the continuum states accessed by a particular transition since this depends naturally not only upon the final state but upon the full transition matrix element. An important question is the connection of this structure with the structure of resonant states just below threshold.

In Section 3.1 a route to ionization by electron impact in the MO model as arising from diabatic crossing of sequences of adiabatic states which themselves show avoided crossings will be suggested, and in Section 3.2 the diabatic MO structure is shown to be reflected in the symmetry properties of the dominant states expected above threshold. This diabatic route implicitly locates the ECM on the saddle point and therefore implies the Wannier configuration $r_1 = -r_2$ to be dominant in leading to three-body breakup. The proven correspondence between the MO modes and the important classical modes of motion allows a semiclassical interpretation to be given of the MO route in terms of the motion of wave packets on the three-body potential surface. Since it has been emphasized repeatedly that the R and μ motions couple strongly and are relatively decoupled from the bending vibration λ motion, it is clear that the decisive motion will be in the $\theta_{12} \approx \pi$ subspace. The corresponding potential surface in the $(r_1 r_2)$-plane is shown in Fig. 9.

In the case of electron impact an initial wave packet would move in from infinity along one of the valleys, i.e., $r_1 = \infty$, r_2 small. It is clear that on the grounds of symmetry it is necessary to discuss motion in one valley only. On reaching the bottom of the valley the packet can be simply reflected back or move out along the other valley. This is elastic scattering and corresponds to motion remaining on the initially populated $1s\sigma_g$ and $2p\sigma_u$ MO only. Inelastic scattering occurs presumably when some portion of the wave packet moves out not along the valleys but along the Wannier ridge $r_1 \approx r_2$. Such symmetric stretch motion corresponds to the diabatic crossing of sets of adiabatic curves. Adiabatic states correspond to resonances and, as shown in Fig. 7, this involves asymmetric stretch motion locally perpendicular to the Wannier ridge. As the wave packet moves out there is a finite probability that these states will be occupied and the packet, or some portion of it, perform AS motion. Two-fragment breakup can occur when, during this motion the packet reaches one valley or the other and instead of being reflected back along the AS trajectory moves out along the valley $r_1 \to \infty$, r_2 small. In the MO picture this corresponds to failure to cross one or another of the avoided crossings on the outgoing trajectory, leading to the asymptotic channel $He^+(n_1 n_2 m) + e^-$, where $n_1 n_2$ are SA limit parabolic quantum numbers. Only that part of the wave packet that retains motion along the ridge can lead to three-body breakup. The dynamics of double photoionization is essentially the same except that in this case the wave packet is prepared in a localized state near the bottom of the well r_1, r_2 small and can propagate either along the valley, leading to single photoionization, or along the ridge, with the possibility of double photoionization if the diabatic crossings are negotiated successfully. The natural conclusion of this is that, unlike the situation for the two-body spectrum, "bound" state structures do not extrapolate smoothly through the three-body threshold. Rather, motion leading to three-body breakup is orthogonal to motion in long-lived resonant states. However, the precise nature of resonant motion very close to threshold is unknown at this time. If the foregoing, admittedly speculative, analysis is correct it implies a strongly correlated motion just above threshold in the Wannier mode $\mathbf{r}_1 = -\mathbf{r}_2$. As will be shown in Section 3.4, experiment appears to confirm this. In turn this implies that any wave function above threshold must explicitly contain the electron–electron repulsion involving the coordinate $\mathbf{R} = \mathbf{r}_1 - \mathbf{r}_2$.

3.1. Ionization Mechanisms near Threshold

One can now pose the question of the mechanism of ionization in the MO picture. As already indicated the movement of a wave packet along the symmetric stretch line of Fig. 7 corresponds to a diabatic traversal of the avoided crossings of adiabatic MO in Fig. 3. The mechanism of ionization near the threshold is governed by two basic processes which affect the character of the ionized state and the cross section for ionization in different ways. The primary process is the establishment of a state in which the CM of the electrons sits on the saddle point of the two-center potential of Fig. 4. This is the prerequisite for three-body dissociation in the Wannier picture. The secondary process is the preservation of this configuration as the two electrons recede, leaving the nucleus symmetrically disposed between them. This three-body breakup configuration is unstable against decay into a two-body breakup configuration in which one electron forms a bound state with the nucleus and the other electron moves off to infinity.

The primary process occurs by two quite separate sequences of transitions involving different molecular symmetry species. This primary process governs the overall magnitude by which ionization occurs and is expected to be decisive in the determination of the symmetry properties (i.e., angular and spin distributions) of ionized electrons. The more complicated atomic case is discussed first. Since the hydrogen atom is assumed to be in the ground $(1s)^2 S^e$ state before the collision, the incident channel can be written $1s\sigma_g$ or $2p\sigma_u$ MO. These two MO will be occupied initially according to the spin and orbital state of the total system. The $1s\sigma_g$ MO can support the gerade two-electron states $^1S^e$, $^3P^0$, $^1D^e$, etc., with $\pi(-1)^S = 1$ and $L + S =$ even, while the $2p\sigma_u$ MO can support the ungerade $^3S^e$, $^1P^0$, $^3D^e$, etc., states with $\pi(-1)^S = 1$ and $L + S =$ odd. Hence these two sets of states have quite different behaviors in their pathways to ionization. The states built on the $1s\sigma_g$ MO have no λ, μ, or ϕ nodes associated with the CM motion. The $1s\sigma_g$ MO is coupled rotationally with all π_g MO and radially with other σ_g MO, thereby generating λ, μ, or ϕ nodes. However, the mechanism for retaining high CM density on the saddle (and hence three-body character) is to follow a diabatic MO that is itself nodeless. This diabatic MO connects the $1s\sigma_g$ MO with the series $3d\sigma_g$, $5g\sigma_g$... etc., having no λ nodes and an increasing even number of μ nodes and corresponds to an extrapolation of the $1s\sigma_g$ MO through the sequence of avoided crossings of Fig. 3a.

Clearly the behavior of the electron ECM at the first broad $1s\sigma_g - 3d\sigma_g$ avoided crossings around 8 a.u. (see fig. 3a) is the primary process which is the precursor of ionization. The simplest pathway is that where the ECM follows the $1s\sigma_g$ MO adiabatically to small interelectronic distances and then on the outgoing half of the collision follows the σ_g diabatic MO through the avoided crossings, riding the saddle to ionization. An alternative pathway is to jump the avoided crossing on the way in to occupy the $3d\sigma_g$ MO and to remain on this MO, hence connecting with the diabatic MO on the way out. The total probability for moving out on the diabatic curve is then $2P(1 - P)$, where P is the probability for traversing the $1s\sigma_g - 3d\sigma_g$ crossing diabatically. This is the primary process involving a double-curve crossing. The secondary process occurs on the outgoing part of the collision as the two electrons recede from each other. Then the ECM must stay on the diabatic MO as the $3d\sigma_g$, $5g\sigma_g$, $7h\sigma_g$, etc., series of crossings is traversed from small to large R. Failure to cross a particular avoided crossing will result in two-fragment dissociation corresponding to excitation of the excited state of the hydrogen atom connecting adiabatically to the $3d\sigma_g$, $5g\sigma_g$, $7h\sigma_g$,... MO.

Of the $^1S^e$, $^1D^e$, etc. singlet states supported by $1s\sigma_g$ ECM motion, the $^1S^e$ should have the maximum probability of crossing since it corresponds classically to a zero-impact parameter collision. However an incident electron at the threshold for ionization has momentum $k = 1$ and therefore classical angular momentum $L \sim bk$. Hence, as the crossing occurs around $R = 4.5$ a.u., one expects states with L values up to about 5 to be involved in threshold ionization.

One other aspect of the $1s\sigma_g$ primary process should be noted. On the second pathway, i.e., initial jumping of the avoided MO crossing to occupy the $3d\sigma_g$ MO inside the crossing, there will be the possibility of coupling to the $3d$ degenerate manifold of states when the two electrons are at their minimum distance of approach. Of these states only the $3d\delta_g$ generates a three-body breakup diabatic MO, i.e., is a precursor MO. This MO is not directly coupled to the $3d\sigma_g$ and any coupling is only possible for $L \geqslant 2$. Hence it is most likely that this pathway will lead to strong dissipation of probability among two-body breakup channels. To summarize, for gerade states based on the $1s\sigma_g$ MO the preferred mechanism of ionization is the following of the adiabatic $1s\sigma_g$ MO down to small interelectronic distances on the inward half of the collision and the following of the diabatic saddle-riding MO out to infinite interelectronic separation on the outward half. For the reasons given above, this should occur with maximum efficiency for the $^1S^e$ state but perhaps with only slightly diminished probability for other low L gerade states.

The same mechanism operates for the triplet $^3P^0$, $^3F^0$ ... sequence built upon the $1s\sigma_g$ MO for the ECM motion, and these states should lead to threshold ionization in the same way. One notes only that for triplets L must be greater than or equal to unity for this primary process to operate.

The question now arises, can the states of symmetry $^3S^e$, $^1P^0$, $^3D^e$,... built on the entrance channel $2p\sigma_u$ ECM MO also lead to three-body breakup? The answer is yes, but the primary process is quite different from that operating for the $1s\sigma_g$ states. The $2p\sigma_u$ MO has $n_\lambda = 0$, $n_\mu = 1$, $n_\phi = 0$. Hence it has a node exactly at the saddle point and states built on this MO will have a low propensity to lead to three-body breakup. Nevertheless there is an efficient pathway for ionization, based upon the rotational coupling of the $2p\sigma_u$ and $2p\pi_u$ MO, which is extremely well-known in ion–atom collisions.[17] At zero interelectronic separation the ECM $2p\sigma_u$ MO is degenerate with the $2p\pi_u$ and can couple to it very efficiently by rotational coupling. This rotation is important since the $2p\pi_u$ has $n_\lambda = 0$, $n_\mu = 0$, $n_\phi = 1$ and is the generator of a diabatic three-body breakup MO connecting the series $2p\pi_u$, $4f\pi_u$, $6h\pi_u$, etc. through its avoided crossings. This sequence of adiabatic MO showing successive avoided crossing is seen in Fig. 3b. The $2p\sigma_u$ MO has a node at the saddle but a sudden rotation of the interelectronic axis through 90° will place the ECM in a $2p\pi_u$ MO having an antinode on the saddle as do all π_u MO of the saddle sequence of Fig. 3a. Hence, the second mechanism for ionization is the initial occupation of the $2p\sigma_u$ MO down to small interelectronic distances followed by a rotational coupling to the $2p\pi_u$ MO, the lowest lying of the MO sequence shown in Fig. 3a. Then on the outgoing half of the trajectory the sequence of avoided crossings among π_u MO energy curves must be negotiated successively to remain on the diabatic π_u MO leading to three-body breakup. However, since the ECM motion now has a projection of its angular momentum along the electronic axis equal to one unit, the angular momentum of the ECM motion itself must be equal at least to unity. Therefore this primary process does not operate for the $^3S^e$-state.

Here the likely mechanisms for electron impact ionization of the hydrogen atom near threshold have been discussed only qualitatively within the MO model. As yet no calculations have been carried out. A similar mechanism of threshold ionization via diabatic coupling of MO has been proposed in the case of proton impact, however, and shown to provide reasonable estimates of cross sections.[31] In this case the mass of the proton implies that all total angular momentum states participate at threshold.

3.2. Symmetries of Electron Impact Ionization at Threshold

Electron impact ionization at threshold has been much discussed in terms of the Wannier mechanism.[27-30] In this mechanism it is assumed that the region $\mathbf{r}_2 = -\mathbf{r}_1$ is the important region for two particles to emerge at threshold. After establishment of energy threshold laws, attention was given to the angular distribution,[32,33] in terms of the most likely states of given $LS\pi$ to appear at threshold. These analyses were made of the six-dimensional wave function $\Psi_{LS\pi}(\mathbf{r}_1, \mathbf{r}_2)$ by considering the nodal structure in the Wannier region via expansion in the six spherical coordinates of $\mathbf{r}_1, \mathbf{r}_2$ and their associated one-electron quantum numbers. In particular, following the first analysis by Greene and Rau,[32] Stauffer[33] made important statements concerning the nodal structure of given $LS\pi$ states. It can be shown from a transformation to molecular coordinates $\mathbf{r}, \mathbf{R}$, that all of the symmetry properties with respect to the Wannier region can be expressed in terms of the three-dimensional $\mathbf{r}$ coordinate describing the motion of the ECM with respect to the nucleus. The nodal structure of continuum wave functions appears directly due to the separation in prolate spheroidal coordinates λ, μ, ϕ. That is, the nodal structure in the Wannier region can be expressed in terms of the 'three-dimensional' quantum numbers n_λ, n_μ, m.

The angular distribution in the continuum is described by a linear combination of wave functions of fixed L, M, S and parity π. More importantly they are eigenstates of the product operator of parity P and permutation P_{12}.

$$PP_{12}\Psi_{LSt}(\mathbf{r}, \mathbf{R}) = \Psi_{LSt}(-\mathbf{r}, \mathbf{R}) = (-1)^t \Psi_{LSt}(\mathbf{r}, \mathbf{R}) \qquad (5.3.1)$$

where $(-1)^t = \pi(-1)^S$.

A state of fixed $LMSt$ or equivalently $LMS\pi$ can be expanded as in Eq. (5.2.3) and Eq. (5.2.6). In the MO picture the Wannier point $\mathbf{r}_2 = -\mathbf{r}_1$ is expressed by the condition

$$\mathbf{r} = 0 \qquad (5.3.2)$$

For fixed $\mathbf{R}$ this origin of ECM coordinates coincides with the saddle point of the two-center potential:

$$r_2 = r_1 \Leftrightarrow \mu = 0 \tag{5.3.3}$$

$$\hat{r}_2 = -\hat{r}_1 \Leftrightarrow \lambda = 1 \tag{5.3.4}$$

i.e., by the origin in prolate spheroidal coordinates. This is important since μ motion is along the unstable saddle direction and connected to decay while λ motion is in the stable direction.

From the expression of the Wannier saddle according to (5.3.3) and (5.3.4) nodal properties can be derived from a consideration of the three internal dimensions $\mathbf{r}$ only. For example from (5.3.1) one sees that at $\mathbf{r} = 0$,

$$[1 + (-1)^{t+1}]\Psi(0, \mathbf{R}) = 0 \tag{5.3.5}$$

This is the first result; it shows that only gerade states with t even or $\pi = (-1)^S$ can be finite at the Wannier saddle. From (5.3.5) one sees that the origin of this rule is the MO g, u parity of the adiabatic function $\Phi_i^t(r, R)$ of Eq. (5.2.6).

The condition $r_2 = r_1$, given by (5.3.3), can also be expressed in MO quantum numbers since $\Phi_i^t = 0$ at $\mu = 0$ when n_μ is odd. More precisely the wave function Φ_i^t has a node or antinode at $r_1 = r_2$ according as $(-1)^{n_\mu} = -1$ or $+1$. Then one can generalize the result of Klar and Schlecht[34] to note that states built solely on $A = +1$ or solely on $A = -1$ have the Wannier exponent 1.127 or 3.881, respectively.

From (5.2.3) it is seen that $^3S^e$ states are built solely on σ_u MO and $^1P^e$ states solely on π_g MO. Both have $A = -1$ which explains the origin of the higher exponent for $^3S^e$ and $^1P^e$ states. All states of $L \geqslant 2$ can be built on MOs of $A = +1$ and therefore can have the lower exponent.

The origin of the other spheroidal coordinate $\lambda = 1$ expresses the condition $\hat{r}_2 = -\hat{r}_1$ or equivalently $\Theta_{12} = 180°$, where Θ_{12} is the interelectronic angle. Since only $m = 0$ MO are finite at $\lambda = 1$ this leads to the result that all states (regardless of $LS\pi$) built on MO with $m > 0$ are zero at $\Theta_{12} = 180°$.

This result can also be extracted from the results of Feagin,[30] who expanded the total wave function in cylindrical coordinates about the point $\mathbf{r} = 0$. At this point the stable part of the ECM motion is represented by a two-dimensional oscillator and the condition that the two electrons are collinear corresponds to vanishing 'vibrational' angular momentum, i.e. to $m = 0$. A similar conclusion was also reached by Selles et al.[35] using a body-fixed frame with z-axis along $\hat{r}_1$.

This has the important consequence that experiments measuring electrons at relative angle 180° measure only $m = 0$, $t = 0$ body-frame wave functions. From (5.2.3) one sees that of those built on σ_g, only states with $(-1)^{L+S+1} = \pi(-1)^L = 1$ can be finite at this point. Such data will be presented in Section 3.4.

3.3. Three-Body Continuum Wave Functions

Many different wave functions have been suggested with which the transition matrix element for electron impact ionization (the so-called $e^- - 2e^-$ process) can be described. The derivation has proceeded mostly via distorted-wave theory[36] and has involved some form of distortion both in initial and final states. The majority of applications, although not all, has been to experiments in which the two electrons have large (in atomic units) relative momenta after the collision. It appears that only one form of the transition matrix element has been able to explain qualitatively all major features of final-state momentum distributions for all experimental geometries and at all incident energies. This is not to say of course that in specific cases other forms have not achieved greater accuracy. The relevant T-matrix element is one in which the exact final scattering state Ψ_f^- of the three-body continuum in the exact T-matrix element

$$T = \langle \Psi_f^- | V_i | \Phi_i \rangle \tag{5.3.6}$$

is approximated by a product of three two-body on-shell Coulomb wave functions, i.e.,

$$T \approx \langle \Psi_{3c}^- | V_i | \Phi_i \rangle \tag{5.3.7}$$

where Φ_i is the initial state with the incident electron asymptotically far from the hydrogen atom and V_i is the perturbation of this initial state by the incident electron. The correlated three-Coulomb wave function (abbreviated 3c) is explicitly[37]

$$\Psi_{3c}^- \simeq (2\pi)^{-3/2} \exp(i\mathbf{k}_a \cdot \mathbf{r}_a) \exp\left(\frac{-\pi\alpha_a}{2}\right) \Gamma(1 - i\alpha_a) {}_1F_1[i\alpha_a; 1; -i(k_a r_a + \mathbf{k}_a \cdot \mathbf{r}_a)]$$

$$\times (2\pi)^{-3/2} \exp(i\mathbf{k}_b \cdot \mathbf{r}_b) \exp\left(\frac{-\pi\alpha_b}{2}\right) \Gamma(1 - i\alpha_b) {}_1F_1[i\alpha_b; 1; -i(k_b r_b + \mathbf{k}_b \cdot \mathbf{r}_b)]$$

$$\times \exp\left(\frac{-\pi\alpha_{ab}}{2}\right) \Gamma(1 - i\alpha_{ab}) {}_1F_1[i\alpha_{ab}; 1; -i(k_{ab} r_{ab} + \mathbf{k}_{ab} \cdot \mathbf{r}_{ab})] \tag{5.3.8}$$

Here to comply with standard practice in the $e^- - 2e^-$ literature the coordinate notation has been changed according to

$$\mathbf{r}_a \equiv \mathbf{r}_1, \mathbf{r}_b = \mathbf{r}_2 \qquad (5.3.9)$$

and

$$\mathbf{r}_{ab} = \mathbf{r}_a - \mathbf{r}_b \equiv \mathbf{R} = \mathbf{r}_1 - \mathbf{r}_2 \qquad (5.3.10)$$

with corresponding notation for the momenta conjugate to these relative coordinates. Note that

$$k_{ab} = \tfrac{1}{2}(\mathbf{k}_a - \mathbf{k}_b) \qquad (5.3.11)$$

The coefficients α are the corresponding Sommerfeld parameters

$$\alpha_a = \frac{-Z}{k_a}, \qquad \alpha_b = \frac{-Z}{k_b} \qquad (5.3.12)$$

$$\alpha_{ab} = \frac{1}{2k_{ab}} = \frac{1}{|\mathbf{k}_a - \mathbf{k}_b|} \qquad (5.3.13)$$

with $Z = 1$ for the hydrogen nucleus.

It is to be emphasized that, as in all distorted-wave theories, (5.3.8) is an ansatz for a distorted wave function. It was designed to treat all Coulomb interactions on an equal footing and to satisfy the exact boundary conditions of the three-body problem for fixed nucleus and two electrons with final momenta $\mathbf{k}_a$ and $\mathbf{k}_b$. That is, Ψ_{3c}^- has the asymptotic form

$$\Psi_{3c}^- \to (2\pi)^{-3} \exp[i(\mathbf{k}_a \cdot \mathbf{r}_a + \mathbf{k}_b \cdot \mathbf{r}_b) + i\phi] \qquad (5.3.14)$$

where

$$\phi = -Zf(\mathbf{k}_a, \mathbf{r}_a) - Zf(\mathbf{k}_b, \mathbf{r}_b) + 2f(\mathbf{k}_{ab}, \mathbf{r}_{ab}) \qquad (5.3.15)$$

and

$$f(\mathbf{k}, \mathbf{r}) = -k^{-1} \ln(kr + \mathbf{k} \cdot \mathbf{r}) \qquad (5.3.16)$$

The final-state wave function (5.3.8) is the one used by Brauner et al.[37] to describe electron-impact ionization of hydrogen and helium. It is equivalent to a form used by Garibotti and Miraglia[48] for heavy-particle impact. In Brauner et al. it was introduced essentially as an ansatz based on a method of Pluvinage.[49] Subsequently it was shown that the ansatz has the desirable feature that it becomes exact as the distance between all three particles tends to infinity.[50] It can also be shown that the ansatz represents the first term of a particular "cluster" expansion of the Møller operator for three-body scattering.[51]

Here a derivation of the T-matrix element of Brauner et al.[37] will be given using distorted-wave theory. This has the advantage that an explicit form for the next-order correction to the T-matrix element can be given.

Standard distorted-wave theory[36] in which the final-state distortion is taken into account gives the T-matrix element

$$T_{fi} = \langle \chi_f^- | V_i | \Phi_i \rangle + \langle \chi_f^- | W_f | \Psi_i^+ - \Phi_i \rangle \tag{5.3.17}$$

Here Φ_i is the initial state of a ground-state hydrogen atom and a free electron with momentum $\mathbf{K}_i$ and Ψ_i^+ is the exact three-body scattering state evolving from it, i.e.,

$$|\Psi_i^+\rangle = (1 + G^+ V_i)|\Phi_i\rangle \tag{5.3.18}$$

where $G^+ = (E - H + i\delta)^{-1}$, H being the total Hamiltonian and V_i that part of it not diagonalized in the initial state Φ_i. The final distorted wave is defined by

$$|\chi_f^-\rangle = (1 + \tilde{G}^- W_f)|\Phi_f\rangle \tag{5.3.19}$$

where $|\Phi_f\rangle$ is the final state of three Coulomb-interacting particles at infinite separation, W_f is that part of the total interaction not diagonalized in $|\chi_f^-\rangle$ and

$$\tilde{G}^- = (E - \tilde{H} - i\delta)^{-1} \tag{5.3.20}$$

with $H = \tilde{H} + W_f$.

As is often the case in distorted-wave applications it is more convenient to specify χ_f^- and solve for W_f rather than the other way around. Here χ_f^- will be taken as the 'three-Coulomb' (3c) wave function used by Brauner et al.,[37] i.e.,

$$\chi_f^- \equiv \Psi_{3c}^- \tag{5.3.21}$$

where Ψ_{3c}^- is given by (5.3.8).

In the following, calculations will be described in which the cross section for electron impact ionization of the hydrogen atom are calculated using the zeroth-order approximation to the T-matrix element, i.e., only the first term in Eq. (5.3.17). Higher-order distorted-wave approximations could be derived by substitution of the form (5.3.18) in Eq. (5.3.17) and suitable expansion of the exact Green operator G^+.

3.4. The Triply-Differential Cross Section at Near-Threshold and Intermediate Energies

In this section the first experimental data on the Triply-Differential Cross Section (TDCS) in coplanar geometry on the fundamental process

$$e^- + H(1s) \rightarrow p^+ + e^- + e^- \qquad (5.3.22)$$

near to threshold will be discussed.

In coplanar geometry the TDCS is given by[38]

$$\frac{d^3\sigma}{d\theta_a d\theta_b dE_b} = (2\pi)^4 \frac{k_a k_b}{K_i} |T(\mathbf{k}_a, \mathbf{k}_b, \mathbf{K}_i)|^2 \qquad (5.3.23)$$

where $\mathbf{K}_i$, $\mathbf{k}_a$, and $\mathbf{k}_b$ are the momenta of the incoming, the scattered, and the ejected electron. The T-matrix element

$$T(\mathbf{k}_a, \mathbf{k}_b, \mathbf{K}_i) = \langle \Psi_f^- | V_i | \Phi_i \rangle \qquad (5.3.24)$$

employs the projectile-target potential

$$V_i = -\frac{Z}{r_a} + \frac{1}{|\mathbf{r}_a - \mathbf{r}_b|} \qquad (5.3.25)$$

where $\mathbf{r}_a$ and $\mathbf{r}_b$ are the electron positions and Z is the nuclear charge ($Z = 1$ for hydrogen). In the following the T-matrix for the scattering from the nucleus

$$T^n = \langle \Psi_f^- | -\frac{Z}{r_a} | \Phi_i \rangle \qquad (5.3.26)$$

and the T-matrix for the scattering from the initially bound electron

$$T^e = \langle \Psi_f^- | \frac{1}{|\mathbf{r}_a - \mathbf{r}_b|} | \Phi_i \rangle \qquad (5.3.27)$$

will be considered separately. Both contributions add coherently to the total T-matrix, of course:

$$T = T^n + T^e \tag{5.3.28}$$

The indistinguishability of the two electrons has to be taken into account. To this end the exchange T-matrix

$$\mathbf{P}_{ab} T(\mathbf{k}_a, \mathbf{k}_b, \mathbf{K}_i) = T(\mathbf{k}_b, \mathbf{k}_a, \mathbf{K}_i)$$

is introduced. Singlet and triplet T-matrices are then

$$T_s = (1 + P_{ab}) T(\mathbf{k}_a, \mathbf{k}_b, \mathbf{K}_i) \tag{5.3.29}$$

and

$$T_t = (1 - P_{ab}) T(\mathbf{k}_a, \mathbf{k}_b, \mathbf{K}_i) \tag{5.3.30}$$

respectively. The TDCS for unpolarized incident electrons is composed of the statistically weighted TDCS for singlet and triplet scattering, i.e.,

$$\frac{d^3\sigma}{d\theta_a d\theta_b dE_b} = (2\pi)^4 \frac{k_a k_b}{K_i} (\tfrac{1}{4}|T_s|^2 + \tfrac{3}{4}|T_t|^2) \tag{5.3.31}$$

The initial state Φ_i in (5.3.21) consists of an incident plane wave and a bound $1s$-state normalized to unity,

$$\Phi_i(\mathbf{r}_a, \mathbf{r}_b) = (2\pi)^{-3/2} e^{i\mathbf{K}_i \cdot \mathbf{r}_a} \phi_{1s}(r_b) \tag{5.3.32}$$

The final two-electron continuum state is described by the symmetric three-body wave function and given explicitly in Eq. (5.3.8).

The final state wave function Ψ_f^- must be normalized to a six-dimensional delta function,

$$\langle \Psi_{f'}^- | \Psi_f^- \rangle \equiv \int d\mathbf{r}_a \int d\mathbf{r}_b \Psi_{f'}^-(\mathbf{k}_a', \mathbf{k}_b'; \mathbf{r}_a, \mathbf{r}_b)^* \Psi_f^-(\mathbf{k}_a, \mathbf{k}_b; \mathbf{r}_a, \mathbf{r}_b)$$

$$= \delta(\mathbf{k}_a - \mathbf{k}_a') \delta(\mathbf{k}_b - \mathbf{k}_b') \tag{5.3.33}$$

in order to obtain the TDCS according to Eq. (5.3.23) on an absolute scale. In the experiments presented TDCS have been measured only relatively. At higher incident energies, accompanied by small values of momentum transfer, the relative measurements were scaled according to an extrapolation to zero momentum transfer and comparison with photoionization cross sections.[38] This prescription is based on the validity of the first Born approximation in the optical limit. Hence it constitutes a good approximation at higher energies. However at intermediate or even lower energies the method must be expected to break down.

Brauner et al.[40] have shown that the symmetric three-body wave function does satisfy the criterion (5.70). This implies that the normalization factors given in Eq. (5.3.8) provide the correct flux at infinity. This normalization has two immediate consequences. The first is that the normalization constant vanishes exponentially at threshold. This behavior emerges from the normalization of that part of the wave function describing the electron–electron interaction. The corresponding normalization factor is given by

$$\left|\Gamma\left(1 - \frac{i}{2k_{ab}}\right)\right|^2 e^{-(\pi/2k_{ab})} \simeq \frac{\pi}{k_{ab}} e^{-(\pi/k_{ab})}, \quad \text{as} \quad k_a b \to 0 \quad (5.3.34)$$

The threshold limit is understood as $E = \frac{1}{2}(k_a^2 + k_b^2) \to 0$. Writing

$$\frac{1}{2}k_a^2 = E \sin^2 \beta$$
$$\frac{1}{2}k_b^2 = E \cos^2 \beta \quad (5.3.35)$$

with

$$\tan \beta = k_a/k_b \quad (5.3.36)$$

one finds

$$\exp\left(-\frac{\pi}{k_{ab}}\right) = \exp\left[-\frac{\pi\sqrt{2}}{\sqrt{[E(1 - \hat{\mathbf{k}}_a \cdot \hat{\mathbf{k}}_b) \sin 2\beta]}}\right] \quad (5.3.37)$$

This behavior, of course, is in conflict with the prediction of Wannier. One should not expect to derive the Wannier threshold law from this theoretical ansatz although the wave function describes the long-range Coulomb interaction correctly. This is because the Wannier phenomenon emerges from the strong configuration interaction between two-electron continuum states and two-electron states where one electron is bound in a high Rydberg state and the other one is free, as discussed earlier. This conclusion is based on the fact that within Wannier's classical analysis the threshold law results from trajectories diverging from the saddle which describe the competition between double vs. single escape. The second consequence of the normalization is that the form of Eq. (5.3.34) leads correctly to the prediction that two electrons cannot escape with the same (finite) energy in the same direction. The corresponding factor in Eq. (5.3.8) produces a zero in the TDCS for this configuration.

It is also interesting that the angular dependence in the exponential factor of Eq. (5.37) can be written in terms of $\gamma = \pi - (\theta_a + \theta_b)$ to give, for equal energy-sharing, a factor

$$\exp[-\sqrt{2\pi}E^{-1/2}(1 + \cos\gamma)^{-1/2}] \simeq \exp(-\pi E^{-1/2})$$
$$\times \exp\left[-\frac{\pi}{8}E^{-1/2}\gamma^2\right] \quad (5.3.38)$$

for small γ. This angular factor is of the Gaussian form derived by several authors; however, they differ in predicting the width of the Gaussian. The normalization factor given in (5.3.35) corresponds exactly to Altick's[39] correlation factor. The full width at half-maximum is then given by

$$\Delta\gamma \simeq 2.66 E^{1/4} \text{ rad}$$

Both the experimental as well as the computed data show widths which are not so narrow that $\cos\gamma$ can be replaced by $1 - \frac{1}{2}\gamma^2$. The l.h.s. of Eq. (5.3.38), however, can still be employed to determine a FWHM. The result reads

$$\Delta\gamma = 4\cos^{-1}\left(\frac{\pi}{\pi + E^{1/2}\ln 2}\right) \quad (5.3.39)$$

For example, for $E_i = 15.6$ eV, Eq. (5.3.39) predicts $\Delta\gamma \simeq 77°$.

3.5. Comparison with Experimental Data

It is only recently that measurements of the TDCS on the fundamental system $e^- + H(1s)$ have become available. The main interest here will center on the structures to be seen in such cross sections and to give, as far as possible, a physical interpretation of them.

The Wannier configuration is the most interesting near threshold. Experiments[38] have been performed at impact energies of 15.6 eV and 17.6 eV (i.e., 2 eV and 4 eV excess energy with respect to three-body breakup) with electrons emerging with equal energies in opposite directions, i.e., $\theta_a + \theta_b = 180°$. (Here θ_a and θ_b are the emission angles with respect to the beam direction and are positive measured anticlockwise and clockwise, respectively). The first question to ask is: in this process which leaves the ECM at rest after the collision, is there memory of the initial beam direction? The results shown in Fig. 10 indicate that the answer is yes. A

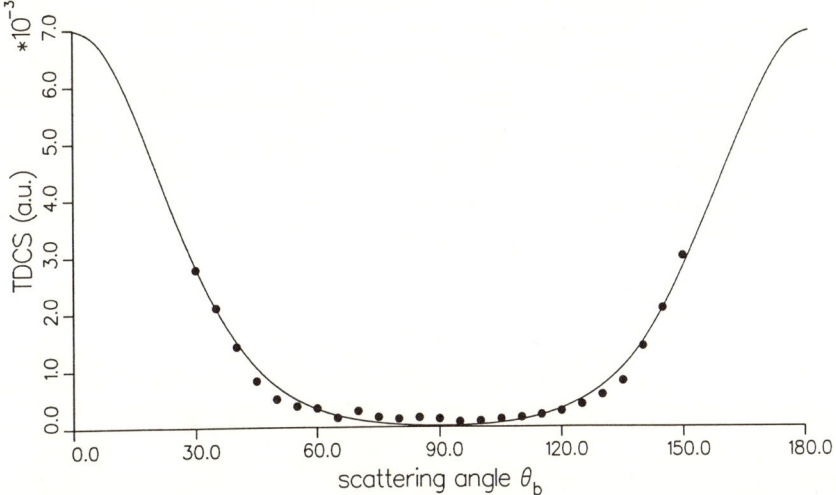

FIGURE 10. The angular distribution of electrons emitted with equal energy 1 eV in the reaction H(e⁻,2e⁻)H⁺. The angle $\theta_a + \theta_b = 180°$ is fixed and $\theta_b = 0$ denotes the beam direction. The theory is normalized to experiment.

pronounced minimum for the relative momentum $\mathbf{k}_a - \mathbf{k}_b$ to be oriented perpendicular to the beam direction is seen in the data. The shape of the calculated cross section is in very close agreement with experiment. One notes that this implies a very small contribution of total S-wave in the final state and a dominance of odd-L components, most probably P-wave.

The calculation involving three Coulomb waves is not particularly transparent in isolating ionization mechanisms. Rather surprisingly it turns out that the major features of the angular distribution are a simple consequence of the probability for momentum transfer in the Coulomb field. This is seen by replacing the Coulomb-correlated wave function by a simple product of plane waves, i.e.,

$$\Psi_f^- = (2\pi)^{-3} \exp(i\mathbf{k}_a \cdot \mathbf{r}_a + i\mathbf{k}_b \cdot \mathbf{r}_b) \qquad (5.3.40)$$

Considering first the electron–electron interaction the T-matrix element (5.3.27) is readily evaluated to give

$$T^e = \frac{1}{2\pi^2} \frac{1}{|\mathbf{K}_i - \mathbf{k}_a|^2} \tilde{\phi}_i(\mathbf{k}_a + \mathbf{k}_b - \mathbf{K}_i) \qquad (5.3.41)$$

where $\tilde{\phi}_i$ is the initial-state momentum wave function. Then the separate singlet and triplet T-matrix elements are given by

$$T^e_{s,t} = \frac{1}{2\pi^2} \tilde{\phi}_i(\mathbf{k}_a + \mathbf{k}_b - \mathbf{K}_i) \left[\frac{1}{|\mathbf{K}_i - \mathbf{k}_a|^2} \pm \frac{1}{|\mathbf{K}_i - \mathbf{k}_b|^2} \right] \quad (5.3.42)$$

where the index s, t refers to the $+$ sign ($-$ sign); $\tilde{\phi}_i$ is the momentum space initial wave function, so that for an initial $1s$ state the cross sections are then proportional to

$$|T^e_s|^2 = \frac{2Z^2}{\pi^6} \frac{1}{(Z^2 + |\mathbf{k}_a + \mathbf{k}_b - \mathbf{K}_i|^2)^4}$$
$$\times \left[\frac{2K_i^2 + k_a^2 + k_b^2 - 2\mathbf{K}_i \cdot (\mathbf{k}_a + \mathbf{k}_b)}{(K_i^2 + k_a^2 - 2\mathbf{K}_i \cdot \mathbf{k}_a)(K_i^2 + k_b^2 - 2\mathbf{K}_i \cdot \mathbf{k}_b)} \right]^2 \quad (5.3.43)$$

and

$$|T^e_t|^2 = \frac{2Z^2}{\pi^6} \frac{1}{(Z^2 + |\mathbf{k}_a + \mathbf{k}_b - \mathbf{K}_i|^2)^4}$$
$$\times \left[\frac{k_b^2 - k_a^2 - 2\mathbf{K}_i \cdot (\mathbf{k}_b - \mathbf{k}_a)}{(K_i^2 + k_a^2 - 2\mathbf{K}_i \cdot \mathbf{k}_a)(K_i^2 + k_b^2 - 2\mathbf{K}_i \cdot \mathbf{k}_b)} \right]^2 \quad (5.3.44)$$

Note that energy conservation requires $k_a^2 + k_b^2 = K_i^2 - Z^2$. In the case of emission of the two electrons with the same energy and at $180°$ we put $\mathbf{k}_b = -\mathbf{k}_a$

$$|T^e_s|^2 = \frac{2Z^2}{\pi^6} \frac{1}{(Z^2 + K_i^2)^4} \left[\frac{K_i^2 + k_a^2}{(K_i^2 + k_a^2)^2 - 4|\mathbf{K}_i \cdot \mathbf{k}_a|^2} \right]^2 \quad (5.3.45)$$

and

$$|T^e_t|^2 = \frac{2Z^2}{\pi^6} \frac{1}{(Z^2 + K_i^2)^4} \left[\frac{\mathbf{K}_i \cdot \mathbf{k}_a}{(K_i^2 + k_a^2)^2 - 4|\mathbf{K}_i \cdot \mathbf{k}_a|^2} \right]^2 \quad (5.3.46)$$

It is readily seen that the angular distribution predicted by (5.3.45) and (5.3.46) has a maximum for $\hat{\mathbf{K}}_i \cdot \hat{\mathbf{k}}_a = \pm 1$, i.e., in the forward and backward directions and a minimum at $90°$, i.e., perpendicular to the beam direction (here the triplet contribution is identically zero). The angular distribution is symmetric around $90°$. One sees that, remarkably, the major qualitative features of the angular distribution in this symmetric geometry, particularly the large deviation from isotropy, are explained by the Fourier transform of the electron–electron Coulomb force.

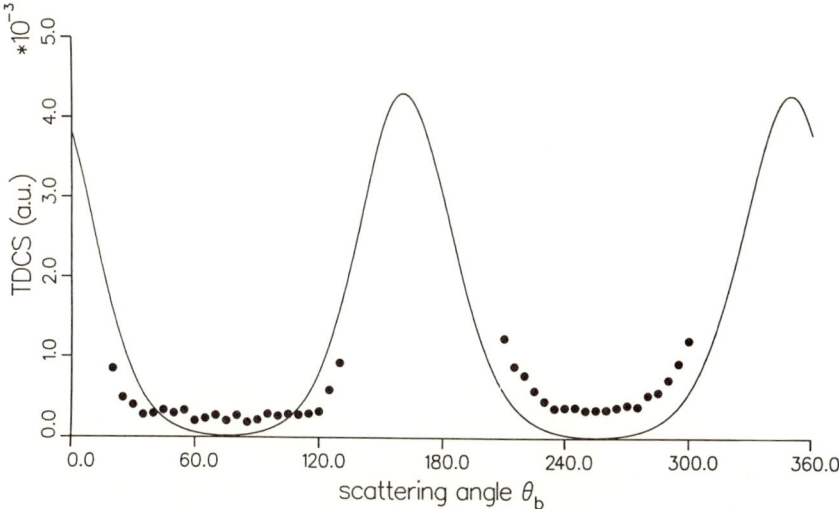

FIGURE 11. The same as Fig. 10 but for $\theta_a + \theta_b = 150°$.

It can also be deduced readily from Eqs. (5.3.40) and (5.3.41) that this feature of maximum cross section for $\mathbf{k}_a - \mathbf{k}_b$ oriented along the direction $\hat{\mathbf{K}}_i$ and minimum for orientation perpendicular to $\hat{\mathbf{K}}_i$ is a general feature not restricted to $\mathbf{k}_a = -\mathbf{k}_b$. This is illustrated in Fig. 11, where the TDCS is shown for fixed relative angle $\theta_a + \theta_b = 150°$ at 15.6 eV. Again the two electrons have finally the same energy. A good qualitative agreement is obtained although unfortunately the experimental data do not cover the regions where pronounced maxima are predicted by theory. Perhaps more interesting from the threshold point of view are the data for equal energy sharing of the total excess energy of 2 eV, but where the angle between the two electrons is varied. These data are shown in Fig. 12. the cross section is plotted as a function of $\theta_a + \theta_b$ for fixed θ_a. The strong influence of the electron–electron repulsion in orientating the electrons is clearly evident in that the cross section peaks in the region of 180° relative angle. Nevertheless the peak is not exactly at 180° and occurs at larger or smaller angles for $\theta_a = 30°$ or 150° respectively. There is also an intriguing subsidiary small peak at around $\theta_b = 90°$ which has so far defied explanation. Strangely the $\theta_a = 90°$ data, as does the calculation, show two peaks and a pronounced *minimum* at $\theta_a + \theta_b \simeq 180°$. The origin of this dip is quite readily explained and is due to a further effect that modifies the single-peak structure expected from a consideration of electron repulsion alone. This arises simply from the statistics of the symmetric geometry $E_a = E_b$ and $\theta_a = \theta_b$. In this case the triplet cross section vanishes, i.e., only a symmetric configuration space wave function is possible. This causes a dip in the total cross section for

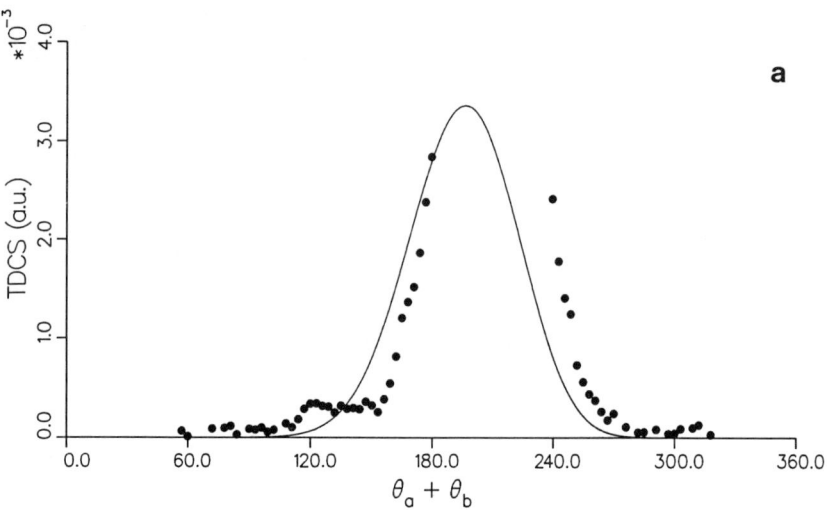

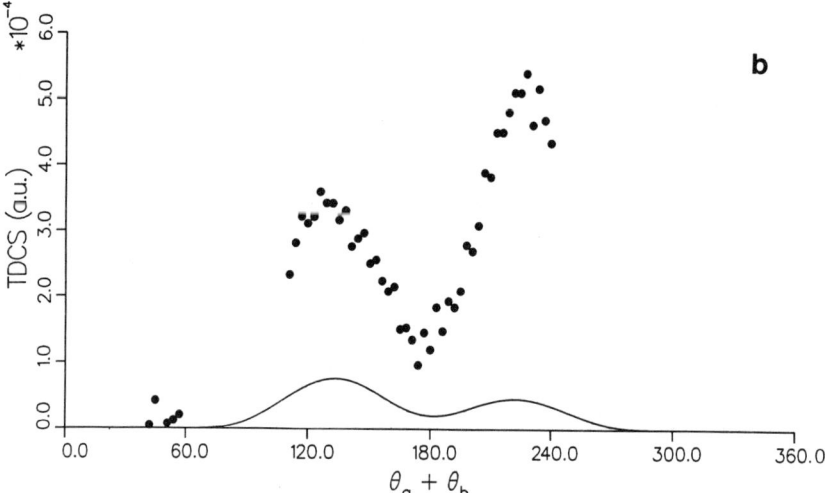

FIGURE 12. The same as Fig. 10 but for $\theta_a + \theta_b$ variable and θ_a fixed at (a) 30° and (b) 90°.

$\theta_a + \theta_b = 2\theta_a$. For $\theta_a = 90°$ this minimum *coincides* with the maximum expected for $\theta_a + \theta_b \simeq 180°$ and causes the peak to split into two.

The calculations shown in Fig. 12 again are not in good quantitative agreement with the data although the main qualitative features are well-described. The calculations of Brauner *et al.*[40] show that, remarkably perhaps, at the intermediate impact energy of 27.2 eV, the agreement of the theory using a three-Coulomb-wave approximation is much worse than at threshold. The intermediate energy range remains a significant problem for theory.

3.6. Double Photoionization

To date experiments measuring the TDCS for double photoionization, either of H^- or He are scarce although such experiments are planned. Nevertheless information on the angular distribution can be gleaned from experiments in which only one of the electrons with energy E_a is detected at a given angle θ_a with respect to the photon polarization vector. This doubly-differential cross section (DDCS) can be written[39]

$$\frac{d\sigma}{\sin\theta_a d\theta_a dE_a} = \frac{1}{2}\frac{d\sigma}{dE_a}[1 + \beta P_2(\cos\theta_a)] \qquad (5.3.47)$$

Developing the ideas of Wannier, Greene[42] made the important prediction that if only one of the electrons is detected in a photoabsorption experiment its emission will be predominantly perpendicular to the linear polarization vector $\hat{\varepsilon}$, i.e., $\beta = -1$ in the cross section of Eq. (5.3.44) for the ejection of the detected electron a with energy E_a at angle θ_a to the polarization direction. According to Greene this result should hold as the threshold is approached both from above and below. Further calculations using Wannier theory by Selles et al.[35] support Greene's prediction that $\beta \to -1$ as $E \to 0$. Experiments[43] appear not to support the Greene prediction. Single photoionization of helium where the second electron is left bound in $He^+(n)$ with $5 \leqslant n \leqslant 10$ show $\beta \approx -0.5$. Double photoionization experiments in which one electron is detected show $\beta \approx -0.45$ for total energy as low as 0.25 eV. This result has been claimed as the first evidence that predictions of the Wannier theory fail near threshold. However, calculations using the wave function of Eq. (5.3.8) suggest that there is no discrepancy between theory and experiment and that Greene's $\beta \to -1$ prediction is not disproved.

The main result of the analysis[44] is to indicate a crucial dependence of the asymmetry parameter β upon the precise way in which the threshold $E \to 0$ is approached. The limit $E_b \to 0$ with E_a fixed shows a limiting value covering the complete range $-1 \leqslant \beta \leqslant 2$, depending on the value of E_a. Only in the case that the ratio $R = E_a/E_b$ is kept fixed as $E \to 0$ does β approach the limiting value -1.

The double photoionization cross section of helium at frequency ω leading to two free electrons with momenta $\mathbf{k}_a$ and $\mathbf{k}_b$ can be calculated in the form of Eq. (5.3.31) from the expression

$$\frac{d\sigma}{d\Omega_a dE_a} = \int \frac{d\sigma}{d\Omega_a d\Omega_b dE_a} d\Omega_b \tag{5.3.48}$$

with[42]

$$\frac{d\sigma}{d\Omega_a d\Omega_b dE_a} = \frac{4\pi^2\alpha}{\omega} k_a k_b |\langle \psi_f | \hat{\boldsymbol{\varepsilon}} \cdot (\boldsymbol{\nabla}_a + \boldsymbol{\nabla}_b) | \psi_i \rangle|^2 \tag{5.3.49}$$

where $E_b = \hbar\omega + E_i - E_a$ and α the fine-structure constant. The total initial binding energy is E_i. For ψ_i the Hylleraas wave function

$$\psi_i = n \exp[-c_1(r_a + r_b) + c_2 r_{ba})] \tag{5.3.50}$$

where $r_{ba} = |\mathbf{r}_b - \mathbf{r}_a|$, $n = 1.470$, $c_1 = 1.858$ and $c_2 = 0.255$ has been used. Again, the correlation in the final state is accounted for by using the symmetric three-particle Coulomb wave function of Eq. (5.3.8) with $Z = 2$ for helium, for the wave function ψ_f.

In Fig. 13 the calculated β values are shown covering ten decades of energy E_b for fixed $E_a = 0.25$ eV. This is the arrangement used by Hall et al.[43] Since the experimental data were obtained with a finite resolution of ≈ 60 meV two curves were calculated. The first is the calculation assuming infinite resolution. The second is a calculation of β in which the cross sections have been averaged over a finite energy resolution weighted by a Gaussian distribution with full width 60 meV. One sees a good general agreement with experiment. Significantly the point of lowest energy E_b ("zero" energy E_b in the experiment) could be placed anywhere below the extent 60 meV of the resolution. In Fig. 13 it has been placed at the median value of 30 meV. This does not affect the agreement with experiment since for finite resolution β is constant for $E_b < 10$ meV.

The variation in β with the ratio R is readily understood. For fixed low $E_a (\ll |E_i|)$ the limit $R \to 0$ corresponds to the unobserved electron b absorbing almost all the photon energy and angular momentum. Since there is no residual angular momentum transfer to the He$^+$ core, the electron a ionizes by "shake-off" into a final s-state. Hence $\beta = 0$, corresponding to isotropy. This is the high-energy E_b limit shown in Fig. 13. The region $R \approx 1$ always gives the minimum in β, i.e., as E_a increases the minimum in the β curve shifts to higher E_b. In addition, the smaller the total energy, the lower the minimum value of β.

The limit $R \to \infty$, i.e., E_a fixed, $E_b \to 0$ depends in general upon E_a and covers the complete range $-1 \leqslant \beta \leqslant 2$. Two limits can be established,

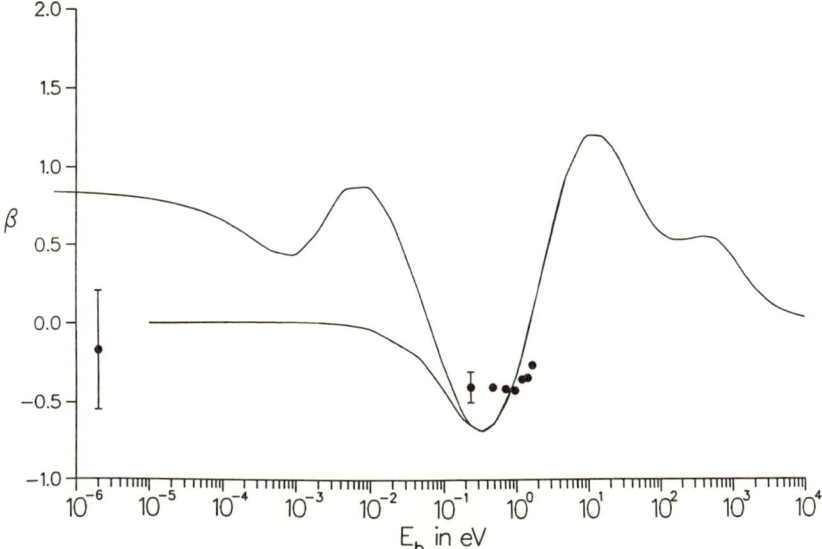

FIGURE 13. The asymmetry parameter β for $E_a = 0.25\,\text{eV}$ and variable E_b. Upper curve, calculation at infinite resolution; lower curve, calculation at 60 meV resolution. Data from Hall et al.[40]

however. For $E_a \to \infty$, $E_b \to 0$ we have $\beta = 2$ since then electron a absorbs almost all the energy and angular momentum, the unobserved electron b shakes off into an s-state and the situation is similar to one-electron photoionization, i.e., $\beta = 2$. For $E_b \to 0$ with E_a fixed but progressively smaller, the limiting value of β will be less than 2 and approach -1 monotonically only as $E_a \to 0$. That is, if experiments are performed with E_a fixed, no matter how small, a value of $\beta > -1$ will be measured as $E_b \to 0$. However if R is *kept fixed* as $E \to 0$ a value of $\beta = -1$ will be the limit. The result is illustrated in Fig. 14, where $\beta(E)$ is shown for $R = 1$ and $R = 2$. In both cases there is a monotonic decrease to $\beta = -1$ as $E \to 0$. For $R \approx 1$, equal energy sharing, the approach to this limit is the most rapid. It corresponds to an experiment performed in such a way that the β measured is always in the minimum of the curve of Fig. 13. The shape of the curves in Fig. 14 is in good qualitative agreement with similar curves shown in Selles et al.[35] obtained on the basis of Wannier theory.

The calculations with the correlated final wave function confirm Greene's result. The result $\beta \to -1$ for R finite but arbitrary as $E \to 0$ is also in accord with the Wannier philosophy that the threshold behavior is independent of the sharing of the total energy E between the electrons. Note, however, that the steepness of the approach to -1 depends strongly on R. As $R \to \infty$ the sudden turn down of the β curve toward -1 occurs for

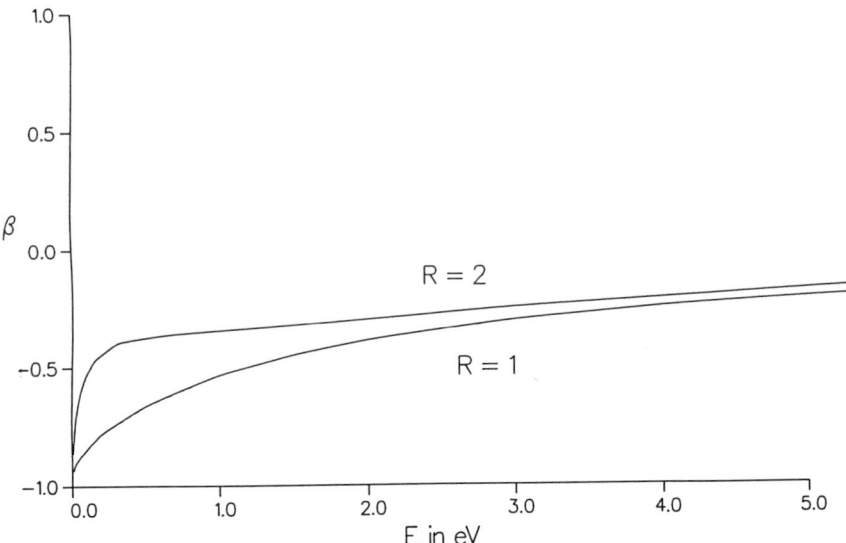

FIGURE 14. The asymmetry parameter as a function of total energy $E = E_a + E_b$ for $R = 1$ and $R = 2$.

lower and lower total energy. This suggests that with finite resolution the experimental test of the prediction of Greene will be extremely difficult. Essentially threshold must be approached such that the resolution is maintained small compared to the smaller of E_a and E_b. Certainly the further understanding of the way in which threshold is approached and the connection with resonance properties below threshold is an outstanding problem. In this respect measurements of the TDCS, giving the mutual angular distribution of two slow electrons as a function of the energy sharing between them, are needed urgently to test recent theoretical predictions.[45] Very recently the first TDCS for double photoionization of helium has been measured[46] for the particular case of equal energy sharing. The angular distributions show good agreement with calculations using the 3c wave function.[45]

3.7. The High-Energy Continuum

For particles interacting through short-range forces it can be shown that the cross section for dissociation of an initially bound pair by impact of a third particle converges to the first Born cross section in the limit of high velocity and energy of the incident particle. For Coulomb forces this cannot be proved. Total cross sections for ionization appear to converge to the first Born result but generally TDCS are not in agreement with first Born predictions. Generally it can be said that for ionization of atomic hydrogen the convergence to the first Born cross section is better the greater the relative velocity of the projectile and the ionized electron after the collision. For cases in which the momenta of the projectile and the ionized electron are similar in the final state there is a strong interaction between them and the first Born approximation, which neglects this interaction, is never valid. This effect of 'final-state' interaction is most dramatic in the case of electron impact and gives rise to significant structure in cross sections. Since the first Born approximation considers only single scattering events, the final state interaction can be thought of as arising from multiple-scattering between the electron and the projectile. Clearly, in the partial summation of the Born series that is represented by the three-Coulomb-wave ansatz of Eq. (5.3.8), other multiple-scattering terms in the three two-body potentials are included. Some of these give rise to identifiable structures in TDCS also.

3.7.1. The First Born Approximation. Before discussing the role of multiple scattering it is illustrative to examine first the structures that are present in the first Born approximation. In this approximation one includes infinite-order interaction between target electron and target nucleus in both initial and final states but only a first-order interaction between the projectile and the target electron. Neglecting exchange in the case of electrons as projectiles the T-matrix element reads

$$T_{fi}^{b1} = \langle \mathbf{k}_a, \psi_{\mathbf{k}_b}^- | V_{eP} | \mathbf{K}_i, \varphi_i \rangle \qquad (5.3.51)$$

where $\mathbf{K}_i$ is the initial momentum of the projectile with respect to the target center-of-mass, φ_i the initial wave function of the target electron, V_{eP} the interaction of the target electron (e) with projectile (P), and $\psi_{\mathbf{k}_b}^-$ the wave function of the ionized electron. In the final state the ejected electron has momentum $\mathbf{k}_b$ with respect to the target nucleus and the projectile has momentum $\mathbf{k}_a$ with respect to the center-of-mass of the electron and the target nucleus. If the ratio of electron mass to nuclear mass is neglected, the center-of-mass of the target may be placed on the target nucleus and the matrix element (5.3.51) cast in terms of the momentum transfer $\mathbf{K} = \mathbf{K}_i - \mathbf{k}_a$ from the projectile to the target as

$$T_{fi}^{b1} = -\frac{1}{2\pi^2} \cdot \frac{Z_P}{K^2} \cdot \langle \psi_{\mathbf{k}_b}^- | \exp(-i\mathbf{K} \cdot \mathbf{r}_b) | \varphi_i \rangle \qquad (5.3.52)$$

The cross-section differential in $\mathbf{k}_a$ and $\mathbf{k}_b$ can be expressed in the form

$$\frac{d\sigma}{d\mathbf{k}_a d\mathbf{k}_b} = \frac{\mu(2\pi)^4}{k_i}|T_{f_i}|^2 \delta\left(\frac{K_i}{2\mu} + \varepsilon_i - \frac{k_a^2}{2\mu} - \frac{k_b^2}{2}\right) \quad (5.3.53)$$

where μ is the projectile-target nucleus reduced mass and ε_i the initial binding energy of the target electron.

The factor Z_P^2/K^4 in $|T_{f_i}^{b1}|^2$ is essentially the Coulomb differential cross section for scattering of the projectile off the target electron considered to be initially free and stationary in the laboratory frame. The energy conservation condition in such a scattering is obtained from (5.3.53) by setting $\varepsilon_i = 0$. The momentum conservation condition arises from the form factor in (5.3.52) in the limit that the electron is free, i.e., the interaction with the target nucleus is switched off. In this case the final state $\psi_{\mathbf{k}_b}^-$ is a plane wave and the form factor in (5.3.52) is given by

$$\langle \psi_{\mathbf{k}_b}^-|\exp(-i\mathbf{K}\cdot\mathbf{r}_b)|\varphi_i\rangle \approx (2\pi)^{3/2}\tilde{\phi}_i(\mathbf{k}_b - \mathbf{K}) \quad (5.3.54)$$

where $\tilde{\phi}_i$ is the momentum-space wave function of the target initial state. In the limit $Z_T \to 0$ it can be shown[52] that, for a 1s hydrogenic initial state

$$\tilde{\phi}_i(\mathbf{k}_b - \mathbf{K}) = 2\sqrt{2\pi}[Z_T^2 + (\mathbf{k}_b - \mathbf{K})^2]^{-2}$$
$$\approx \varphi_i(0)\delta(\mathbf{k}_b - \mathbf{K}) \quad (5.3.55)$$

which illustrates the emergence of the momentum-conservation condition for scattering of a free electron $\mathbf{k}_b - \mathbf{K} = \mathbf{k}_b + \mathbf{k}_a - \mathbf{K}_i = 0$. The two conditions of energy and momentum conservation lead to the kinematic restrictions of a binary encounter between the projectile and the target electron. The simultaneous fulfillment of both conditions expressed by the δ-functions in Eqs. (5.3.53) and (5.3.55) in the limit $Z_T \to 0$, i.e., $\varepsilon_i \to 0$ is equivalent to the condition $\mathbf{k}_a \cdot \mathbf{k}_b = 0$ for electron or positron impact or to $k_b^2 = 2(\mathbf{k}_a/M_a)\cdot\mathbf{k}_b$ for heavy-particle impact, where M_a is the particle mass. For finite Z_T, $\tilde{\phi}_i(\mathbf{k}_b - \mathbf{K})$ still peaks at $\mathbf{k}_b = \mathbf{K}$ but instead of a δ-function has the form of a distribution of width proportional to Z_T in momentum space. This gives rise to the "binary ridge" structures seen[53] in the cross-section differential in the ejected electron energy $E_b = \frac{1}{2}k_b^2$ (i.e., integrated over all angles of $\hat{\mathbf{k}}_a$ and $\hat{\mathbf{k}}_b$). Note that the binding in the initial state implies that $\mathbf{k}_b - \mathbf{K}$ is not necessarily zero. However, $\mathbf{Q} = \mathbf{k}_b - \mathbf{K}$ is just the momentum taken up by the target nucleus by virtue of initial binding of the target electron. This recoil of the target nucleus broadens the binary δ-function into a peak of width proportional to Z_T and the initial binding energy ε_i in (5.90) shifts the peak maximum.

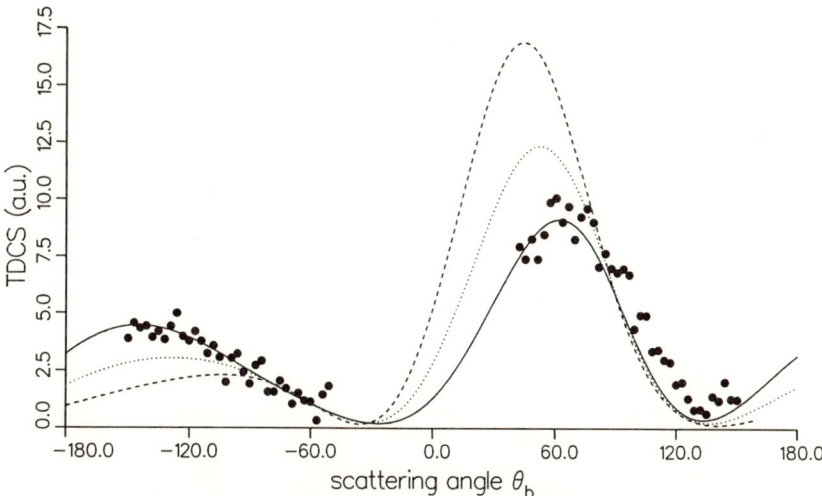

FIGURE 15. The TDCS for the process H(e, 2e)H$^+$ at 250 eV impact energy. E_b is fixed at 5 eV and θ_a at 3°. The continuous curve is for electron impact, the long-dashed curve for positron impact, and the short-dashed curve is the first Born approximation.

The peak is very clearly seen in the triply differential cross section for the $e^- - 2e^-$ ionization process in asymmetric geometry shown in Fig. 16. Thus, the simplest form of structure arising in ionization cross sections for fast-particle impact is a peak due to effective two-body scattering between target electron and projectile. This peak is observable when the influence of the target nucleus is small, i.e., the ejected electron can be represented approximately by a plane wave. Further well-known structure arises when, instead of a plane wave for the ionized electron, the Coulomb interaction of the slow electron with the nucleus is "switched on" in the final state, i.e., a Coulomb wave is used. This is the first Born approximation. Then the typical triple-differential cross section shown in Fig. 15 is obtained, where, in addition to the binary peak, a new peak appears in the backward direction. This 'recoil' peak is due directly to the final-state interaction of electron and nucleus and the aptness of its name emerges from the following argument. If the final state of the ionized electron is written

$$\langle \psi_{\mathbf{k}_b}^- | = \langle \mathbf{k}_b | (1 + G_{eT}^- V_{eT})^\dagger \tag{5.3.56}$$

where G_{eT} is the Green operator containing the full kinetic energy but only the potential V_{eT} between target nucleus and electron, then (5.3.51) can be written

$$T_{f_i}^{b1} = \langle \mathbf{k}_a, \mathbf{k}_b | V_{ep} + V_{eT} G_{eT}^+ V_{eP} | \mathbf{k}_i, \varphi_i \rangle \tag{5.3.57}$$

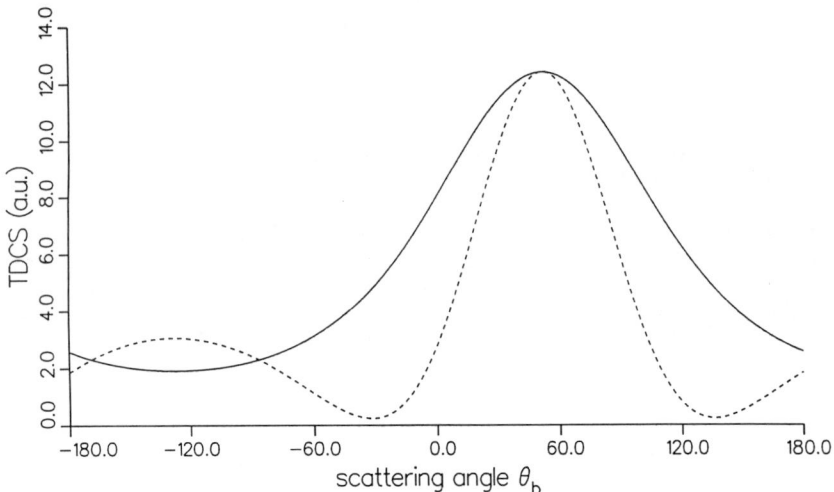

FIGURE 16. The dashed line is the first Born approximation shown in Fig. 15. For conparison, the TDCS for plane waves in the final state (continuous curve) has been normalized to it at the maximum of the binary peak.

The first term is just the binary encounter approximation giving the binary peak. The second term corresponds to the electron in bound state $|\varphi_i\rangle$ being first scattered by the projectile (interaction V_{eP}) and then scattering (recoiling) an infinite number of times off the massive target nucleus. It is this second term that gives the recoil peak, as can be seen by comparing the two curves of Fig. 16.

In detail the modulus squared of the form factor in (5.3.52) can be written for the ionization of a hydrogenic atom of charge Z_T:

$$|\langle\psi_{\mathbf{k}_b}^-|\exp(-i\mathbf{K}\cdot\mathbf{r}_b)|\varphi_{1s}\rangle|^2$$
$$= 8\pi\left(\frac{Z_T}{k_b}\right)\exp\frac{\{-(2Z_T/k_b)\tan^{-1}[2Z_T k_b(K^2 + Z_T^2 - k_b^2)]\}}{[1 - \exp(-2\pi Z_T/k_b)][(K^2 + Z_T^2 - k_b^2)^2 + 4Z_T^2 k_b^2]}$$
$$\times F(\mathbf{K}, \mathbf{k}_b)|\tilde{\varphi}_{1s}(\mathbf{K} - \mathbf{k}_b)|^2 \qquad (5.3.58)$$

where

$$F(\mathbf{K}, \mathbf{k}_b) = (K^2 - \mathbf{K}\cdot\mathbf{k}_b)^2 + \frac{Z_T^2}{k_b^2}(\mathbf{K}\cdot\mathbf{k}_b)^2 \qquad (5.3.59)$$

One notes that apart from the Fourier transform of the initial state contained in the binary encounter approximation, the factor F is the angle-dependent factor giving rise to the recoil peak and modifying the binary peak. The remaining factor is angle-independent but depends strongly upon the parameter Z_T/k_b, characterizing the strength of the final-state Coulomb interaction. In the limit $k_b \ll Z_T$ and $k_b \gg K$, i.e., ionization of electrons with vanishingly small velocity, the form factor (5.3.58) becomes

$$\lim_{k_b \to 0} |\langle \psi_{\mathbf{k}_b}^- | \exp(-i\mathbf{K} \cdot \mathbf{r}) | \varphi_{1s} \rangle|^2 = 8\pi \left(\frac{Z_T}{k_b}\right) \exp\left(\frac{-4Z_T^2}{K^2 + Z_T^2}\right)$$
$$\times [K^4 + (\mathbf{K} \cdot \hat{\mathbf{k}}_b)^2](K^2 + Z_T^2)^{-4} \qquad (5.3.60)$$

With the direction of momentum transfer as the axis one has $\hat{\mathbf{K}} \cdot \hat{\mathbf{k}}_b = \cos \Theta$. Then one sees from (5.3.60) that the first Born angular distribution is of the form $K^2 + \cos^2 \Theta$, i.e., is symmetric with respect to this axis, with binary and recoil peaks of equal heights and 180° apart. The peaks are at $\pm 90°$ with respect to the position of the minimum. It is to be emphasized that this angular distribution, of general form $a + bP_2(\cos \Theta)$, arising in the limit $k_b \to 0$, K, arbitrary, is not to be confused with the well-known forward-scattering, fast-collision optical limit $K \to 0$, k_b, arbitrary. This latter limit gives rise to a similar angular distribution due to the dipole nature of the Coulomb interaction in the optical limit. In the $k_b \to 0$ limit all multipoles act coherently. Hence, the distinction between binary and recoil peaks no longer has meaning. Rather the double-peak shape of the angular distribution arises from an extrapolation into the continuum of the shape of the coherently excited *nlm* manifold of states for large principal quantum number n.[46] In this very asymmetric situation the three-body continuum approximates a two-fragment breakup in which the influence of the fast electron is small.

3.7.2. Beyond the First Born Approximation. Much experimental data[54] on hydrogen has been obtained in asymmetric geometry, i.e., where the incident electron has energy much in excess of the 13.6 eV binding energy of the ground-state electron which itself is ionized with a final energy small with respect to the final energy of the incident electron. In this situation exchange scattering is negligible and the TDCS in coplanar geometry shows in all cases the two prominent features, the binary and recoil peaks, reproduced by the first Born approximation. Nevertheless the first Born calculation fails to predict the precise positions and relative heights of the two peaks correctly. For example, the binary and recoil peaks are not collinearly aligned along the direction of momentum transfer as demanded by the first Born approximation. Much effort has gone into incorporating multiple-scattering effects into theories of impact ionization. Cross sections calculated by Brauner *et al.*[37] with the symmetric three-

Coulomb wave function show good agreement with the shape and positions of binary and recoil peaks for a variety of impact energies and momentum transfers. This, and the strong departure from the first Born TDCS, indicates the remarkable effect of final-state interactions even when the two electrons have large relative momenta. Again this is evidence of the long-range Coulomb forces operating.

A good example is shown in Fig. 15. This is the TDCS plotted as a function of the angle θ_b of the slow electron with respect to the beam direction. The incident electron has 250 eV and scatters through an angle of 3°. The slow electron is emitted with an energy of 5 eV. Also shown are the TDCS in the first Born approximation and that for positron impact for which there are as yet no experimental data. The first Born result is quadratic in the charge of the incident particle and therefore identical for electron and positron impact. Although the incident electron has a velocity more than four times the Bohr velocity (i.e., the orbital velocity of an electron in the 1s state of the hydrogen atom) one notes that the ratio of binary to recoil peak heights is predicted very badly by the first Born approximation, whereas the multiple-scattering theory gives excellent agreement. It is also seen that the results for positron impact depart from the first Born TDCS in the opposite direction as compared to electron impact. This is readily understood as being due to the attraction/repulsion between the projectile and the ionized electron in the final state. When projectile and electron both move in the forward direction (binary peak) the cross section is enhanced for positrons as projectiles and decreased for electrons. When the projectile and the ionized electron move in opposite directions the converse is true. Again this is a very clear example of final-state Coulomb interactions neglected in the first Born approximation.

Further numerical results of Brauner et al.[37] indicate that there is a slow but uniform convergence to the first Born result as the incident energy E_i tends to infinity. However even at an incident energy of 2 keV there is still a difference of a few percent between the multiple-scattering and first Born results in the binary peak region.

The discussion in Section 3.4 illustrated the extremely strong tendency of the two electrons to orient themselves at 180° to each other when they emerge with equal, but low energies. The results is asymmetric geometry have shown similar strong electron correlation effects. The electron–electron correlation is also a dominant feature, even at very high energies when coincidence measurements are made of electrons emerging at very small relative angles ($\theta_a + \theta_b \leqslant 10°$) and with roughly equal energies. Then the electron–electron repulsion leads to a strong dip in the cross section with a zero cross section for equal energy sharing when $\theta_a + \theta_b = 0$. Such a dip was predicted theoretically[55] and has been confirmed recently by experiments[56] on the electron-impact ionization of helium. The results are shown

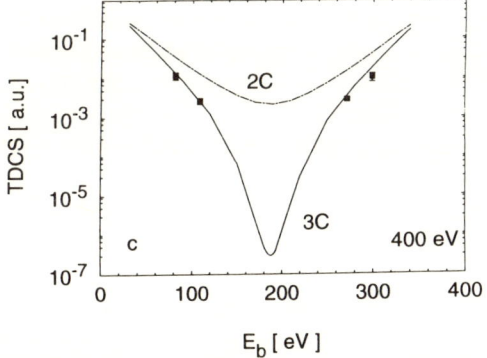

FIGURE 17. The TDCS for 400-eV electron impact ionization of helium with $\theta_a = \theta_b = 4°$. The curve labeled 3c is calculated with Ψ_{3c} of Eq. (5.3.8) as final state; that labeled 2c with a wave function neglecting electron repulsion.

in Fig. 17 for an impact energy of 400 eV. The electrons were measured at $\pm 4°$ to the beam direction, i.e., relative angle $\theta_a + \theta_b = 8°$ and the cross section is plotted as a function the energy of one electron, which, of course, fixes the energy of the other electron.

The theoretical treatment[55] of this ionization process shows that a zero in the cross section for vanishing relative momentum is only obtained when the repulsion between the two electrons is taken into account explicitly. This is the case for cross sections calculated with Ψ_{3c} as the final state [Eq. (5.3.8)]. However, the first Born approximation gives a finite cross section for zero relative electron momentum in the final state since the electron repulsion is neglected. Nevertheless, on kinematic grounds it does predict a minimum in the cross section as a function of energy when the electrons have equal energy (for fixed relative angle). For finite, but small, relative momentum both the first Born approximation and the cross section calculated with the correlated continuum wave function Ψ_{3c} have minima at equal energy sharing, but the latter predicts a much deeper minimum due to the explicit inclusion of electron repulsion. That this repulsive effect is necessary is illustrated in Fig. 17, which shows that much better agreement with experiment is obtained when the repulsion is taken into account than when it is neglected. One should note, however, that the approximation in which the electron–electron repulsion is neglected in Fig. 17 (curve labeled 2C) is *not* the first Born approximation. In this approximation, one electron is placed in a plane wave, the other in a Coulomb wave.

Such an asymmetric approximation is plainly unsuited to describe equal-energy two-electron states. Rather in Fig. 17 the curve labeled 2C has been calculated from (5.3.23) and (5.3.24) using a wave function Ψ_{2c} as the final state. This wave function is simply a product of two electron–nucleus eigenstates with momenta $\mathbf{k}_a$ and $\mathbf{k}_b$, i.e., in Ψ_{3c} of Eq. (5.3.8) the electron–electron eigenstate of relative momentum $\mathbf{k}_{ab}$ is replaced by unity. In particular the dominant part of the electron–electron normalization factor appearing in (5.3.34) is neglected in Ψ_{2c}. It is precisely this factor that gives the zero in the cross section calculated with Ψ_{3c} as $k_{ab} \to 0$.

4. CONCLUSIONS

In the preceding, the present state of knowledge of the dynamics of three-body Coulomb interacting systems has been considered. Below the three-body breakup threshold, apart from the bound spectrum of singly-excited states, there exist infinities of doubly-excited resonant states which dominate the two-fragment breakup channels. Considerable progress has been made in recent years in unraveling the dynamics of these resonant states and assigning new schemes of classification in terms of new approximate quantum numbers. Although much effort has gone into a description in terms of hyperspherical coordinates, prominence has been given here to the use of the adiabatic approximation using the relative vector of the two particles of like charge as adiabatic parameter. This strategy unifies atomic and molecular problems and leads directly to new MO quantum numbers, as the adiabatic two-center Coulomb problem is separable.

The adiabatic MO picture also allows a rather direct connection to be made with recent solutions of the classical three-body Coulomb problem. An expansion of motion around the saddle point of the two-center Coulomb potential leads directly to "bending," "asymmetric stretch," and "symmetric stretch" motion of the two particles of like charge with respect to that of opposite charge. These motions are exactly those emerging from classical mechanics, and a semiclassical quantization leads to quantum numbers in close correspondence to those of the MO model.

The theory of three-body breakup threshold processes is still rather unclear. Although fragmentary information concerning the dominant mechanisms of ionization and the symmetries of preferred configurations emerge from both MO treatments and Wannier approaches, there is no well-founded theory of the population of threshold states by particle impact or through direct two-electron excitation by photons. Most studies using Wannier theory, including recent developments, concentrate on properties of the final state alone, although the probability of populating a specific final-state configuration clearly depends upon the full T-matrix.

Although the description of *precisely* the threshold region may be difficult, there are indications that several eV above threshold the situation may be rather simpler, particularly when the two emerging electrons have roughly equal energies. Here there have been many distorted-wave theories put forward but again prominence has been given to a form of final-state wave function that emphasizes the relative vector of the two electrons explicitly. This enables one to study directly the effect of electron correlation in the final state. It is found that structure in the TDCS near threshold emerges mainly from two effects. One is the preponderance of the relative momentum after the collision to be aligned along the initial beam direction (a strong "memory" effect, even close to threshold). The other is the strong electron–electron repulsion tending to align the electrons at 180° to each other when they have rather similar energies. This effect also dominates the TDCS for double photoionization of helium.

As the total energy of the electron pair is increased in the continuum, either in electron or photon impact, the influence of electron correlation becomes less noticeable except when the relative momentum of the two electrons is much smaller than their total momentum relative to the nucleus, i.e., when they emerge "close together." In the latter case a strong dip in the cross section occurs[55,56] when the relative momentum approaches zero. Even when the relative momentum is comparable to the total, i.e., when the energy sharing between the two electrons is very asymmetric, there are still strong deviations of the shape of the TDCS from the first Born approximation result indicating that electron correlation is not negligible. Generally the binary peak is lower and the recoil peak higher than in the first Born TDCS. This may explain why *total* ionization cross sections appear to converge to the first Born result already at energies of the order of 500 eV, where the TDCS still shows significant differences between first Born results and cross sections calculated with Ψ_{3c} as the final state. The agreement of the total cross sections is fortuitous, due to a convenient cancellation of errors. Even at impact energies of the order of 2 keV there are departures of the order of 5% from the first Born TDCS.

The one major result of all this work is that electron correlation is significant throughout the complete spectrum of energies of three-body atomic systems. It is equally responsible for the existence of long-lived resonant states above the lowest two-body fragmentation threshold and for significant structures in multiply differential cross sections for electron and photon impact ionization.

ACKNOWLEDGMENTS

I am deeply grateful to all present and former colleagues and students in the Freiburg group who have contributed to this work; in particular to J. Berakdar, M. Brauner, J. T. Broad, H. Klar, F. Maulbetsch, K. Richter, J.-M. Rost, and D. Wintgen. We all appreciate the generous support of the Deutsche Forschungsgemeinschaft, partly under the auspices of SFB 276.

REFERENCES

1. G. H. Wannier, *Phys. Rev.* **90**, 817 (1953).
2. I. Langmuir, *Phys. Rev.* **17**, 339 (1921).
3. J. H. van Vleck, *Philos. Mag.* **44**, 842 (1922).
4. M. Born *The Mechanics of the Atom* (Ungar, New York, 1927).
5. I. Percival and J. G. Leopold, *J. Phys. B* **13**, 1037 (1980).
6. M. C. Gutzwiller, *J. Math. Phys.* **12**, 343 (1971).
7. E. B. Bogomolny, *Nonlinearity* **5**, 805 (1992).
8. D. Wintgen, K. Richter, and G. Tanner, *Chaos* **2**, 19 (1992).
9. H. A. Bethe, *Ann. Phys. Leipzig* **5**, 325 (1930).
10. E. A. Hylleraas, *Zeit. Phys.* **65**, 209 (1930).
11. G. Hunter and H. O. Pritchard, *J. Chem. Phys.* **46**, 2146 (1967).
12. J. M. Feagin and J. S. Briggs, *Phys. Rev. Lett.* **14**, 627 (1986); *Phys. Rev. A* **37**, 4599 (1988).
13. J. M. Rost and J. S. Briggs, *J. Phys. B* **24**, 4293 (1991).
14. U. Fano, *Rep. Prog. Phys.* **46**, 97 (1983).
15. J. M. Rost, R. Gersbacher, K. Richter, J. S. Briggs, and D. Wintgen, *J. Phys. B* **24**, 2455 (1991).
16. K. Helfrich, *Zeit. Phys.* **D13**, 295 (1989).
17. J. S. Briggs, *Rep. Prog. Phys.* **39**, 217 (1976).
18. C. Wulfman and K. Sukeyuki, *Chem. Phys. Lett.* **23**, 367 (1973).
19. O. Sinanoglu and D. R. Herrick, *J. Chem. Phys.* **62**, 886 (1973).
20. D. R. Herrick, *Adv. Chem. Phys.* **52**, 1 (1983).
21. C. D. Lin, *Phys. Rev. A* **29**, 1019 (1984).
22. D. R. Herrick and M. E. Kellman, *Phys. Rev. A* **21**, 418 (1980).
23. A. Vollweiter, J. M. Rost, and J. S. Briggs, *J. Phys. B* **24**, L155 (1991).
24. C. A. Coulson and A. Joseph, *Int. J. Quant. Chem.* **1**, 337 (1967).
25. G. S. Ezra, K. Richter, G. Tanner, and D. Wintgen, *J. Phys. B* **24**, L413 (1991).
26. U. Fano and A. R. P. Rau, *Atomic Collisions and Spectra* (Academic Press, Orlando).
27. R. Peterkop, *J. Phys. B* **4**, 513 (1971).
28. A. R. P. Rau, *Phys. Rev. A* **4**, 207 (1971).
29. H. Klar, *Zeit. Phys. A* **307**, 75 (1982).
30. J. M. Feagin, *J. Phys. B* **17**, 2433 (1984).
31. D. I. Abramov, S. Yu. Ovchinnikov, and E. A. Solovev, *JETP* **47**, 424 (1988).
32. C. H. Greene and A. R. P. Rau, *Phys. Rev. Lett.* **48**, 533 (1982).
33. A. D. Stauffer, *Phys. Lett.* **91A**, 114 (1982).
34. H. Klar and W. Schlecht, *J. Phys. B* **9**, 1699 (1976).
35. P. Selles, J. Mazeau, and A. Huetz, *J. Phys. B* **20**, 5183 (1987); **23**, 2613 (1990).
36. C. J. Joachain *Quantum Collision Theory* (North Holland, Amsterdam), Ch. 17.
37. M. Brauner, J. S. Briggs, and H. Klar, *J. Phys. B* **22**, 2265 (1989).
38. H. Ehrhardt, K. Jung, G. Knoth, and P. Schlemmer, *Phys. Lett.* **110A**, 92 (1985).
39. P. L. Altick, *J. Phys. B* **18**, 1841 (1984).
40. M. Brauner, J. S. Briggs, H. Klar, J. T. Broad, T. Rösel, K. Jung, and H. Ehrhardt, *J. Phys. B* **24**, 657 (1991).
41. H. Klar and M. Fehr, *Zeit. Phys. D* **23**, 295 (1992).
42. C. H. Greene, *J. Phys. B* **20**, L357 (1987).
43. R. I. Hall, A. G. McConkey, L. Avaldi, K. Ellis, M. A. MacDonald, G. Dawber, and G. C. King, *J. Phys. B* **25**, 1195 (1992).
44. F. Maulbetsch and J. S. Briggs, *Phys. Rev. Lett.* **68**, 2004 (1992).
45. F. Maulbetsch and J. S. Briggs, *J. Phys. B* **26**, 1679 (1993).

46. O. Schwarzkopf, B. Krässig, J. Elmiger, and V. Schmidt, *Phys. Rev. Lett.* **79**, 3008 (1993).
47. J. S. Briggs and M. Day, *J. Phys. B* **13**, 4797 (1980).
48. C. R. Garibotti and J. E. Miraglia, *Phys. Rev. A* **21**, 527 (1980).
49. P. Pluvinage, *Ann. Phys. Paris* **5**, 145 (1950); *J. Phys. Radium* **12**, 789 (1951).
50. H. Klar, *Zeit. Phys. D* **16**, 231 (1990).
51. J. S. Briggs, *Phys. Rev. A* **41**, 539 (1990).
52. K. Dettmann and G. Leibfried, *Zeit Phys.* **218**, 1 (1969).
53. M. Inokuti, *Rev. Mod. Phys.* **43**, 297 (1971).
54. H. Ehrhardt, K. Jung, G. Knoth, and P. Schlemmer, *Zeit. Phys. D* **1**, 3 (1986).
55. M. Brauner and J. S. Briggs, *J. Phys. B* **19**, L325 (1986).
56. G. Pan, P. Hvelplund, H. Knudsen, Y. Yamazaki, M. Brauner, and J. S. Briggs, *Phys. Rev. A* **47**, 1531 (1993).

INDEX

Abelian (ε) limit, 176, 182
Adiabatic approximation, 284, 287, 297
Adiabatic method, 285
Adiabatic motion, 301
Adiabatic potential curves, 260
AGS equations, for arrangement operators
    for break-up operators, 56
    charged-particles, 46
    for Coulomb-modified short-range, 54
    neutral particles, 39, 42
Amplitude, transition, 191
Anomalous term, 179, 180, 183, 185, 192
Approximation
    adiabatic, 287, 302, 336
    binary encounter, 332
    Born, 283
    Born–Oppenheimer, 284, 285
    disorted wave, 317
    finite dimensional, 199, 202
    first Born, 319, 329, 331, 333–335
    parameters, 196
Asymmetry parameter
    β, 264, 325–328
    q, 258
Asymptotic behavior, 177, 179, 187, 195
Asymptotic channel, 191
Asymptotic condition, 8
Asymptotic situation, 187, 189
Asymptotic states: *see* States, asymptotic
Atomic/molecular physics, 216
Atomic/molecular system, 170

Bipolar harmonic, 143
Boundary condition(s), 101, 105, 110, 113, 114, 123, 129, 130, 132–135, 138, 156, 174, 187, 194
Branching ratio, 255, 260, 273

Breakup, 98, 123, 124, 130, 132, 135, 155, 158, 159, 170, 189, 192, 212, 213, 216, 307–309, 311, 333, 336

Chain rule, 188
Channel(s)
    diabatic, 260
    Hamiltonian, 36
    interaction, 37
    rearrangement, 123, 131, 145, 146, 155, 156, 160, 190
    resolvent, 360
    scattering, 124
    state, 36, 43
Chaotic motion, 193, 406
Charge dependence, 154
Close coupling, 248
Collective variables, 296
Coulomb effects on analytical properties of particle-transfer amplitudes, 88
Convergence, 180, 197, 200
    of particle-wave series, 7
    strong, 205, 206
    uniform, 206
    weak, 206
Cook–Hack method, 178
Coordinates
    hyperspherical, 124, 132, 242, 245, 251, 273, 275, 285, 336
    Jacobi, 121, 150, 188, 190
    prolate spheroidal, 251, 275
Core, 203–205
Correlation, 221, 223, 224, 226, 245, 251, 259, 326, 334, 337
    diagram, 290
    electron, 225, 246, 254, 272

Coulomb
  energy, 152, 153, 156
  function, 114, 177
  interaction, 110, 120, 144, 146, 151
  modified scattering amplitude, 117
  parameter, 108, 133, 135, 157, 160
  plus short-range potential, 173, 177
  problem
    few-body, 281
    three-body, 170, 282, 283
  scattering, 172
  state, 179, 180
  See also Forces, Green's function, Potential, S-matrix
Crank–Nicholson algorithm, 195
Crossing
  avoided, 310, 311
  diabatic, 308, 309
Cross section, 176, 214, 221, 235, 236, 238, 240, 242, 245–247, 260, 263, 270, 273, 323
  Coulomb differential, 330
  differential, 330
    triply (TDCS), 317, 318, 320, 323, 325, 331, 333, 337
  double photoionization, 325
  first Born, 329
  photo detachment, 254, 258
  resonant, 240
  at threshold, 238, 241

Degeneracies, 225
Degenerate, 226
Diabatic molecular orbital (MO), 310
Dipole, 226, 241, 243, 252
  amplitude, 263
  field, 226
  moments, 225
  strength, 259
  term, 246
Distorted wave theories, 315, 336
Distortion
  Coulomb, 192
  logarithmic, 178
Dollard amplitudes
  two-fragment reactions
    short-range and Coulomb potentials, 57
    Coulomb-modified short-range, 57
  break-up processes
    short-range potentials, 59
    short-range plus screened Coulomb potentials, 60

Dollar amplitudes (*cont.*)
    Coulomb-modified short-range, 60
  See also Wave operators
Dollard's modified
  approximate wave operators, 205
  asymptotic Hamiltonian, 204
Doppler
  shift, 228, 267, 268
  shifted laser spectroscopy, 229
Double photoionization, 282, 325, 328, 364

Effective range, 213
  functions, 99, 108, 114, 116, 158, 159
Effective-two-body LS equations
  arrangement scattering amplitudes, 64, 66
  break-up amplitudes, 65
Electron
  detachment, 267
  –electron
    Coulomb force, 322
    interaction, 319, 321
    repulsion, 335
Energy(ies)
  CM photon, 242
  conservation, violation of 198, 209
  H^- beam, 270
  (laser) photon, 231, 257, 261, 262, 268, 272
  ponderamotive, 265–267
  quiver, 264–267
  resolution, 233–235, 253, 254, 257, 273
  shift, 242–244
  spread, 234
Exchange
  t-matrix, 318

Faddeev
  calculation, 214
  equation, 101, 125–128, 132, 146, 152
  reformulation, 100
  results, 170
Faddeev–Noble
  equation, 129, 138, 139, 152, 157
  method, 142
Fano function, 240, 247, 248, 255, 258, 260, 263
Field-induced tunneling, 260
Fields
  electric, 225, 227, 230, 232, 236, 239, 242, 245, 254, 256, 257, 260, 266, 272
  magnetic, 227, 230, 236

INDEX 343

Final-state
  interaction, 331
  Coulomb interaction, 334
Floquet, 267
Forces
  Coulomb, 99, 100, 108, 117, 282
  polarization, 156
  tensor, 102, 142
  three-body, 103, 127
  three-nucleon, 100, 101, 103, 104, 152, 157
Foil(s), thin, 227
  experiments, 222
  thickness, 271
  transmission studies, 270, 272
Form factor, 109
Frame transformation, 239, 240
Free-electron continuum, 247
Fusion: See Muon

Grand angular momentum, 251, 274
Green's function, 171
  Coulomb, 185

H (beam), 230
H$^-$
  beam, 228, 233, 239, 242, 249, 251, 253, 256, 258, 261, 264, 270, 272
  ion, 221, 265, 270, 284
^{3}H, 97, 100, 103, 144, 151, 154
^{3}He, 97–100, 144, 151, 154, 189
He-atom, 284, 303, 326, 337
Hilbert space, 172, 175
  states, 174, 175
Hyperfine interaction, 225
Hyperspherical
  calculations, 296
  treatment, 260
  See also Coordinates
Interference, continuum, 259
Intertwining relation(s), 25, 38, 175, 184
Ionization, 92, 309, 310, 328, 332, 333
  auto-, 250, 276
  double-, 207
  electron impact-, 282, 312, 314–316, 333–335
  mechanisms, 309, 321
  process, 335
Isospin, 99, 142, 147
  impurity, 148, 152

Jacobi See Coordinates

K-Matrix, 107, 113, 115, 119

Lambda (Λ) doubling, 297
LAMPF, 228, 231, 234, 240, 243, 254, 257, 259
  high-resolution atomic beam, 232
Laser beams, 227, 231, 233–235, 265
Laser intensity, 229
Laser pulse, 234, 242
Laser, H$^-$-interaction region, 260
Lifetimes, 228, 229
Lippmann–Schwinger equation
  for T-matrix, 27, 34
  for wave function, 25

Møller operators
  three-fragment
    neutral particles, 44
    charged particles, 58
    Coulomb-modified short-range, 58, 60
  two particles, 24, 28, 37
  two fragment,
    neutral particles, 43, 45
    charged particles, 58
    Coulomb-modified short-range, 47
  See also Wave operator
Multiphoton
  detachment, 267–269
  effects, 264
  experiments, 222, 264, 265, 272, 273
Muon
  catalyzed-fushion, 101, 162
  internal conversion, 163

Nodal lines, 302
Nodal structure, 302, 312
Normalization, 107, 149, 307, 318, 319, 354
Numerical results
  Coulomb-modified short-range p-d versus n-d elastic 73, 82
    effective-range parameters, 75
    energy dependence of phase shifts, 75
    l-dependence of phase shifts, 75
    deuteron-nucleus elastic scattering, 88
  p-d verus n-d
    break-up cross sections, 85
    elastic differential cross sections, 73, 77, 80

On-shell limit, 183
On-shell matrix element, 185

On-shell relation, 186

Partial wave, 114, 146
  S matrix
    for short-range potential, 7
    for screened Coulomb potential, 15
    for unscreened Coulomb potential, 9, 10
    shoft-range and Coulomb potential
      total, 13
    Coulomb-modified short-range, 11
Partial-wave T-matrix
  for screened Coulomb potential, 15, 17
  short-range and Coulomb potential
    total, 13
    Coulomb-modified short-range, 19, 21, 33
  for short-range potential, 7
  for unscreened Coulomb potential, 9, 10
Path integral, 306
Pauli principle, 99, 126, 143, 146, 165
p-d scattering, 100, 109, 110, 151, 156, 159, 172, 190, 192, 212, 213, 216
p-d system, 99, 162, 164
Peak
  binary, 331–333, 337
  recoil, 331–333, 337
Periodic orbit(s), 302, 304–306
Perturbed
  anomalous term, 196
  Dollard-type wave operator, 196
  Hamiltonian, 198
  Møller wave operator, 196, 197
  S-matrix, 196, 197, 206
Phase shift(s), 107, 111, 176, 184
  Aharonov–Bohm, 239
  Coulomb, 186
  Coulomb-modified, 99, 111, 115
  for Coulomb potential, 8
  for screened Coulomb potential, 15
  short-range plus Coulomb potential
    total, 12, 19
    Coulomb-modified short-range, 12
  for short-range potential, 7
  short-range plus screened Coulomb potential
    Coulomb-modified short-range, 13
    total, 13
Photo detachment, 221, 222, 224, 225, 237, 239, 240–242, 247, 252–254, 261, 272
  electron, 222, 235, 264
Photon, 228, 235, 256
  beam, 231

Photon (*cont.*)
  energy, 231, 242
  probes, 226
Polarizability, electric, 136
Polarization, 236, 237, 239, 264, 268, 269; see *also* Potential
Positron, 283
Potential, 222, 251, 254
  adiabatic, 252
  barrier, 240, 242, 243, 263
  center-of-mass Coulomb, 47
  centrifugal, 254
  channel, 46, 47
  Coulomb, 128, 131, 137, 139, 147, 149, 177
    screened, 46
    unscreened, 8
  Coulomb-like, 187–189, 191, 206
  curves, 252, 253, 263
  effective
    for arrangement scattering, 62, 64
    for break-up processes, 65, 66
  one-pion exchange (OPEP), 102
  optical, long-range behavior of, 89
  polarization, 79, 101, 109, 135, 137–139, 141
  separable, 62
  short-range, 5, 46
  two-body
    Graz, 207–209, 211
    N-N, 212
    paris, 207
    p-p, 207, 209–211
    separable, 207
    Yamaguchi, 211
Propagator, 173, 174, 176, 199

Quantization
  semiclassical, 282, 303, 305
  WKB, 306
Quantum defect, 259
Quantum number, 199, 275, 286, 290, 291, 336
  correlation, 274
  electric, 298, 300
  Herrick, 298–301
  MO, 301
  principal, 226, 232, 242, 274
  two-electron, 299

Radial wave function
  asymptotic behavior
    for screened Coulomb potential, 15

Radial wave function (*cont.*)
  asymptotic behavior (*cont.*)
    for short-range potential, 6
    for unscreened Coulomb potential, 8
  for Coulomb potential
    screened, 19
    unscreened, 8
  irregular, 12
  logarithmic derivative, 6
  physical, 6, 9, 12
  regular, 5, 6, 12
  short-range plus Coulomb potential
    screened, 15
    unscreened, 11
  for short-range potential, 5
Rearrangement, 125, 126
  scattering, 124
  See also Channels
Regularity condition, 5
Relativistic effects, 252
Relativistic H^- beam, 222
Relativistic kinematics, 227
Renormalization factor, 16, 51, 57
Renormalization phase factor, 184
Renormalized, 176, 185, 203
  channel, 39
  Coulomb, 29
  Coulomb-distorted channel, 47
  free, 29
  full, 25, 29, 40
Resolvent equation, 42
Resonance(s), 237, 248, 257–259, 262, 281, 302
  doubly excited, 261
  Feshbach, 222, 250, 252–256, 258, 260, 272
  Losurdo–Stark, 256, 257
  position, 221
  quenching, 262
  shape, 222, 240, 241, 250, 253, 255–258, 272
  spectrum, 251
  width, 221, 254, 255, 259, 260, 262, 263
Riccati
  equation, 110
  Bessel function, 113
Rydberg sequence, 248

Saddle
  behavior, 293
  motion, 294
  point, 293, 294, 298, 300, 312

Saddle (*cont.*)
  potential, 302
  Wannier, 313
Scale(s), 143, 157, 163, 318
Scaling, 211
Scattering, 170
  amplitude, 176
    Coulomb, 178
  length, 99, 114, 116, 117, 153, 155–158, 218
  p-p, 207, 212
  Rayleigh, 135–137
  theory
    mathematical, 173
    stationary state, 171, 184
    time-independent, 169, 172, 174
  time, lattice, 199, 209
Schrödinger equation, 100, 105, 110, 121, 123, 126, 173, 195, 251, 267
Screened potential, 183
Screened Møller wave operator, 183
Screening, 183, 187, 189
Selection rules, 276
Singular, 185, 199, 200
  behavior, 197
  factors, 185
Singularity, 185
  kinematical, 39
S-matrix, 172, 174–176, 184, 185, 195, 197–200, 209
  Coulomb, 186
  relative error, 198, 208
  for three-particle scattering
    two-fragment reactions, 38
    break-up processes, 60
  for two-particle scattering, 20, 26
Sommerfeld parameter, 315
Spline technique, 150
Stark
  effect, 244, 291
  splitting, 243
  state, 257, 262
State(s)
  adiabatic, 308
  asymptotic, 172–174, 180, 182, 193, 198
  Coulomb scattering, 180, 181
  doubly excited, 222, 226, 246, 247, 251, 252, 264, 273, 296
  Losurdo–Stark, 271
  resonant, 283, 285, 296, 308
  S, 144
  S', 144, 154, 165

Strong operator limit, 174, 182, 183
Strong resolvent convergence, 172, 200
Supercomputers, 170, 171
Symmetry
  charge, 151
    breaking, 153
    dependence, 154
  chiral, 101, 106
  group, 195

Three-body scattering state, 43
  asymptotic behavior, 41
Threshold, 241, 252, 260, 269
  behavior, 327
  detachment, 222, 235, 236, 246, 264, 273
  higher, 240, 242
  law, 282, 308
  lowest, 237
  $n = 2$, 252, 253, 272
  shift, 242–245
  three-body, 283
    breakup, 213, 282–284, 293, 294, 307, 309, 336
  two-fragment, 282, 337
  Wannier, 319
  Wigner, law, 235, 240, 241, 245, 269, 282, 307
Time-dependent
  approach, 191
  formulation, 170, 185, 187
  framework, 172
  methods, 170, 172
  scattering theory, 216
Time evolution, 174, 193–195, 200, 205
  operator, 173, 200
t-matrix, 107, 115, 171, 172, 184, 185, 314, 316
  Coulomb, 183, 184
Transition amplitude
  for arrangement processes
    neutral particles, 62
    charged particles, 48, 52
  for break-up processes, 44
Transition operator
  break-up processes
    Coulomb-modified short-range, 56
    short-range potentials, 44
    short-range plus screened Coulomb potentials, 56
  two-fragment reactions
    short-range potentials, 39, 41
    short-range plus screened Coulomb potentials, 48

Transition operator (*cont.*)
  for two-particle scattering
    Coulomb, 29, 30
    Coulomb-modified short-range, 29, 30
    full, 26
Triad equations, 125
Two-body scattering amplitude
  Coulomb-modified short-range plus screened Coulomb potential, 13, 22
  for center-of-mass Coulomb potential, 48
  for Coulomb potential, 9, 10
  for screened Coulomb potential, 17
  for short-range potential, 7
  total
    short-range plus Coulomb potential, 13
    short-range plus screened Coulomb potential, 15
Two-body scattering state
  asymptotic behavior
    for Coulomb potential, 8
    for screened Coulomb potential, 15, 17
    for short-range potential, 7
  center-of-mass Coulomb, 48
  for Coulomb potential
    screened, 19
    unscreened, 9
  short-range plus Coulomb potential
    screened, 15
    unscreened, 11
  for short-range potential, 7
Two-center Hamiltonian
  Coulomb, 287
  electronic, 288
Two-potential formula
  for three-particle scattering, 48
  with effective-two-body theory, 65, 69
  for two-particle scattering, 29

Uncertainty principle, 229, 234, 255
Unscreening property
  center-of-mass Coulomb scattering
    state, 53
    amplitude, 53
  three-particle scattering
    break-up amplitude, 56
    two-fragment scattering amplitude, 51
  two-particle scattering
    Coulomb-modified short-range scattering amplitude, 21, 31, 32
    Coulomb scattering amplitude, 31
    full scattering amplitude, 31

Unscreening property (*cont.*)
   two-particle scattering (*cont.*)
      full screened phase shift, 19
      partial-wave S-matrix, 20
      partial-wave T-matrix, 20, 31
      radial wave function, 20, 31
      scattering state, 20, 31
      screened Coulomb-modified short-range phase shift, 33
      screened Coulomb phase shift, 33

Variable phase
   for Coulomb potential, 8
   for short-range potential, 6
   for screened Coulomb potential, 14
Variational Kohn estimates, 149, 157
Variational Kohn results, 119
Variational method, 224
Variational techniques, 118, 120, 223

Wannier, 245, 246, 307, 312, 319, 325
   exponent, 313

Wannier (*cont.*)
   theory, 325, 327
   theshold law, 319
Wave
   function, 312, 314, 316, 318, 329, 330
      correlated three-Coulomb, 314, 321
      molecular orbital (MO), 286, 292
      reduced, 106
      symmetric three-Coulomb, 326, 333, 334
      symmetrized, 249
      three-dimensional, Coulomb, 10
      two-electron, 223, 224
   operator, 169, 172, 175, 177, 190–193, 195, 197, 200
      Coulomb-modified, 179
      Dollard, 178, 179, 181, 183, 187, 188, 192, 205, 206, 216
      Møller, 173, 178, 180, 181, 183, 188, 191
   packet(s), 193, 207, 209, 308, 309
Wolfenstein–Gershtein effect, 164

Zero-screening limit, 17, 19, 21, 30, 51, 52, 56